Theory and Applied Principles of Condensed Matter Physics

Theory and Applied Principles of Condensed Matter Physics

Edited by
Raymond Stevens

C WILLFORD PRESS

www.willfordpress.com

Published by Willford Press,
118-35 Queens Blvd., Suite 400,
Forest Hills, NY 11375, USA

ISBN: 978-1-68285-629-1

Cataloging-in-Publication Data

Theory and applied principles of condensed matter physics / edited by Raymond Stevens.
    p. cm.
Includes bibliographical references and index.
ISBN 978-1-68285-629-1
1. Condensed matter. 2. Physics. I. Stevens, Raymond.
QC173.454 .T44 2019
530.41--dc23

For information on all Willford Press publications
visit our website at www.willfordpress.com

WILLFORD PRESS

Contents

Permissions

List of Contributors

Index

Preface

Condensed matter physics is the field of physics that studies the macroscopic and microscopic physical properties of condensed phases of matter. These include solids and liquids, as well as superconducting, ferromagnetic and antiferromagnetic phases, Bose-Einstein condensates, etc. The theoretical understanding of condensed matter especially relating to the electronic properties of solids is furthered by various models such as Drude model, Band structure and density functional theory. Modern condensed matter physics involves the use of numerical computation to understand high temperature superconductivity, topological phases and gauge symmetries. This book brings forth some of the most innovative concepts and elucidates the unexplored aspects of condensed matter physics. The aim of this book is to present the researches that have transformed this discipline and aided its advancement. Coherent flow of topics, student-friendly language and extensive use of examples make this book an invaluable source of knowledge.

After months of intensive research and writing, this book is the end result of all who devoted their time and efforts in the initiation and progress of this book. It will surely be a source of reference in enhancing the required knowledge of the new developments in the area. During the course of developing this book, certain measures such as accuracy, authenticity and research focused analytical studies were given preference in order to produce a comprehensive book in the area of study.

This book would not have been possible without the efforts of the authors and the publisher. I extend my sincere thanks to them. Secondly, I express my gratitude to my family and well-wishers. And most importantly, I thank my students for constantly expressing their willingness and curiosity in enhancing their knowledge in the field, which encourages me to take up further research projects for the advancement of the area.

Editor

Ordering and order-disorder phase transition in the (1 × 1) monolayer chemisorbed on the (111) face of an fcc crystal[*]

A. Patrykiejew, T. Staszewski

Department for the Modeling of Physico-Chemical Processes, Maria Curie-Skłodowska University, 20-031 Lublin, Poland

In this paper we have considered a simple lattice gas model of chemisorbed monolayer which allows for the harmonic fluctuations of the bond length between the adsorbate atom and the surface site. The model also involves a short-ranged attractive potential acting between the adsorbed atoms as well as the surface periodic corrugation potential. It has been assumed that the adsorbed atoms are bonded to the uppermost layer of the substrate atoms. In particular, using Monte Carlo simulation method we have focused on the orderings appearing in the dense monolayer formed on the (111) face of an fcc solid. Within the lattice gas limit, the chemisorbed layer forms a (1 × 1) structure. On the other hand, when the bonds are allowed to fluctuate, three other different ordered phases have been found to be stable in the ground state. One of them has been found to be stable at finite temperatures and to undergo a phase transition to the disordered state. The remaining two ordered states have been found to be stable in the ground state only. At finite temperatures, the ordering has been demonstrated to be destroyed due to large entropic effects.

Key words: *chemisorption, phase transitions, computer simulation, nanoscopic systems*

1. Introduction

The ever-growing technological importance of the on-demand tailored nanomaterials, makes it of great importance to understand and master their formation, structure and thermodynamics as well as electronic properties. Nanostructures of reduced geometry, like finite adsorbed islands, have become a field of intensive research in the last decades [1–3]. Owing to the development of powerful experimental methods, like the scanning tunneling microscopy (STM), it is now possible to study the inner structure of even small adsorbate clusters and to determine the arrangement of individual atoms within such nanoscopic islands [4–8].

Order-disorder phenomena in strongly adsorbed and chemisorbed monolayers have been a subject of active research for many years now [9–18] and those studies have vastly relied on Monte Carlo simulations carried out in a general framework of lattice gas models [9, 10, 15].

At present, theoretical studies of chemisorption are usually based on *ab initio* quantum mechanical calculations, and employ various methods, like the density functional theory [19–21]. Although quantum mechanical calculations can be carried out only for rather small clusters, they nonetheless allow for a precise determination of interaction potentials and the structure of different ordered phases in chemisorbed layers. On the other hand, ab-initio approaches cannot be still efficiently used to study cooperative phenomena and phase transitions in particular. Such studies require rather large systems consisting of $10^3 - 10^4$ atoms and are beyond the reach of *ab initio* molecular dynamics [22, 23] and Car-Parrinello [24, 25] methods. Therefore, one has to use classical simulation methods with appropriately tuned interaction potentials [26, 27].

---

[*]Dedicated to our friend Stefan Sokołowski on the occasion of his 65th birthday

In the case of weakly adsorbed layers, made of simple gases (e.g., Ar, Kr, Xe, N_2, CH_4) on various solid substrates (e.g., graphite, boron nitride and metal crystals), the adsorbate-adsorbate and the adsorbate-substrate interactions can be quite well described by the Lennard-Jones potential [28, 29]. This potential does not properly describe the interactions operating in the systems involving metals and/or semiconductors. For metal-metal interactions, the embedded atom method (EAM) provides an approach allowing for the development of potentials with numerous applications in theoretical and computer simulation studies of single metals and metallic alloys [30–33]. For covalently bonded systems, the potentials proposed by Stillinger and Weber [34, 35], by Tersoff [36, 37] and by Brenner [38] are often used.

In this work, we consider a simple lattice-like model of finite chemisorbed monolayer films that involves the pair adsorbate-adsorbate interaction, the harmonic potential that accounts for possible displacement of adsorbed atoms from the active centers to which the chemisorbed atoms are bonded, and the corrugation potential resulting from the lattice structure of the crystalline substrate [28, 39].

The idea of including translational degrees of freedom (elastic interaction) into lattice gas models is not new. Such models have been used to study the phase behavior and unmixing transition in Si-Ge alloys [40], Ising ferro- and antiferromagnets [41–43], adsorbed monolayers [44] and frustrated materials [45].

Although the results presented in this paper should be considered as preliminary and have been obtained for a very simple version of the model, nevertheless they show novel forms of ordering that may appear in chemisorbed layers.

The model can be readily extended to take into account three-body interactions, the possibility of the appearance of different types of active sites at the substrate surface, as well as different symmetry properties of both the adsorbing surface and the adsorbed layer. Thus, it holds promise of having predictive power for a broad range of questions.

2. The model

We assume that the potential representing the pair interaction between adsorbate atoms takes the following form:

$$u(r) = \begin{cases} 4\varepsilon \left[(\sigma/r)^{12} - (\sigma/r)^6 \right] \exp\left[-\sigma/(r - R_c) \right], & r < R_c, \\ 0, & r \geq R_c, \end{cases} \tag{1}$$

i.e., it is a standard Lennard-Jones (12,6) potential supplemented by the cut-off function, with R_c being the parameter determining the range of interaction. Equation (1) can be treated as a special case of the pair interaction term that appears in the Stillinger-Weber potential [34].

We also assume that all adsorbed atoms are bonded to adsorption centers, which form a two-dimensional lattice of the symmetry imposed by the substrate structure. Throughout this work, we have considered the surface lattice of triangular symmetry and have taken the surface lattice constant a as a unit of length. A further assumption is that $\sigma < a$ so that all lattice sites are accessible to adsorption. In this work, we are not interested in the adsorption process but only in the structure and properties of films in which all sites are occupied by the adsorbed atoms. Thus, each surface site, located at $\mathbf{r}_{0,i}$, is occupied by an adatom. Nonetheless, the bonds between adatoms and surface sites have been assumed not to be completely rigid, but permitted to undergo small harmonic deformations. Therefore, the adatom energy changes with the displacement from the lattice site, $\mathbf{u} = \mathbf{r} - \mathbf{r}_{0,i}$, as follows:

$$u_{\text{har}}(\mathbf{u}) = \frac{1}{2} f \mathbf{u}^2, \tag{2}$$

where f is the force constant of the harmonic potential.

In what follows we use the reduced units. All the energy-like parameters are given in units of ε, e.g., the potential energy is given as $u^* = u/\varepsilon$ and the force constant is given by $f^* = f/\varepsilon$. All the distances are given in units of the surface lattice constant a. Thus, $r^* = r/a$, $R_c^* = R_c/a$ and $\sigma^* = \sigma/a$.

Due to the lattice nature of the crystalline surface, the surface potential experienced by adatoms exhibits periodic variations and can be expressed as follows:

$$v(\mathbf{r}) = v_0 \sum_k \cos[\mathbf{q}_k \mathbf{r}]. \tag{3}$$

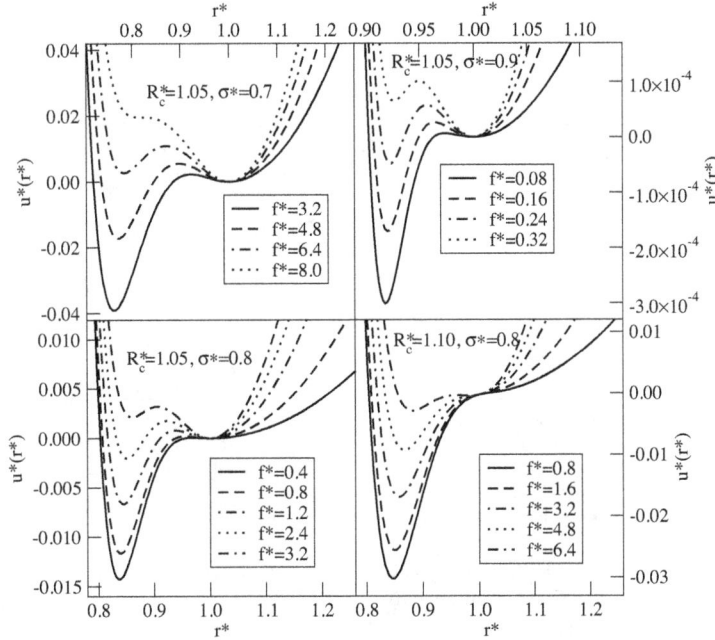

Figure 1. The examples of the pair interaction potential $u(r) + u_{\text{har}}(|\mathbf{u}|)$ for the systems with different σ^*, R_c^* and the force constant f^*.

In the above, v_0 determines the amplitude of the corrugation potential and the sum runs over the reciprocal lattice vectors of the surface lattice ($\mathbf{q}_k$).

When $v_0 = 0$, the interaction energy of the pair of atoms adsorbed over the neighboring sites depends only upon their distance r, and the displacements $u_i = |\mathbf{u}_i|$, $(i = 1, 2)$ of both interacting atoms should be the same with the displacement vectors $\mathbf{u}_1 = -\mathbf{u}_2$.

Figure 1 gives some examples of the pair interaction potential, being the sum of $u(r)$ and $u_{\text{har}}(\mathbf{u})$, obtained for different sets of the parameters: $\sigma^* = \sigma/a$, $R_c^* = R_c/a$ and $f^* = f/\varepsilon$. Due to the addition of a harmonic term, the pair interaction energy becomes positive for large r and possibly exhibits two minima. It should be emphasized that the pair interaction potential is qualitatively very similar to that obtained from *ab initio* quantum mechanical calculations obtained for Pb adsorbed on Si within the generalized gradient approximation (GGA) [46].

Here, we assume that the adsorbed layer has been formed on the (111) face of a perfect fcc crystal and forms a simple (1 × 1) structure when the bonds are strictly rigid ($f^* = \infty$).

3. The ground state behavior

The ground state behavior of the model depends on the assumed lattice symmetry, the pair potential cutoff distance (R_c^*), the diameter of adatoms (σ^*), the value of the force constant f^*, and the amplitude of the surface potential v_0^*.

The first series of calculations aiming at the determination of stable ground state structures has been performed for the systems without the corrugation potential ($v_0^* = 0$), assuming that $\sigma^* = 0.80$ and $R_c^* = 1.05$, and using finite clusters of different size and shape. The calculations have been carried out over a wide range of f^* between 0.4 and 3.0. Four different ordered structures, depicted in figure 2, have been found. The stability of each of them is determined by the magnitude of the force constant f^*. Figure 3 shows the ground state phase diagram obtained for the systems with $\sigma^* = 0.80$ and $R_c^* = 1.05$ and different values of f^*.

In order to determine the structure of the system in the ground state, we have calculated the energy for all possible structures as a function of the displacement. Of course, the structure (1 × 1) corresponds to zero displacement. The energy of the structures S and T is characterized by a single displacement,

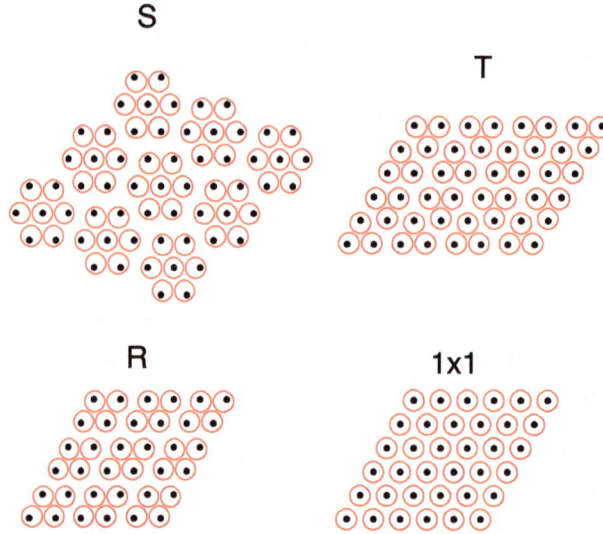

Figure 2. (Color online) The four different ordered structures found for the systems characterized by different force constant. Small filled circles are the locations of surface active sites while big circles represent the positions of adatoms.

while the energy of the structure R is determined by two displacements in two orthogonal directions along the diagonals of the rhombus. The stable state of a system is the one in which the energy reaches its minimum. We have determined the displacements that minimize the energy for each structure, as well as determined which of the structures is stable for a given set of parameters.

For sufficiently high values of f^* ($f^* > f^*_{T-(1\times1)}$), a simple (1×1) structure is a stable state. In this case, the adatoms are not displaced from the positions of adsorption centers, i.e., the average displacement **u** is equal to zero. When f^* decreases, the adatoms exhibit a gradually increasing tendency to be displaced from the positions given by the vectors $\mathbf{r}_{0,i}$ and this leads to the formation of the structures T, R and S, depicted in figure 2. The structure T consists of equilateral triangles of adatoms arranged in such a

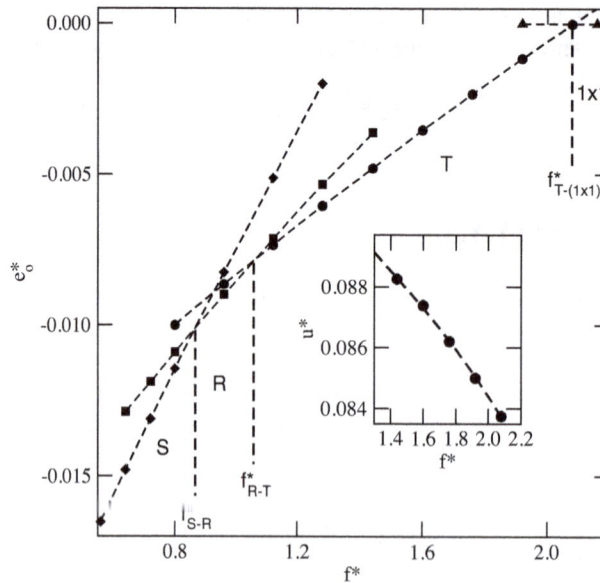

Figure 3. The ground state phase diagram, in the plane energy (e^*_0) — force constant (f^*) for the systems with $\sigma^* = 0.8$ and $R^*_c = 1.05$. The inset shows the changes of adatom displacement with f^* in the ordered structure T.

way that each adatom has a displacement from the surface lattice site of the same length. Of course, the magnitude of this displacement gradually increases when f^* becomes lower (see the inset to figure 3). The structure T is stable as long as f^* stays between $f^*_{T-(1\times1)}$ and f^*_{R-T}. The structure R develops for the lower values of f^*, between f^*_{S-R} and f^*_{R-T}. In this structure, the groups of four adatoms form rhombic clusters, characterized by two different displacements of adatoms from the surface lattice (along the diagonals). In perfectly ordered T and R states, the unit cell vectors are aligned with the symmetry axes of the triangular lattice. When f^* falls below f^*_{S-R}, the structure labeled as S (star-like) becomes stable. This structure consists of clusters made of seven atoms each and arranged in such a way that the central atom is not displaced from the adsorption site, while its six nearest neighbors are equally displaced towards the central atom. The central atoms form a $\sqrt{7} \times \sqrt{7}$ lattice rotated by 19.1° with respect to the original triangular lattice of adsorption sites. Of course, in the R and S structures, the atomic displacements also gradually increase when the force constant f^* becomes lower, just the same as in the case of T structure. For the values of f^* well below f^*_{S-R}, the tendency of adatoms to be displaced from lattice sites becomes high enough to lead to the formation of structures consisting of clusters larger than in the S structure and this point will be addressed later on in section 5. Of course, when f^* approaches zero, the entire adsorbed island forms a close packed triangular lattice, but in the case of chemisorption, this situation can be excluded from consideration.

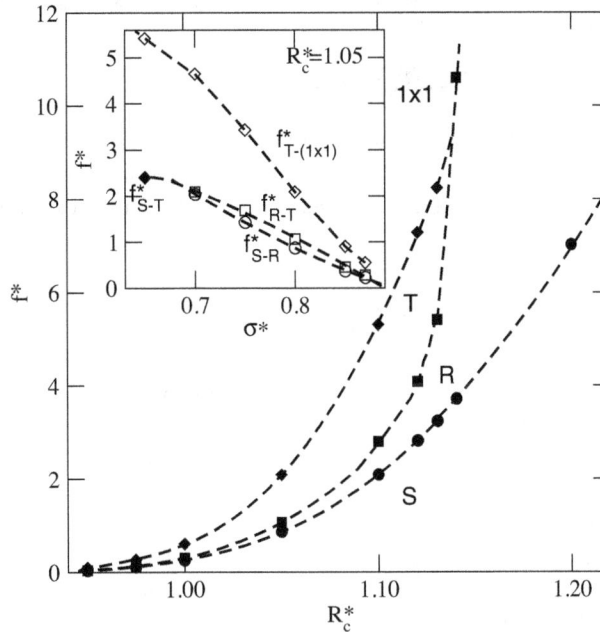

Figure 4. The changes of the regions of stability of different ordered phases with the cut-off distance for the systems with $\sigma^* = 0.8$.

It should be noted that although our ground state calculations have been carried out for finite systems, the results have not been affected by finite size effects. The assumed range of the interaction potential [equation (1)] is very short, and due to rather large displacements of adatoms from lattice sites, the system energy in the ground state is not influenced by the presence of free boundaries. For the potentials with a longer interaction range, given by R_c^*, finite size effects become of importance. In such systems, the stability regions of different ordered states are also affected by finite size and boundary effects. The change of the interaction range affects the strength of the pair potential, cf. the lower panels of figure 1, and hence the values of f^* that delimit the regions of stability of different ordered states. This is illustrated by the results given in figure 4. The results show that the T structure exists only for rather short-ranged potentials with R_c^* lower than about 1.139, while the R and S structures are stable even for the potentials of a longer range. However, with the increase of R_c^*, the results are subjected to gradually increasing finite size effects. The ground state calculations for the clusters of perfectly ordered S structure and of different size have shown that for $f^* \leqslant 2.4$, the S structure occurs only in finite clusters. Upon the increase of the cluster size, the displacement of adatoms corresponding to the minimum of potential energy (u_{min}) goes to zero. The ground state properties of the model also depend upon the value of σ^*. Of course, for σ^*, at which the minimum of the potential occurs at the distance approaching a, only the (1×1) structure is stable. Upon the decrease of σ^*, the location of the potential minimum shifts towards shorter distances between the interacting atoms (cf. figure 1) allowing for rather large atomic displacements. This gives rise to the appearance of S, R and T structures. We have performed ground state calculations for a fixed value of $R_c^* = 1.05$ and for different σ^* between 0.65 and 0.87. The results of the calculations are summarized in the inset to figure 4. It appears that the R structure occurs only for σ^* larger than about 0.68. For still smaller atoms, only the S, T and (1×1) structures are stable in the ground state. Due to the lowering of σ^*, the energy of the pair interaction increases so that for a given f^* the adatoms are more likely to exhibit larger displacements from the lattice sites. This explains why the values of f_{S-R}^*, f_{R-T}^* and $f_{T-(1\times1)}^*$ increase when σ^* becomes lower.

In the case of an (111) fcc surface considered here, the effects of the corrugation potential on the ground state properties can be readily anticipated. The adsorption centers are located right above the surface atoms, i.e., over the maxima of the external field, and the atomic displacements in all S, R and T structures are directed towards the positions of a lower surface potential $v(\mathbf{r})$. Therefore, an increase of the corrugation potential amplitude (v_0^*) has the same effect as the lowering of the force constant (f^*). In fact, the calculations performed have demonstrated that the values of f_{S-R}^*, f_{R-T}^* and $f_{R-(1\times1)}^*$ linearly increase with v_0^* as follows:

$$f_{\alpha-\beta}^*(v_0^*) = f_{\alpha-\beta}^*(v_0^* = 0) + C v_0^* \qquad (4)$$

with $\alpha - \beta$ being S–R, R–T or T–(1×1), and the magnitude of the constant C depending only upon σ^* and R_c^*. In other words, due to the turning on of the corrugation potential at zero temperature ($T = 0$), the limits of stability of all differently ordered structures uniformly shift towards higher values of f^*.

4. Monte Carlo simulation method

The above presented model has been studied using a standard Metropolis Monte Carlo simulation method in the canonical ensemble [47]. We have monitored the system energy and the contributions to the energy due to the pair interaction represented by the potential [$u(r)$], the harmonic interaction [$u_{har}(\mathbf{u})$] and due to the corrugation potential [$v(\mathbf{r})$]. Furthermore, we have calculated the heat capacity from the energy fluctuations.

In order to monitor the formation of the ordered structures, we have calculated the probability distribution of atomic displacements $p_d(|\mathbf{u}|)$ and the probability distribution of the number of nearest neighbors $p_{NN}(n)$. Since all surface lattice sites are covered by adatoms, we have assumed that the nearest neighbors are only those atoms located on the nearest adsorption sites with the inter-atomic distance smaller than unity. Thus, only the atoms displaced from lattice sites have been counted.

In the case of the ordered T structure, the distribution $p_d(|\mathbf{u}|)$ should be unimodal and $p_{NN}(n)$ should be close to unity for $n = 2$, while approaching zero for other values of n. The structure S should give bimodal distribution $p_d(|\mathbf{u}|)$; with one maximum at $|\mathbf{u}| \approx 0$ (for the central atoms) and the second maximum at larger value of $|\mathbf{u}|$ (for the nearest neighbors of the central atom). The distribution $p_{NN}(n)$ should be equal to about 1/7 for $n = 6$, about 6/7 for $n = 3$ and approach zero for other values of n. Finally, the structure R should also be characterized by bimodal distribution $p_d(|\mathbf{u}|)$ with the maxima at the positions corresponding to the two displacements in the rhombus, while the distribution $p_{NN}(n)$ should approach the value of 0.5 for $n = 2$ and 3 and should be approximately zero for other values of n.

The calculations have been performed using two different shapes of the simulation box. One series of calculations has been carried out using simple rhombic cells oriented along the symmetry axes of the triangular lattice and of different side length, L (labeled as R-cells). Note that the ordered structures T and R require even values of L to accommodate them within the cell. In the case of the ordered structure S, we have also used rhombic simulation cells, but appropriately rotated and of the side length adjusted to accommodate the S structure (labeled as S-cells).

5. Results and discussion

The primary goal here has been to study the mechanism of disordering of the structures S, R and T. Under the assumption of strong bonding between adatoms and surface sites, the desorption has been prohibited so that the disordering can only involve the destruction of the order specific to S, R and T structures. We have performed Monte Carlo simulations for several systems using finite R- and S-cells of different size. Taking into account that one of our goals was to study truly nanoscopic systems, we have not applied periodic boundary conditions.

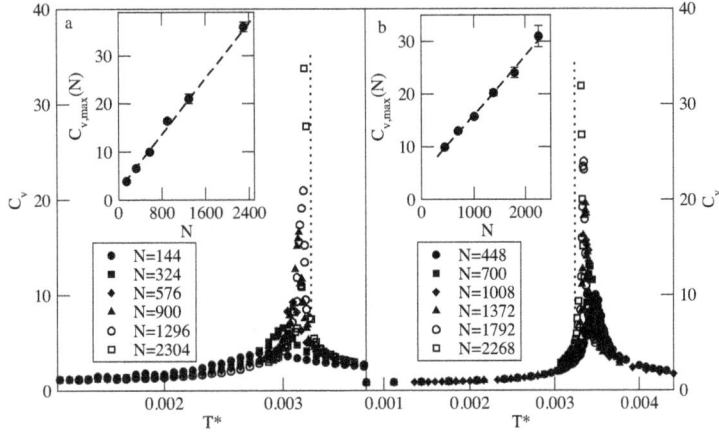

Figure 5. Main panels show the heat capacity curves for the system with $\sigma^* = 0.8$, $R_c^* = 1.05$ and $f^* = 0.4$ for different sizes of the adsorbed island (given in the figure). The left-hand panel (a) gives the results for the simple rhombic simulation cell and the right-hand panel (b) corresponds to the simulation cell accommodating the S structure. The insets show the scaling plots of the heat capacity maximum versus system size.

At first, we have considered the systems with $\sigma^* = 0.8$ and $R_c = 1.05$ and the values of f^* for which the ordered S structure is stable in the ground state, i.e., with $f^* < f_{S-R}^* = 0.864$. Parts (a) and (b) of figure 5 show the heat capacity curves obtained for the system with $f^* = 0.4$. Part (a) of figure 5 gives the results obtained using the R-cells of different size, while part (b) gives the results obtained for finite S-cells, also of different size. Both series of calculations have demonstrated that the heat capacity exhibits sharp maxima of the height and location depending upon the system size, indicating the presence of a phase transition. The insets to parts (a) and (b) clearly show that the maximum value of the heat capacity changes linearly with the number of atoms in the system N, i.e., with the system volume, since all lattice sites are occupied by the adsorbate atoms. According to the finite-size scaling theory of phase transitions [47, 48], this sort of behavior is characteristic of the first order transition. The theory also predicts the transition temperature in finite systems $T_{tr}(N)$ to be shifted with respect to the transition temperature in the infinite system, $T_{tr}(\infty)$, and the following relation should hold:

$$T_{tr}(N) = T_{tr}(\infty) + W/N, \tag{5}$$

where the constant W is proportional to $1/(E_+ - E_-)$ with E_- and E_+ being the internal energies of the coexisting phases at $T \to T_{tr}^-$ and $T \to T_{tr}^+$, respectively [49]. The difference $E_+ - E_-$ is positive and hence the temperature $T_{tr}^*(N)$ should increase with $1/N$. This does occur when the S-cells accommodating the ordered S structure are used (see figure 6). On the other hand, the temperature $T_{tr}^*(N)$ decreases with $1/N$ when the R-cells are used. In this case, the ordering is not perfect. The regions close to the island boundaries do not show the ordering characteristic of the S structure, even at very low temperatures. At higher temperatures, the disorder easily propagates into the system interior and considerably affects the estimated transition temperature. Of course, these effects of propagated disorder gradually decrease when the simulation cell size becomes larger, so that the transition temperature gradually increases with the system size. This is very well illustrated by the curves showing the changes of the system energy with temperature and obtained for different system sizes (see the inset to figure 6). In the case of simulation

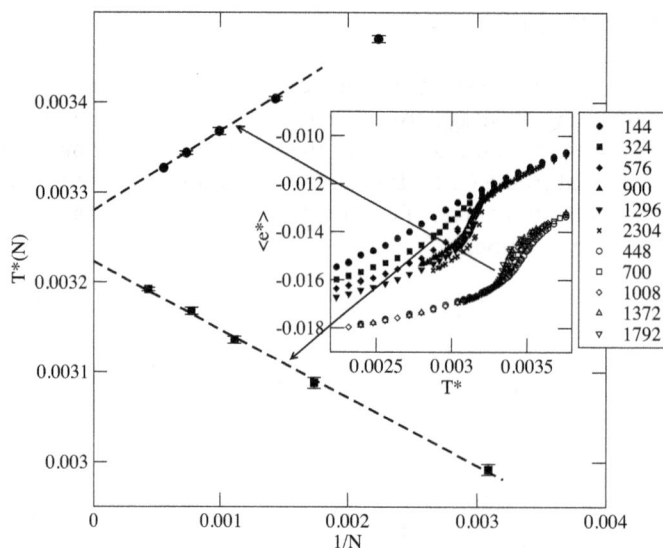

Figure 6. The changes of the temperature at which the heat capacity reaches its maximum plotted against the number of atoms in the adsorbed island for the systems with $\sigma^* = 0.8$, $R_c^* = 1.05$ and $f^* = 0.4$. Filled circles correspond to the simulation cell perfectly accommodating the S structure, while the filled squares, to the simple rhombic box. The inset shows the changes of the average potential energy $\langle e^* \rangle$ with temperature T for both cases and different size of the adsorbed island.

with S-cells, the energy at low and high temperatures does not depend upon the system size at all. Only at the temperatures close to the phase transition, the size effects set in. In fact, the results obtained from the runs with and without periodic boundary conditions applied have given the same results. On the other hand, the energy curves obtained for R-cells demonstrate large finite size effects already at low temperatures. This is a consequence of the already mentioned disordering close to the island boundaries. Therefore, only the simulation with S-cells gives the correct results and the estimated transition temperature $T_{tr}(\infty)$ is equal to about 0.00328.

The nature of the transition becomes quite clear when one considers the behavior of the probability distributions $p_d(|\mathbf{u}|)$ and $p_{NN}(n)$ given in figure 7. The distributions recorded below the transition temperature show a perfect ordering into the S structure, while those obtained at the temperatures above the transition point indicate that the disordering leads to the destruction of the S structure. In particular, the

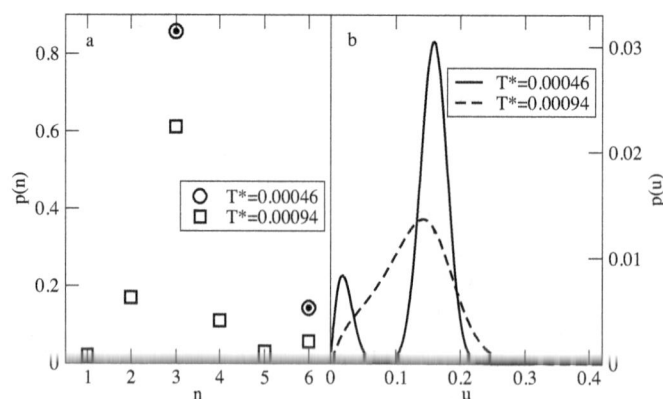

Figure 7. The probability distribution functions $p(n)$ and $p(|\mathbf{u}|)$ for the system with $\sigma^* = 0.8$, $R_c^* = 1.05$, and $f^* = 0.4$ at the temperature below and above the order-disorder transition point obtained for the adsorbed island containing 1372 atoms. The solid circles in the left-hand panel show the probabilities of an atom to have 3 and 6 nearest neighbors in the perfect S structure.

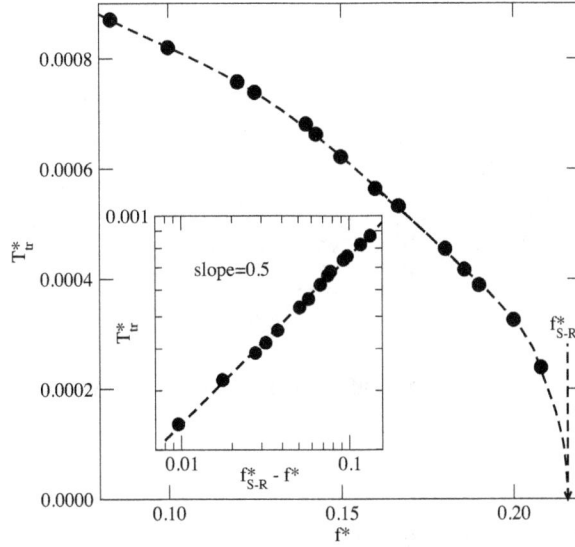

Figure 8. The main figure shows the changes of the transition temperature for the S structure versus f^* obtained for the systems with $\sigma^* = 0.8$ and $R_c^* = 1.05$. The inset shows the scaling plot of the transition temperature versus $f_{S-R}^* - f^*$.

distribution $p_d(|\mathbf{u}|)$ looses its bimodal shape and the distribution $p_{NN}(n)$ shows quite large contributions due to different numbers of nearest-neighbors, other than 3 or 6.

We have carried out simulations for a series of systems with different values of the force constant, f^*, and estimated the transition temperatures. The results of our calculations have been summarized in figure 8, which illustrates the changes of $T^*(\infty)$ with f^*. As expected, the transition temperature (T_{tr}^*) gradually decreases towards zero when f^* grows towards f_{S-R}^*,

The log–log plot of T_{tr}^* versus $\Delta f^* = f_{S-R}^* - f^*$ (given in the inset to figure 8) demonstrates that there are three regions of Δf^* over which the system exhibits different behavior. For small values of Δf^*, i.e., for the values of f^* close to f_{S-R}^*, the first-order character of the transition becomes more pronounced,

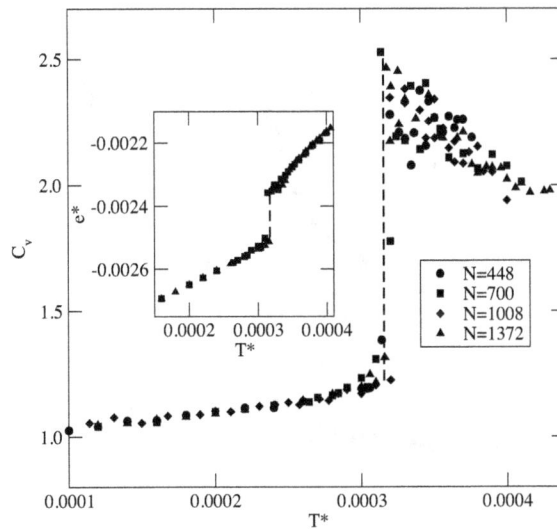

Figure 9. The main figure shows the heat capacity curves for the system with $\sigma^* = 0.8$, $R_c^* = 1.05$ and $f^* = 0.8$ and different size of the ordered into the S structure adsorbed islands. The inset shows the changes of the potential energy versus temperature for the same system. The vertical dashed line marks the location of the order-disorder transition.

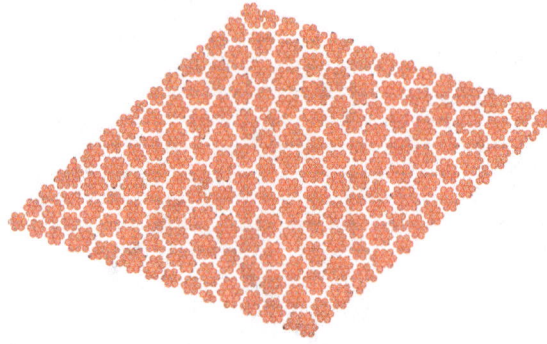

Figure 10. (Color online) A snapshot of the configuration obtained for the system with $\sigma^* = 0.8$, $R_c^* = 1.05$ and $f^* = 0.16$ at $T^* = 0.0024$.

for example when $f^* = 0.72$, and 0.8 the behavior of the heat capacity and of the potential energy is different than in the previously discussed case of $f^* = 0.4$, as well as for other values of f^* up to 0.64. Both quantities exhibit discontinuities at the transition temperature (see figure 9) without showing any systematic finite size effects.

For intermediate values of Δf^* between about 0.64 and 1.6, the transition is only weakly first-order and exhibits the behavior already described while discussing the system with $f^* = 0.4$.

For sufficiently large Δf^*, i.e., for sufficiently small f^*, the structure of adsorbed layer changes, since adatoms enjoy much more freedom to displace from the surface lattice sites. Consequently, the attractive pair potential wins over the harmonic forces and enhances the tendency towards clustering. Hence, adatoms group into lager clusters than those in the S structure. This is well illustrated by the snapshot given in figure 10, recorded at $T^* = 0.0024$ for the system with $f^* = 0.16$ and consisting of 1792 atoms. One can see that these clusters may contain a variable number of atoms and be of different shape. We have studied this system using R- and S-cells with periodic boundary conditions applied. The heat capacity and potential energy curves obtained for systems of different size have demonstrated (see figure 11) that the disordering occurs over a rather narrow temperature range, though the transition region is considerably smeared and the heat capacities obtained for systems of rather large size do not reach such high values as observed in the systems ordering into the S structure. Nonetheless, the heat capacity maxima

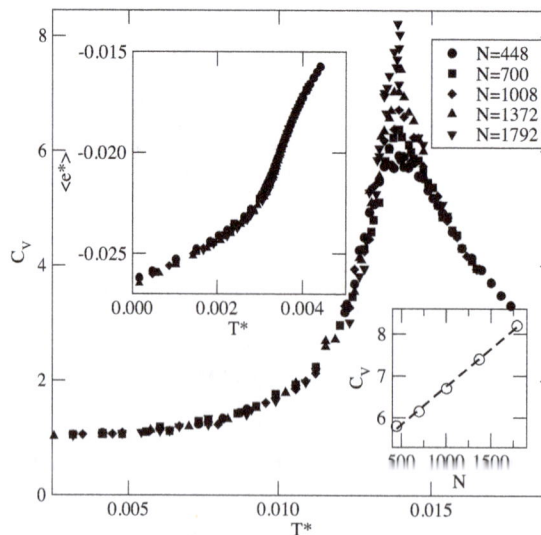

Figure 11. The temperature changes of the heat capacity (main figure) and of the potential energy (the upper inset) for the system with $\sigma^* = 0.8$, $R_c^* = 1.05$ and $f^* = 0.16$. The lower inset shows the changes of the heat capacity maximum with the system size.

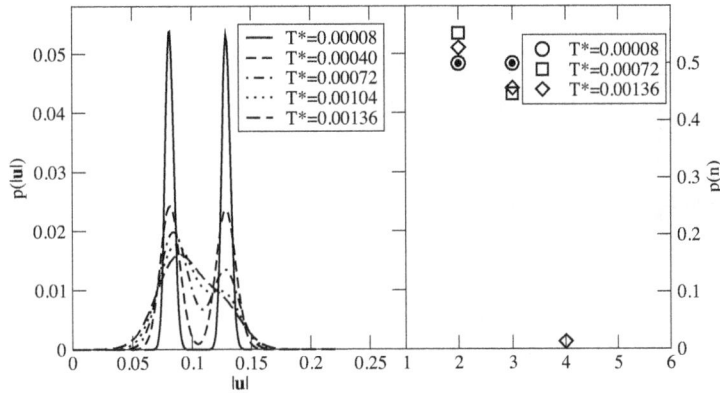

Figure 12. The probability distribution functions $p(n)$ and $p(|\mathbf{u}|)$ for the system with $\sigma^* = 0.8$, $R_c^* = 1.05$ and $f^* = 0.96$ at different temperatures obtained for the adsorbed island containing 1372 atoms. The solid circles in the left-hand panel show the probabilities of an atom to have 2 and 3 nearest neighbors in the perfect R structure.

increases linearly with N, suggesting that the system undergoes a weakly first-order phase transition. The transition occurs between the structure consisting of a large number of small clusters and the disordered state with more-or-less uniform distribution of atoms.

As soon as the value of the force constant becomes higher than f_{S-R}^*, the systems are expected to order into the R structure at low temperatures. The simulation has demonstrated that this does occur. Figure 12 presents the examples of distribution functions $p_d(|\mathbf{u}|)$ and $p_{NN}(n)$ obtained at different temperatures for the system with $f^* = 0.96$. It is well seen that only at very low temperatures does the ordering into the R structure occur. Upon an increase of temperature, the bimodal distribution $p_d(|\mathbf{u}|)$ gets distorted and then disappears at all. This is accompanied by the changes in the distribution $p_{NN}(n)$. In particular, the probability for an atom to have two nearest neighbors increases while the probability of having three nearest neighbors decreases. This indicates that triangular clusters made of three atoms appear at elevated temperatures. This has been confirmed by the inspection of snapshots, which demonstrated the formation of isolated triangles, as well as a loss of alignment of rhombic clusters. In particular, the triangles have been observed to appear close to the regions containing differently oriented rhombic clusters. Upon an increase of temperature, the degree of disordering also gradually increases. The heat capacity and potential energy curves (not shown here) have clearly demonstrated that the R structure disorders continuously and does not undergo any order-disorder phase transition. In particular, neither the potential energy nor the heat capacity show any finite size effects. In a perfectly ordered R phase, all rhombic clusters made of four atoms each are aligned as shown in figure 2. Our simulation has shown that this alignment is destroyed as soon as the temperature is raised above 0. The energy cost associated with changing the orientations of rhombic clusters and the formation of isolated triangles is very low, while the entropic effects are very large and destroy the alignment completely.

Qualitatively similar results have been obtained for the systems characterized by the values of f^* higher than f_{R-T}^*, i.e., when the ordering into the T structure occurs in the ground state. Similarly to the R structure, the disordering of the T structure gradually occurs and is accompanied by the loss of alignment of triangular clusters and by the appearance of rhombic clusters at nonzero temperatures. In fact, already the simulation performed at a very low temperature of $T^* = 0.00008$ and using the starting configuration being a perfect T structure has demonstrated a certain degree of disordering. Figure 13 shows the probability distributions $p_d(|\mathbf{u}|)$ and $p_{NN}(n)$ recorded at very low temperatures that demonstrate the above mentioned partial disordering. In particular, the distribution $p_d(|\mathbf{u}|)$ exhibits a small second maximum at large displacements and the distribution $p_{NN}(n)$ shows a non-zero probability of an atom to have three nearest neighbors already at $T^* = 0.00008$.

We have also performed calculations for the systems with higher potential cutoff, with $R_c^* = 1.14$, as well as for the systems with different size of adsorbed atoms, with $\sigma^* = 0.85$ and 0.7, but apart from the obvious quantitative differences resulting from the changes in the pair potential properties, the results

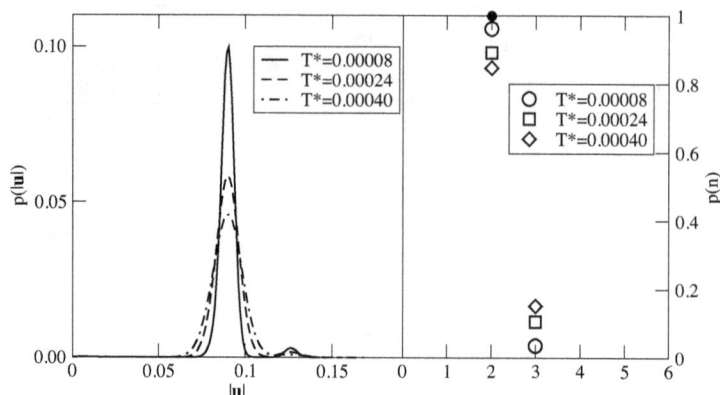

Figure 13. The probability distribution functions $p(n)$ and $p(|\mathbf{u}|)$ for the system with $\sigma^* = 0.8$, $R_c^* = 1.05$ and $f^* = 1.12$ at different temperatures obtained for the adsorbed island containing 1372 atoms. The solid circles in the left-hand panel show the probabilities of an atom to have 2 nearest neighbors in the perfect T structure.

have been qualitatively the same as for the above discussed systems.

As already discussed in section 3, the corrugation potential is not expected to lead to important qualitative changes in the behavior of the systems considered here. Indeed, our calculations carried out for several systems have shown that the increase of v_0^* is somehow equivalent to the decrease of f^*. It has been already shown that for $v_0^* = 0.0$ and the values of f^* corresponding to the stability region of the S structure, a gradual decrease of f^* leads to an increase of the order-disorder transition temperature (cf. figure 8) and the stability of the S structure lowers only for sufficiently low values of f^*. Thus, one expects that for a fixed value of f^*, an increase of the corrugation potential amplitude (v_0^*) should also lead to a gradual increase of the transition temperature. The calculations carried out for finite clusters consisting of 700 atoms with $\sigma^* = 0.8$, $R_c^* = 1.05$, different values of f^* between 0.4 and 0.96 and for several values of v_0^* have demonstrated a gradual increase of the transition temperature over a certain range of v_0 (see

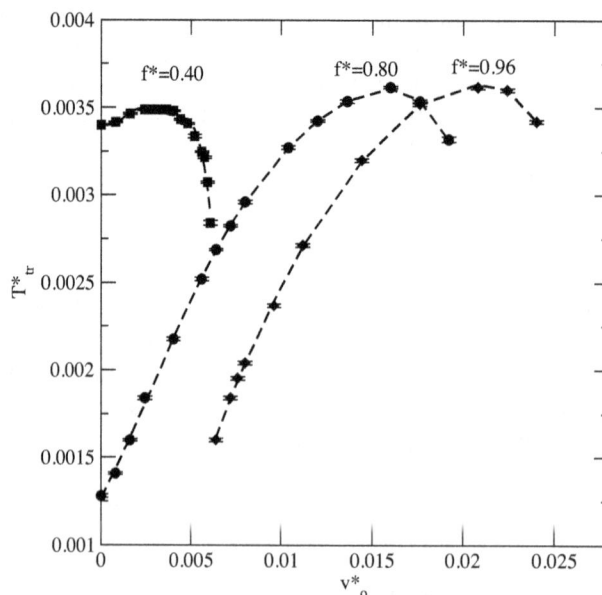

Figure 14. The order-disorder transition temperature of the S phase versus the amplitude of the corrugation potential obtained for the systems with $R_c^* = 1.05$, $\sigma = 0.8$ and different values of f^* (given in the figure).

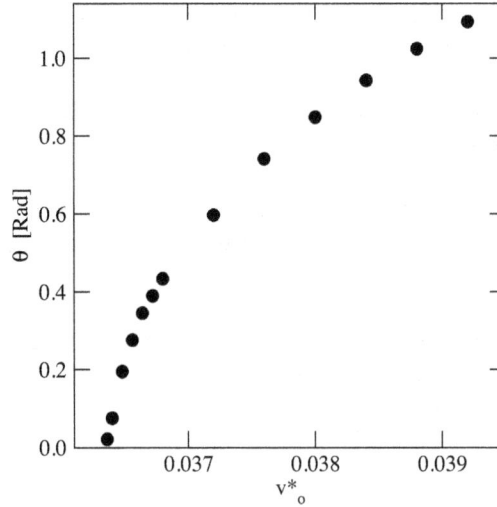

Figure 15. The rotation angle of a single 7-atom cluster versus the amplitude of the surface potential in the ground state of the system with $R_c^* = 1.05$, $\sigma = 0.8$ and $f^* = 0.4$.

figure 14). However, when the amplitude v_0^* becomes high enough, the transition temperature decreases rather sharply. This behavior is quite similar to the one observed for the systems without the corrugation potential and sufficiently low values of F^*.

It is interesting to note that in the case of $f^* = 0.96$, for which a uniform system ($v_0^* = 0$) orders into the R structure in the ground state, the S structure appears to be stable when v_0^* exceeds a certain threshold value (marked by a vertical dashed line in figure 14). This clearly shows that an increase of v_0^* is equivalent to the decrease of f^*.

Taking into account the properties of the corrugation potential generated over the (111) face of an fcc crystal, one expects the 7-atom clusters constituting the S structure to be slightly rotated around the central atom. This rotation should result from the fact that the displacement vectors in the uniform case are not directed towards the surface potential minima, but rather towards the saddle points. The rotation of the entire cluster does not cost any energy of the pair interaction, but affects the harmonic and surface interactions. The harmonic interaction hinders the cluster rotation, so that nonzero rotational angles may occur only when the energy gain due to the corrugation potential is high enough. The ground state calculations for the systems characterized by rather low $f^* = 0.4$ have shown, however, that even for the amplitude of the surface potential well above the upper limit of the S structure stability and equal to about 0.0062, the cluster rotation does not occur. The rotation of a single 7-atom cluster has been found to be present only when v_0^* exceeds the value of about 0.0364 (see figure 15). In the case of large systems consisting of many clusters, the clusters consisting of more than 7 atoms begin to develop already for v_0^* larger than about 0.0062.

6. Final remarks

In this paper, we have discussed the ordering appearing in dense chemisorbed monolayer films formed on the (111) fcc lattice. It has been demonstrated that when the pair potential is short ranged and the bonds between the chemisorbed atoms and the surface atoms are not rigid, the film may form different ordered phases. The ordering strongly depends upon the elasticity of the bonds. When the bonds are very stiff, the film orders into a simple (1×1) phase. Upon a decrease of the bond stiffness, the adatoms exhibit a gradually increasing tendency to be displaced from surface sites. Since the pair potential is attractive at small distances, the adatoms tend to form small clusters. In the ground state, we have found the ordered states in which such clusters consist of three (T), four (R) and seven (S) atoms (R) (cf. figure 2).

The structures T and R have been demonstrated to be stable only in the ground state. At finite temperatures, large entropic effects destroy the orderings. On the other hand, the structure S has been found to be stable at finite temperatures. Our Monte Carlo simulation results, supported by finite-size scaling analysis, have shown that the S phase is disordered via the first-order transition. The transition temperature depends on the bond elasticity.

For sufficiently low values of the harmonic potential force constant, the adatoms enjoy a rather large freedom to displace from adsorption sites and tend to form still larger clusters.

Although the model considered here is very simple, nevertheless it exhibits interesting physics, novel phase behavior and new types of ordering. In this work, we have considered only the case of a fully filled triangular lattice, but the model can be also used to study the orderings on the surfaces of different symmetry of adsorption sites. It can be also readily extended to take into account three body forces [34] and orientation-dependent interactions [50].

Monte Carlo methods can be also used to study the adsorption processes and possible different orderings appearing for lower coverages. This requires simulations in the grand canonical ensemble, commonly used in the studies of adsorption phenomena [14].

References

1. Springer Handbook of Nanotechnology, 3rd Edn., Bhushan B. (Ed.), Springer, Heidelberg, 2010; doi:10.1007/978-3-642-02525-9.
2. Einax M., Dieterich W., Maass P., Rev. Mod. Phys., 2013, **85**, 921; doi:10.1103/RevModPhys.85.921.
3. Tegenkamp C., J. Phys.: Condens. Matter, 2009, **21**, 013002; doi:10.1088/0953-8984/21/1/013002.
4. Equilibrium Structure and Properties of Surfaces and Interfaces, NATO ASI Series, Series B, Vol. 300, Gonis A., Stock G.M. (Eds.), Plenum Press, New York, 1992; doi:10.1007/978-1-4615-3394-8.
5. Custance O., Gómez-Rodríguez J.M., Baró A.M., Juré L., Mallet P., Veuillen J.-Y., Surf. Sci., 2001, **482–485**, 1399; doi:10.1016/S0039-6028(01)00774-9.
6. Stepanovsky S., Yakes M., Yeh V., Hupalo M., Tringides M.C., Surf. Sci., 2006, **600**, 1417; doi:10.1016/j.susc.2005.12.041.
7. Hwang I.-S., Chang S.-H., Fang C.-K., Chen L.-J., Tsong T.T., Phys. Rev. Lett., 2004, **93**, 106101; doi:10.1103/PhysRevLett.93.106101.
8. Chen Q., Liu J., Zhou X., Shang J., Zhang Y., Shao X., Wang Y., Li J., Chen W., Xu G., Wu K., J. Phys. Chem. C, 2015, **119**, 8626; doi:10.1021/jp5117432.
9. Ching W.Y., Huber D.L., Wang G.-C., Lagally M.G., Surf. Sci., 1978, 77, 550; doi:10.1016/0039-6028(78)90140-1.
10. Williams E.D., Cunningham S.L., Weinberg W.H., J. Chem. Phys., 1978, **68**, 4688; doi:10.1063/1.435579.
11. Ordering in Two-Dimensions, Sinha S.K. (Ed.), North-Holland, Amsterdam, 1980.
12. Phase Transitions in Surface Films, NATO ASI Series, Series B, Vol. 51, Dash J.G., Ruvalds R. (Eds.), Plenum, New York, 1980; doi:10.1007/978-1-4613-3057-8.
13. Bak P., Rep. Prog. Phys., 1982, **45**, 587; doi:10.1088/0034-4885/45/6/001.
14. Patrykiejew A., Sokołowski S., Binder K., Surf. Sci. Rep., 2000, **37**, 207; doi:10.1016/S0167-5729(99)00011-4.
15. Chakarova R., Oner D.E., Zoric I., Kasemo B., Surf. Sci., 2001, **472**, 63; doi:10.1016/S0039-6028(00)00923-7.
16. Roelofs L.D., Cortan A.R., Einstein T.L., Park R.L., J. Vac. Sci. Technol., 1981, **18**, 492; doi:10.1116/1.570774.
17. Einstein T.L., Langmuir, 1991, 7, 2520; doi:10.1021/la00059a021.
18. Tarasenko A.A., Nieto F., Pereyra V., Uebing C., Surf. Sci., 1999, **441**, 329; doi:10.1016/S0039-6028(99)00798-0.
19. Kohn W., Sham L., Phys. Rev., 1965, **140**, A1133; doi:10.1103/PhysRev.140.A1133.
20. Ying S.C., Smith J.R., Kohn W., Phys. Rev. B, 1975, **11**, 1483; doi:10.1103/PhysRevB.11.1483.
21. Su Z., Lu X., Zhang Q., Chem. Phys. Lett., 2003, **375**, 106; doi:10.1016/S0009-2614(03)00841-8.
22. Kresse G., Hafner J., Phys. Rev. B, 1993, **47**, 558; doi:10.1103/PhysRevB.47.558.
23. Kresse G., Hafner J., Phys. Rev. B, 1994, **49**, 14251; doi:10.1103/PhysRevB.49.14251.
24. Car R., Parrinello M., Phys. Rev. Lett., 1985, **55**, 2471; doi:10.1103/PhysRevLett.55.2471.
25. Izvekov S., Philpott M.R., Eglitis R.I., J. Electrochem. Soc., 2000, **147**, 2273; doi:10.1149/1.1393520.
26. Gómez L., Diep H.T., Phys. Rev. Lett., 1995, **74**, 1807; doi:10.1103/PhysRevLett.74.1807.
27. Pinto O.A., López de Mishima B., Dávila M., Ramírez Pastor A.J., Leiva E.P.M., Oviedo O.A., Phys. Rev. E, 2013, **88**, 062407; doi:10.1103/PhysRevE.88.062407.
28. Steele W.A., The Interaction of Gases with Solid Surfaces, Pergamon Press, Oxford, 1974.
29. Bruch L.W., Cole M.W., Zaremba E., Physical Adsorption: Forces and Phenomena, Clarendon Press, Oxford, 1997.
30. Daw M.S., Baskes M.I., Phys. Rev. B, 1984, **29**, 6443; doi:10.1103/PhysRevB.29.6443.
31. Foiles S.M., Phys. Rev. B, 1985, **32**, 3409; doi:10.1103/PhysRevB.32.3409.
32. Johnson R.A., Phys. Rev. B, 1988, **37**, 3924; doi:10.1103/PhysRevB.37.3924.

33. Lee B.J., Baskes M.I., Phys. Rev. B, 2000, **62**, 8564; doi:10.1103/PhysRevB.62.8564.

34. Stillinger F.H., Weber T.A., Phys. Rev. B, 1985, **31**, 5262; doi:10.1103/PhysRevB.31.5262.

35. Ding K., Andersen H.C., Phys. Rev. B, 1986, **34**, 6987; doi:10.1103/PhysRevB.34.6987.

36. Tersoff J., Phys. Rev. Lett., 1988, **61**, 2879; doi:10.1103/PhysRevLett.61.2879.

37. Tersoff J., Phys. Rev. B, 1988, **37**, 6991; doi:10.1103/PhysRevB.37.6991.

38. Brenner W.D., Phys. Rev. B, 1990, **42**, 9458; doi:10.1103/PhysRevB.42.9458.

39. Steele W.A., Surf. Sci., 1973, **36**, 317; doi:10.1016/0039-6028(73)90264-1.

40. Dünweg B., Landau D.P., Phys. Rev. B, 1993, **48**, 14182; doi:10.1103/PhysRevB.48.14182.

41. Tavazza F., Landau D.P., Adler J., Phys. Rev. B, 2004, **70**, 184103; doi:10.1103/PhysRevB.70.184103.

42. Cannavacciuolo L., Landau D.P., Phys. Rev. B, 2005, **71**, 134104; doi:10.1103/PhysRevB.71.134104.

43. Zhu X., Tavazza F., Landau D.P., Phys. Rev. B, 2005, **72**, 104102; doi:10.1103/PhysRevB.72.104102.

44. Presber M., Dünweg B., Landau D.P., Phys. Rev. E, 1998, **58**, 2616; doi:10.1103/PhysRevE.58.2616.

45. Shokef Y., Souslov A., Lubensky T.C., Proc. Natl. Acad. Sci. USA, 2011, **108**, 11804; doi:10.1073/pnas.1014915108.

46. Cudazzo P., Profeta G., Continenza A., Surf. Sci., 2008, **602**, 747; doi:10.1016/j.susc.2007.12.001.

47. Landau D.P., Binder K., A Guide to Monte Carlo Simulation in Statistical Physics, Cambridge University Press, Cambridge, 2000.

48. Finite Size Scaling and Numerical Simulation of Statistical Systems, Privman V. (Ed.), World Scientific, Singapore, 1990.

49. Binder K., Rep. Prog. Phys., 1987, **50**, 783; doi:10.1088/0034-4885/50/7/001.

50. Dongare A.M., Neurock M., Zhigilei V., Phys. Rev. B, 2009, **80**, 184106; doi:10.1103/PhysRevB.80.184106.

Computational studies on the behaviour of anionic and nonionic surfactants at the SiO$_2$ (silicon dioxide)/water interface

E. Núñez-Rojas[1], H. Dominguez[2]*

[1] Departamento de Química, Universidad Autónoma Metropolitana-Iztapalapa, México, D.F. 09340

[2] Instituto de Investigaciones en Materiales, Universidad Nacional Autónoma de México, México, D.F. 04510

Molecular dynamics simulations to study the behaviour of anionic (Sodium Dodecylsulfate, SDS) and nonionic (Monooleate of Sorbitan, SPAN80) surfactants close to a SiO$_2$ (silicon dioxide) surface were carried out. Simulations showed that a water layer was first adsorbed on the surface and then the surfactants were attached on that layer. Moreover, it was observed that water behaviour close to the surface influenced the surfactant adsorption since a semi-spherical micelle was formed on the SiO$_2$ surface with SDS molecules whereas a cylindrical micelle was formed with SPAN80 molecules. Adsorption of the micelles was conducted in terms of structural properties (density profiles and angular distributions) and dynamical behaviour (diffusion coefficients) of the systems. Finally, it was also shown that some water molecules moved inside the solid surface and located at specific sites of the solid surface.

Key words: *computer simulations, SDS surfactant, SPAN80 surfactant, adsorption, Cristobalite*

1. Introduction

Adsorption of surfactant molecules at solid-liquid interfaces has been investigated for years not only for its relevance in science but also for its numerous industrial applications, such as detergency, crude oil refining, treatment of waste water, adsorption on activated charcoal and even in pharmaceutical preparations [1–3].

In particular, self-assembly of surfactant molecules on solid surfaces has shown different issues from those observed at liquid/vapor and at liquid/liquid interfaces. For instance, it has been observed that interactions between hydrophobic tails, repulsions between headgroups and interactions between surfactant molecules with solid surface [4, 5] could change the isotherms at the critic micellar concentration (CMC). Therefore, studies of surfactant aggregation will help us to obtain more physical insights of self-assembly phenomena [6, 7].

From the experimental point of view, the surfactant adsorption on surfaces has been studied by different techniques, such as streaming potential methods [8], calorimetry [9], neutron reflection [10], ellipsometry [11], fluorescence spectroscopy [12] and atomic force microscopy (AFM) [13]. In fact, AFM has proved to be a reliable technique to obtain information on the topology of surfactant aggregation since it allows us to observe how surfactants are formed on surfaces. For instance, CTAB arrays in parallel stripes on a graphite surface [14], SDS forms hemimicelles on a rough gold surface [15] and similar morphologies have been seen for other surfactants on hydrophobic surfaces [16–18].

On the other hand, computer simulations have been very useful to study such complex systems. For instance, Monte Carlo simulations have been used to provide information on structural transitions of sur-

*Corresponding author. E-mail: hectordc@unam.mx. Present address: Department of Physics and Astronomy, University of British Columbia, Vancouver, British Columbia, Canada. On sabbatical leave.

factant aggregation [19] while molecular dynamics simulations have been used to investigate aggregation at atomistic scales [20, 21]. In previous papers we have reported the surfactant behaviour on different surfaces [21–24]. It has been observed how graphite surfaces impose orientational order on the surfactant tails [21, 24] and how different solid faces of a titanium dioxide produced different aggregates on the surface [22, 23]. In the present work we are interested to extend the studies of surfactant aggregation on a hydrophilic SiO_2 (silicon dioxide) surface in order to compare the behaviour of surfactant molecules on different solid substrates.

2. Computational method and model

Molecular dynamics simulations of anionic and nonionic surfactant molecules at silicon dioxide surface (SiO_2), in its Cristobalite form, were carried out for the present study. The surface was constructed using an atomistic model with a surface orientation (001) (figure 1). The parameters used for a solid surface were taken from reference [25].

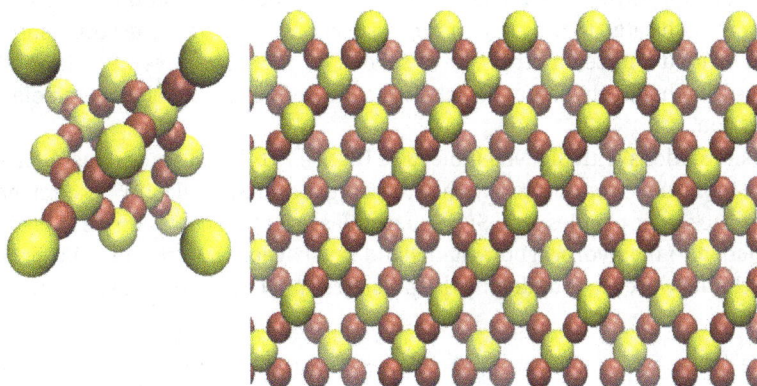

Figure 1. (Color online) The unit crystal cell and a snapshot of the solid surface. Oxygen atoms are in red and silicon atoms are in yellow.

For the anionic surfactant molecule, sodium dodecyl sulfate (SDS), a model of 12 united carbon atoms attached to a headgroup (SO_4) was simulated by using a force field already reported in the literature [21]. For the nonionic surfactant molecule, Monooleate of Sorbitan (SPAN80), there was used, a model with 17 united carbon atoms, three OH^- and one ester groups in the head group. The force field reported in reference [23] was used for this molecule.

In the case of simulations with the anionic molecules, the initial configuration was prepared from a monolayer of 36 molecules in all-trans-configuration with the headgroups initially pointed to the solid surface. Then, 2535 water molecules were added (using the SPC model [26]) to the system and 36 sodium cations (Na^+). In the case of simulations with nonionic molecules, 25 molecules were used with the same number of water molecules.

A simulation box having dimensions $X = Y = 43.7019$ and $Z = 150$ Å was used with the usual periodic boundary conditions. The Z-dimension of the box was long enough to prevent the formation of a second water/solid interface due to the periodicity of the system. Instead, a liquid/vapor interface was present at one end of the box ($z > 0$) whereas at the other end of the box ($z < 0$) beyond the solid there was an empty space. All simulations were carried out in the NVT ensemble with a time step of 0.002 ps using DL-POLY package [27]. Bond lengths were constrained using SHAKE algorithm with a tolerance of 10^{-4}, and the temperature was controlled using the Hoover-Nose thermostat having a relaxation time of 0.2 ps [28] at $T = 298$ K. Long-range electrostatic interactions were handled using the Particle Mesh Ewald method, and the Van der Waals interactions were cut off at 10 Å. Finally, the simulations were run up to 40 ns and configurational energy was monitored as a function of time in order to determine the moment the systems have reached equilibrium. Then, the last 2 ns were collected for analysis.

3. Results

In this section we present calculations of the surfactant molecules at the SiO_2 surface. Studies on the behaviour of the surfactant molecules and how they aggregate at the liquid/solid interface are discussed.

3.1. Density profiles

In order to determine where the surfactant molecules arrayed in the system, mass Z-dependent density profiles for the headgroups and the tails were calculated, i.e., normal to the liquid/solid interface.

From figure 2 we observed that water molecules (dotted line) were not only adsorbed but also absorbed by the solid surface which is located at a position of $Z = -23$ Å in the figure. In fact, the first water profile peak (to the right of the surface) indicated strong adsorption, i.e., a water layer on the surface. The other water peaks (to the left of the surface) suggested that few particles were inside the solid surface. The presence of water molecules inside SiO_2 surfaces has been also observed in real experiments [29, 30]. On the other hand, SDS density profiles showed a strong first peak for the polar group (≈ 4 Å from the surface in figure 2 (a) suggesting that the surfactant was well adsorbed on the surface. Moreover, it was noted that the peak was located to the right of the adsorbed water layer. The result showed that the surfactant molecules formed a micellar structure adsorbed on the water layer on the solid surface. It was also possible to observe that the hydrocarbon chains (solid line) were sited between polar-groups (dashed line) along Z direction.

In the case of simulations with SPAN80 molecules [figure 2 (b)], the water density profile also showed that some of those molecules went deep into the substrate. Besides, unlike SDS system, here strong peaks were observed for the water profiles. The SPAN80 profiles are also shown, the first peak (≈ 5 Å from the surface) corresponded to the hydrocarbon chains and it was very close to the polar group, nevertheless the hydrocarbon chains were surrounded by the polar groups. These profiles suggested the formation of

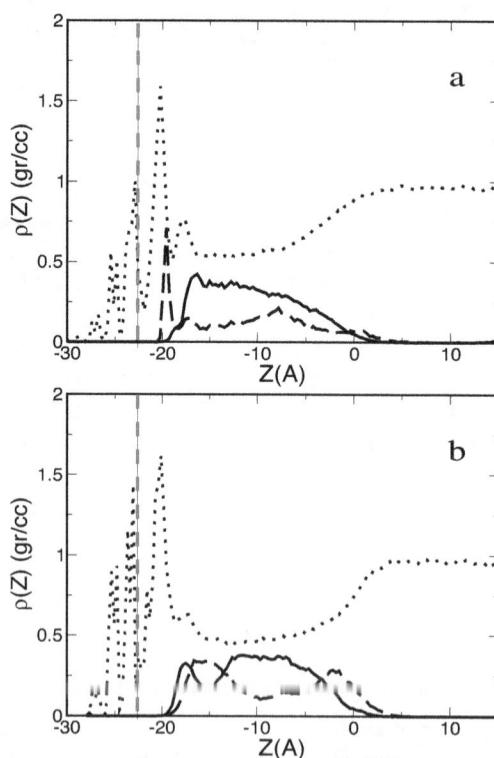

Figure 2. Density profiles calculated along the Z-direction of the SiO_2 solid surface. (a) SDS surfactant and (b) SPAN80 surfactant. Dotted lines represent water, dashed lines show a surfactant headgroup and continuous lines show surfactant tails. The solid surface position is indicated by the red line.

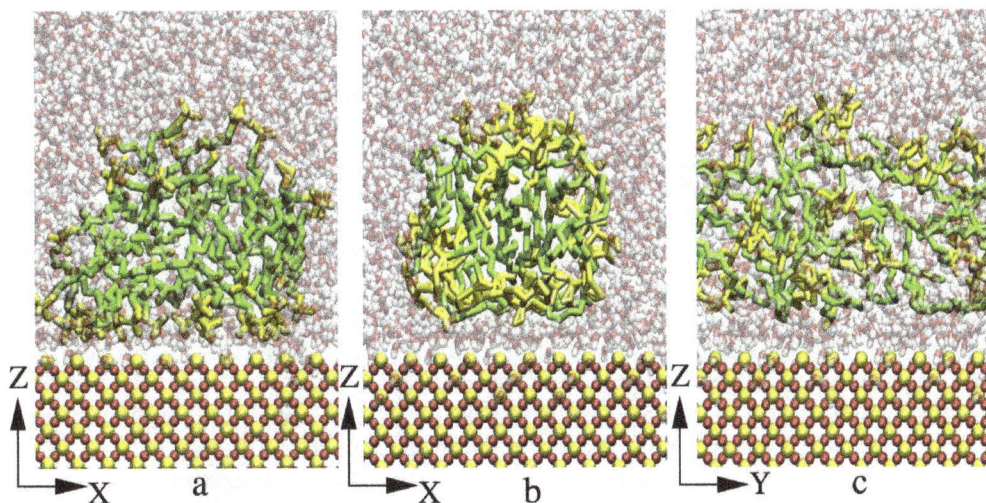

Figure 3. (Color online) Snapshots for SDS (a) and for SPAN80 (b and c) micelles on a SiO$_2$ surface. Spherical and cylindrical shapes are depicted for the SDS and SPAN80, respectively. Tail groups are in green and head groups in yellow.

a micelle on the surface.

In figure 3, snapshots of the final SDS [figure 3 (a)] and SPAN80 [figures 3 (b) and 3 (c)] are shown. In those figures it was possible to observe the micellar structure mentioned above. In the case of the SDS, the micelle structure had a spherical-like shape (snapshot in $X - Z$ and $Y - Z$ looked alike) whereas for SPAN80 the micelle had a cylindrical-like structure [see figures 3 (b) and 3 (c)].

In order to verify the structure of the micelles on the surfaces, particle density profiles in the other

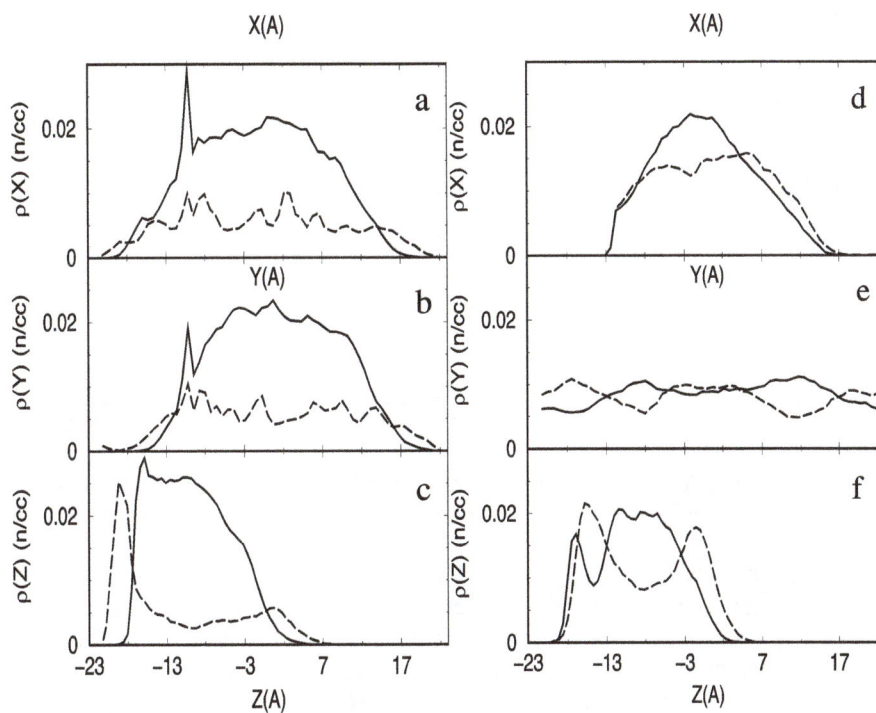

Figure 4. Particle density profiles of the surfactant headgroups (dashed lines) and tails (solid lines) for the SDS (left-hand) and SPAN80 (right-hand) systems. The panels indicate the profiles in the X, Y and Z axis.

two directions, X and Y, were calculated. In figures 4 (a) and 4 (b), head groups density profiles (dashed line) and tail groups (solid line) of SDS molecules are shown. There were observed similarities between the two particle profiles, $\rho(X)$ and $\rho(Y)$, where it was noted that polar groups surrounded the tail groups. In both tail group profiles, a small sharp peak can be seen. In this case, SDS molecules are well adsorbed on the SiO_2 substrate with some surfactants attached on the surface by the head group [see figure 3 (a)]. Then, there is a possibility that a SDS molecule could remain anchored with its tail moving in one region only. This could explain the small sharp peak observed in both profiles, i.e., there are a bit more tail groups in that region. On the other hand, from the Z-particle density profile in figure 4 (c), a big head group peak could be seen close to the solid surface (see also figure 2), i.e., there was an excess of head groups attached to the surface. Moreover, the $\rho(Z)$ indicated that the size of the micelle in the Z-direction was smaller than in the $X-Y$ directions. Therefore, the structure can be described as a sphere deformed along the perpendicular direction of the solid surface.

This calculation was also carried out for the SPAN80 molecules [figures 4 (d)–4 (f)]. In figure 4 (d), the density profile along X-axis is shown where a molecular aggregation is observed. In figure 4 (e), the profiles indicated a uniform distribution of the molecules along the Y-axis whereas in figure 4 (f) it was again possible to observe aggregation of the molecules next to the surface. Here, it can be seen that headgroups partially surrounded the tail groups. Then, these profiles revealed that the molecules structured themselves as a cylindrical micelle along the Y direction. These results suggested that the effect of the solid surface in the micelle formation was minimum.

3.2. Water orientation at the SiO_2 surface

As it was observed above, water density profiles showed strong peaks close to the SiO_2 surface suggesting that those molecules might have some structure close to the solid. Therefore, studies of how water molecules were oriented in the solid were carried out. The analyses were conducted over the molecules in the adsorbed layers only (defined by the peaks of figure 2).

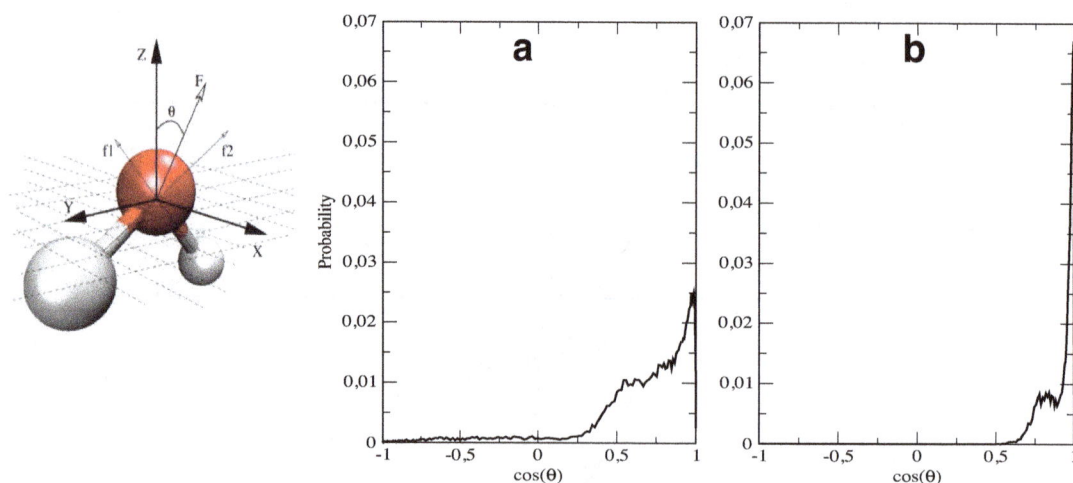

Figure 5. (Color online) Angular probability of the water dipole vector with the vector normal to the interface. (a) For the SDS and (b) for the SPAN80 surfactant systems.

In figure 5, angular distributions of water molecules are shown for the SDS and SPAN80 systems. The orientation of the water molecules was measured as the angle between the bisector vector of the OH bonds and the vector normal to the interface (see figure 5). In both systems, SDS and SPAN80, there was observed a privileged orientation of the water molecules, i.e., they were nearly perpendicular to the surface. In fact, there was noted a larger number of water molecules pointing perpendicular to the surface for the SPAN80 system [high peak in figure 5 (b)] than for the SDS system (small peak). Moreover, the water angle distribution presented a small peak when water interacted with SPAN80 [figure 5 (b)] and a wider distribution when water interacted with SDS.

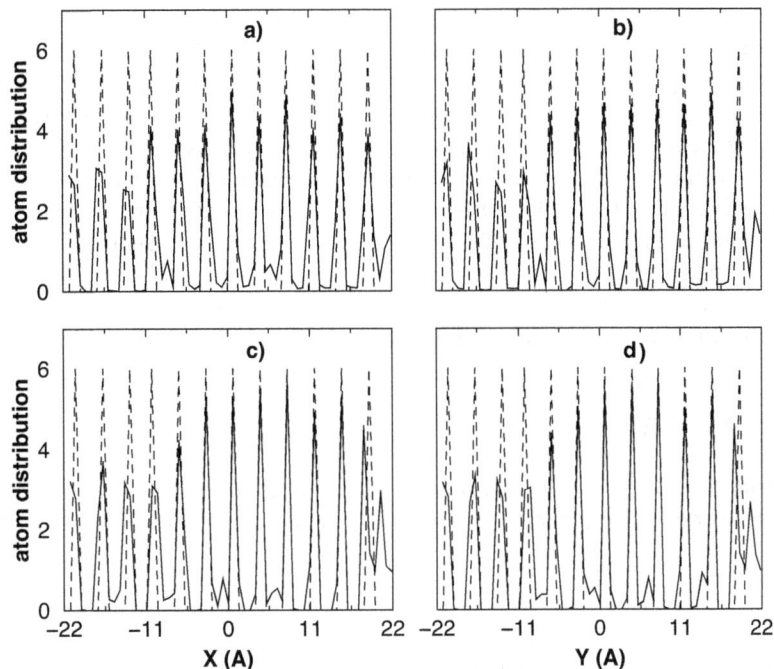

Figure 6. Distribution of oxygen atoms in water (solid lines) and silicon atoms of the SiO_2 surface (dashed lines). Panels (a) and (b) atom positions for the SDS surfactant in the X and Y axis. Panels (c) and (d) atom positions for the SPAN80 surfactant in the X and Y axis.

In figure 6, the distribution (in the first adsorbed water layer) of water molecules in the X and Y directions, i.e., in the $X - Y$ plane, is plotted. Figures 6 (a) and 6 (b) show the water oxygens positions in the solid surface for the SDS system. In the same figure 6, the positions of the silicon atoms of the SiO_2 surface are also shown in the same $X - Y$ directions. In all cases it was observed that water molecules (solid lines) were located only in specific sites. In fact, they were placed on the silicon atoms of the solid surface (indicated by the dashed lines in figure 6). Figures 6 (c) and 6 (d) are the same plots for the SPAN80 system. From the above results we noted that water oxygens were adsorbed above silicon sites.

3.3. Mobility of surfactant molecules

In order to describe the attachment of the surfactant molecules to the solid surface, the diffusion coefficient of the molecular aggregates was also calculated. The diffusion coefficients were calculated in each direction by measuring the square mean displacements of the surfactant atoms and using the Einstein relation. For the SPAN80 surfactant, the diffusions were 0.126×10^{-9} m^2/s, 0.095×10^{-9} m^2/s and 0.085×10^{-9} m^2/s in the X, Y and Z directions, respectively. For the SDS surfactant, the diffusions were 0.125×10^{-9} m^2/s, 0.135×10^{-9} m^2/s and 0.078×10^{-9} m^2/s in the X, Y and Z directions, respectively. In both systems, the diffusions in the perpendicular direction (Z-axis) was lower than in the plane. However, the values for the SPAN80 were higher than for the SDS surfactant, suggesting that the second surfactant had a higher mobility than the first one. It is worthy to mention that these diffusions were much lower than the diffusions found for the same molecules on a Titanium dioxide surface [22, 23].

4. Conclusions and discussion

A series of Molecular Dynamics simulations were carried out in order to describe the behaviour of two different surfactant molecules interacting with a silicon dioxide solid surface. In the case of the anionic surfactant (SDS), there was observed a spherical micelle formation on a layer of water molecules previously adsorbed on the solid. The micelle was described by density profiles and they showed a deformation of the micelle next to the adsorbed layer of water molecules. This deformation can be explained in terms of the SDS charged headgroups interactions with water molecules on the solid surface. On the other hand, the nonionic surfactant (SPAN80) did not show much influence by the solid surface. It formed a cilyndrical micelle next to the adsorbed water layer. In this case, there was observed a thicker water layer between the surfactant and the solid surface. The influence of the surfactant on the surface was characterized by water molecules in the surface. Dipole water orientation, in the solid surface, was more tilted for the SDS molecules than for the SPAN80 suggesting a stronger SDS-surface interaction and consequently more intensive adsorption of those molecules on the surface. Based on the previous results (diffusion coefficients), it was noted that both surfactants showed less affinity with the SiO_2 surface than with other surfaces such as graphite and titanium dioxide.

Acknowledgements

HD acknowledges support from DGAPA-UNAM-Mexico and Conacyt-Mexico for sabbatical scholarships. The authors acknowledge support from DGTIC-UNAM for the supercomputer facilities.

References

1. Ahmaruzzaman M., Sharma D.K., J. Colloid Interface Sci., 2005, **287**, 14; doi:10.1016/j.jcis.2005.01.075.
2. Bhatnagar A., Jain A.K., J. Colloid Interface Sci., 2005, **281**, 49; doi:10.1016/j.jcis.2004.08.076.
3. Polymers in Drug Delivery, Uchegbu I.F., Schtzlein A.G. (Eds.), CRC Press, Boca Raton, 2006.
4. Gaudin A.M., Fuerstenau D.W., Trans. Am. Inst. Min. Metall. Pet. Eng., 1955, **202**, 66.
5. Bhömer M.R., Koopal L.K., Langmuir, 1992, **8**, 2649; doi:10.1021/la00047a014.
6. Sammalkorpi M., Panagiotopoulos A.Z., Haataja M., J. Phys. Chem. B, 2008, **112**, 2915; doi:10.1021/jp077636y.
7. Mu G., Li X., J. Colloid Interface Sci., 2005, **289**, 184; doi:10.1016/j.jcis.2005.03.061.
8. Fuerstenau D.W., J. Phys. Chem., 1956, **60**, 981; doi:10.1021/j150541a039.
9. Király Z., Findenegg G.H., J. Phys. Chem. B, 1998, **102**, 1203; doi:10.1021/jp972218m.
10. Penfold J., Staples E.J., Tucker I., Thompson L.J., Langmuir, 1997, **13**, 6638; doi:10.1021/la970468o.
11. Tiberg F., Joensson B., Lindman B., Langmuir, 1994, **10**, 3714; doi:10.1021/la00022a053.
12. Chandar P., Somasundaran P., Turro N.J., J. Colloid Interface Sci., 1987, **117**, 31; doi:10.1016/0021-9797(87)90165-2.
13. Ducker W.A., Wanless E.J., Langmuir, 1996, **12**, 5915; doi:10.1021/la9605448.
14. Manne S., Cleveland J.P., Gaub H.E., Stucky G.D., Hansma P.K., Langmuir, 1994, **10**, 4409; doi:10.1021/la00024a003.
15. Schniepp H., Shum H., Saville D., Aksay I., J. Phys. Chem. B, 2007, **111**, 8708; doi:10.1021/jp073450n.
16. Patrick H.N., Warr G.G., Manne S., Aksay I., Langmuir, 1997, **13**, 4349; doi:10.1021/la9702547.
17. Jaschke M., Butt H.J., Gaub H.E., Manne S., Langmuir, 1997, **13**, 1381; doi:10.1021/la9607767.
18. Ducker W.A., Wanless E.J., J. Phys. Chem. B, 1996, **100**, 3207; doi:10.1021/jp952439x.
19. Zheng F., Zhang X., Wang W., Dong W., Langmuir, 2006, **22**, 11214; doi:10.1021/la0622424.
20. Shah K., Chiu P., Jain M., Fortes J., Moudgil B., Sinnott S., Langmuir, 2005, **21**, 5337; doi:10.1021/la047145u.
21. Dominguez H., J. Phys. Chem. B, 2007, **111**, 4054; doi:10.1021/jp067768b.
22. Núñez-Rojas E., Dominguez H., J. Colloid Interface Sci., 2011, **364**, 417; doi:10.1016/j.jcis.2011.08.069.
23. Núñez-Rojas E., Dominguez H., Rev. Mex. Fis., 2013, **59**, 530.
24. Aranda-Bravo C.G., Mendez-Bermudez J.G., Dominguez H., J. Mol. Liq., 2014, **200**, 465; doi:10.1016/j.molliq.2014.11.023.
25. Dominguez H., Gama Goicoechea A., Mendoza N., Alejandre J., J. Colloid Interface Sci., 2006, **297**, 370; doi:10.1016/j.jcis.2005.10.020.
26. Berendsen H.J.C., Grigera J.R., Straatsma T.P., J. Phys. Chem. B, 1987, **91**, 6269; doi:10.1021/j100308a038.
27. Forester T.R., Smith W., DL-POLY Package of Molecular Simulation, CCLRC, Daresbury Laboratory, Daresbury, Warrington, 1996.
28. Hoover W.G., Phys. Rev. A, 1985, **31**, 1695; doi:10.1103/PhysRevA.31.1695.
29. Elimbi A., Dika J.M., Djangang C.N., J. Mineral. Mat. Charac. Eng., 2014, **2**, 484; doi:10.4236/jmmce.2014.25049.
30. Keller W.D., Pickett E.E., Am. J. Sci., 1950, **248**, 2640273.

Adsorption from binary solutions on chemically bonded phases*

M. Borówko, T. Staszewski

Department for the Modeling of Physico-Chemical Processes, Maria Curie-Skłodowska University, 20-031 Lublin, Poland

We use density functional theory to investigate adsorption of liquid mixtures on solid surfaces modified with end-grafted chains. The chains are modelled as freely joined spheres. The fluid molecules are spherical. All spherical species interact via the Lennard-Jones (12-6) potential. The Lennard-Jones (9-3) potential describes interactions of solvent molecules with the substrate. We study the relative excess adsorption isotherms, the structure of surface layer and its composition. The impact of the following parameters on adsorption is discussed: the grafting density, the grafted chain length, interactions of solvent molecules with grafted chains and with the substrate, and the presence of active groups in grafted chains. The theoretical results are consistent with experimental observations.

Key words: *density functional theory, adsorption from solutions, polymer-tethered surfaces*

1. Introduction

A great deal of research has focused on adsorption from solutions on solid surfaces modified with end-grafted chains [1]. Understanding the adsorption equilibrium is of enormous importance for a variety of biological and technological processes. One of them is the reversed-phase liquid chromatography (RPLC). This technique is among the most popular methods for separation of sample components.

Various theoretical methods have been used to study the retention in chromatography with chemically bonded phases. Among these are lattice-based analytical theories [2–4], self-consistent field methods [5, 6], density functional theory [7, 8] and computer simulations [9–17]. The retention is driven by the distribution of solute molecules between the mobile phase and the stationary phase (alkyl chains tethered to silica surface). The process presents theoretical challenges due to its complexity, and thus numerous problems related to its mechanism are still unsettled.

The aim of the chromatographic analysis is to achieve the elution of all sample components in reasonable time and with a satisfactory selectivity of the separation. To optimize this process, one commonly uses a mixture of two solvents. Changing the composition of the mobile phase we change the system selectivity. Much effort has been directed toward the theoretical prediction of the solution retention as a function of the mobile phase composition [2, 18–20].

The retention strongly depends on adsorption of solvents at the stationary phase [2]. For this reason, numerous experimental data referring to adsorption of binary solvents commonly used in chromatography have been recently published [20–26]. However, in the literature one can find only a few theoretical articles connected with adsorption from solutions on chemically bonded phases. The bonded phases are usually treated as 'usual' adsorbents, and the theory of adsorption from solutions on solid surfaces is employed to interpret experimental data. The liquid part of a system is formally divided into the surface phase and the bulk phase. The location of the dividing surface is rather arbitrary and several methods

---

*This work is dedicated to Professor Stefan Sokołowski. We are deeply grateful to him for the longstanding and fruitful cooperation.

for its estimation have been considered [13, 20]. Gritti and Gouichon [20] have used the so-called bi-Langmuir equation for excess adsorption isotherms. They assumed that the adsorbent was composed of two patches, one representing the surface covered with grafted chains and the second corresponding to the bare surface of the solid. Their approach has been used to study adsorption on different bonded phases [20, 21, 24]. This phenomenological theory does not give any insight into the structure of the chain layer. Quite recently, the density functional theory has been used to study adsorption from solutions on the polymer-tethered surfaces [27–30]. In this model, the penetration of the solvent molecules into the chain layer is allowed. The densities of all species gradually change with the distance of the solid surface and tend to their bulk values.

In this work, we present the results of the density functional study of the competitive adsorption from binary solutions on chemically bonded phases. We show how the selected parameters affect the composition of the liquid inside the grafted chain layer. Hitherto, this problem has not been analyzed. The local changes in the mixture composition can influence the chromatographic separation. We also consider relative excess adsorption isotherms for the model systems investigated. We discuss the impact of such parameters as: the grafting density, the grafted chain length and interactions of solvents with grafted chains and with the substrate, the presence of active groups in the chains. We want to find general trends rather than to approximate experimental data for concrete systems. Our conclusions are consistent with the results of experiments and computer simulations found in the literature. Short grafted chains are used in popular stationary phases. Therefore, we concentrated on grafted oligomers.

The article is organized as follows. In the next section, we describe the model and the basic aspects of the theoretical approach. The results are presented and analyzed in section 3. Finally, we summarize the conclusions.

2. Model and theory

We study adsorption from a binary solution on a surface modified with end-grafted chains. We employ the computational method used in our previous papers concerning this problem [27–30]. Therefore, we discuss here only the most important aspects of the model. The method was originally proposed by Yu and Wu [31–33]. Numerous research groups used the density functional theory to study the systems involving either free [34–37] or grafted chains [38–42]. We treat the system as a ternary mixture in contact with an impenetrable wall. The mixture consists of tethered polymers (P) and molecules of solvents (labeled as 1 and 2). The grafted polymers (P) are chains of M freely jointed segments. The connectivity of a given chain is enforced by the bonding potential

$$\exp[-\beta V_B(\mathbf{R})] = \prod_{i=1}^{M-1} \delta(|\mathbf{r}_{i+1} - \mathbf{r}_i| - \sigma^{(P)})/4\pi(\sigma^{(P)})^2, \tag{1}$$

where $\mathbf{R}_k \equiv (\mathbf{r}_1, \mathbf{r}_2, \ldots, \mathbf{r}_M)$ is the vector specifying the positions of all segments, $\sigma^{(P)}$ is the segment diameter, the symbol δ denotes the Dirac function, $\sigma^{(P)}$ is the polymer segment diameter and $\beta^{-1} = k_B T$.

The first segment of the chain is bonded with the surface by the potential

$$\exp\left[-\beta v_{s1}^{(P)}(z)\right] = C\delta(z - \sigma^{(P)}/2), \tag{2}$$

where z is a distance from the surface and C is a constant. The surface-binding segments are located at the distance $z = \sigma^{(P)}/2$ from the wall. These segments cannot leave the surface but they can move within the xy-plane.

All of the remaining segments, $i = 2, 3, \ldots, M$ are neutral with respect to the surface and interact with the substrate via the hard wall potential

$$v^{(k)} = \begin{cases} \infty, & z < \sigma^{(P)}/2, \\ 0, & \text{otherwise,} \end{cases} \tag{3}$$

where $v^{(P)} = v_{si}^{(P)}$ ($i \geq 2$).

The solvent molecules are attracted by the surface, according to the Lennard-Jones (9–3) potential

$$v^{(k)} = 4\bar{\varepsilon}_s^{(k)} \left[(z_0^{(k)}/z)^9 - (z_0^{(k)}/z)^3 \right],$$ (4)

where $\bar{\varepsilon}_s^{(k)}$ characterizes the strength of interactions between the kth solvent and the wall ($k = 1, 2$) and $z_0 = \sigma^{(k)}/2$. We also consider the solvents neutral with respect to the substrate, interacting with the surface via the hard-wall potential [equation (3)].

The chain segments and fluid molecules interact via Lennard-Jones potential (12–6)

$$u^{(kl)} = \begin{cases} 4\bar{\varepsilon}^{(kl)} \left[(\sigma^{(kl)}/r)^{12} - (\sigma^{(kl)}/r)^6 \right], & r < r_{\text{cut}}^{(kl)}, \\ 0, & \text{otherwise}, \end{cases}$$ (5)

where $\bar{\varepsilon}^{(kl)}$ is the parameter characterizing interactions between species k and l, $\sigma^{(kl)} = 0.5(\sigma^{(k)} + \sigma^{(l)})$ for $k, l = 1, 2, \text{P}$; r is the distance between the interacting spheres, $r_{\text{cut}}^{(kl)}$ is the cutoff distance. In this work $r_{\text{cut}}^{(kl)} = 3\sigma^{(kl)}$.

We assume that the grafting density, $\rho_\text{P} = N_\text{P}/A_\text{s}$, is fixed where N_P is the number of grafted chains and A_s denotes the area of the surface.

The theory is constructed in terms of the local densities of spherical molecules, $\rho^{(k)}$ ($k = 1, 2$) and the local density of segments of the grafted chains

$$\rho_s^{(P)}(\mathbf{r}) = \sum_{i=1}^{M} \rho_{s,i}^{(P)}(\mathbf{r}) = \sum_{i=1}^{M} \int d\mathbf{R}\, \delta(\mathbf{r} - \mathbf{r}_i) \rho^{(P)}(\mathbf{R}),$$ (6)

where $\rho^{(P)}$ is the local density of the chains and $\rho_{s,i}^{(P)}$ is the density of i-th segments.

Using the procedure prosed by Yu and Wu [31–33] we calculate the density profiles of all components. As usually, the free-energy functional is expressed as the sum $F = F_{\text{id}} + F_{\text{hs}} + F_\text{c} + F_{\text{att}}$. The free energy of an ideal gas, F_{id}, is known exactly [43]. The excess free energy following from hard-sphere interactions, F_{hs}, is calculated from a modified version [32] of the fundamental measure theory of Rosenfeld [44]. The chain connectivity contribution, F_c, follows from the first-order perturbation theory of Wertheim [45]. A reader can find all necessary expressions in reference [46] [equations (7), (10) and (12), respectively]. The attractive interactions between spherical species are expressed using mean-field approximation

$$\begin{aligned} F_{\text{att}} &= \frac{1}{2} \sum_{k=\text{P},1,2} \int d\mathbf{r}_1 d\mathbf{r}_2 \rho_s^{(k)}(\mathbf{r}_1) \rho_s^{(k)}(\mathbf{r}_2) u_{\text{att}}^{(kk)}(\mathbf{r}_{12}) \\ &+ \sum_{\substack{k,l=\text{P},1,2; \\ k<l}} \int d\mathbf{r}_1 d\mathbf{r}_2 \rho_s^{(k)}(\mathbf{r}_1) \rho_s^{(l)}(\mathbf{r}_2) u_{\text{att}}^{(kl)}(\mathbf{r}_{12}), \end{aligned}$$ (7)

where $u_{\text{att}}^{(kl)}$ is the attractive part of Lennard-Jones potential following from the Weeks-Chandler-Anderson scheme [47]

$$u_{\text{att}}^{(kl)}(r) = \begin{cases} -\varepsilon^{(kl)}, & r < 2^{1/6}\sigma^{(kl)}, \\ u^{(kl)}(r), & r \geq 2^{1/6}\sigma^{(kl)}, \end{cases}$$ (8)

and $\rho_s^{(1)} = \rho^{(1)}$ and $\rho_s^{(2)} = \rho^{(2)}$.

The grafting density, ρ_P, is given by

$$\int_0^\infty dz\, \rho_{s,i}^{(P)}(z) = \rho_\text{P}.$$ (9)

The equilibrium density profiles are obtained by minimizing the thermodynamic potential

$$\begin{aligned} Y &= F + \int d\mathbf{R}\, \rho^{(P)}(\mathbf{R}) v^{(P)}(\mathbf{R}) \\ &+ \sum_{k=1,2} \int d\mathbf{r}_k \rho^{(k)}(\mathbf{r}_k)(v^{(k)}(\mathbf{r}_k) - \mu^{(k)}) \end{aligned}$$ (10)

under the constraint (9). In the above, $\mu^{(k)}$ denotes the chemical potential of the component k ($k = 1, 2$). In the considered model, the density distributions vary only with the distance from the surface, $\rho^{(k)}(\mathbf{r}) = \rho^{(k)}(z)$, $k = \mathrm{P}, 1, 2$. Minimization of the thermodynamic potential leads to a set of the Euler-Lagrange equations (equation (17) in reference [46]) which can be solved numerically. Knowing the equilibrium distributions of all components we can calculate different quantities characterizing the adsorption and the structure of the surface layer.

In the case of liquid mixtures, different species compete for room in the surface region, molecules of a given component are displaced by molecules of the other component. As a consequence, the composition of a solution considerably depends on the distance from the wall. To characterize the composition of the solution we define the local volume fraction of the kth component ($k = 1, 2$)

$$x^{(k)} = \frac{\rho^{(k)}(\sigma^{(k)})^3}{\rho^{(1)}(\sigma^{(1)})^3 + \rho^{(2)}(\sigma^{(2)})^3}, \qquad k = 1, 2. \tag{11}$$

Notice that $x^{(k)}$ is not the volume fraction in the whole system but only in the liquid.

For a competitive adsorption from a solution, the relative excess adsorption isotherms are usually used [20, 24, 48, 49]

$$N_k^{\mathrm{e}} = \int \mathrm{d}z \left[x^{(k)}(z) - x_b^{(k)} \right], \tag{12}$$

and $x_b^{(k)}$ is the volume fraction in the bulk mixture. Obviously, $N_1^{\mathrm{e}} = -N_2^{\mathrm{e}}$. The relative excess N_k^{e} can be directly compared with experimental data because it is proportional to the difference in the bulk solution composition after and before the adsorption $\Delta x = x_b^{(k)} - x_0^{(k)}$.

The theory involves numerous molecular parameters. The chain segments and solvent molecules can have different sizes. The bonded phase is characterized by the length of chains, M, and by the grafting density, ρ_{P}. Interactions between molecules of different components are characterized by the energy parameters $\varepsilon^{(kl)}$ ($k, l = \mathrm{P}, 1, 2$) while the parameter $\varepsilon_s^{(k)}$ ($k = 1, 2$) describes interactions with the substrate. Moreover, the densities of components 1 and 2 ($\rho_b^{(1)}, \rho_b^{(2)}$) in the bulk solution should be specified.

We express the energy parameters in units of thermal energy $k_{\mathrm{B}}T$. In this order we define dimensionless energy parameters $\varepsilon^{(kl)} = \bar{\varepsilon}^{(kl)}/k_{\mathrm{B}}T$ and $\varepsilon_s^{(k)} = \bar{\varepsilon}_s^{(k)}/k_{\mathrm{B}}T$. To compare our model with other theories we introduce the Flory-Huggins type parameters $\chi^{(kl)} = -[\varepsilon^{(kl)} - 0.5(\varepsilon^{(kk)} + \varepsilon^{(ll)})]$. The parameter $\chi^{(12)}$ characterizes the nature of the bulk fluid. The parameter $\chi^{(\mathrm{P}k)}$ describes the compatibility of the k-th component with respect to the chain segments.

3. Results and discussion

An adsorption equilibrium is the result of a complex interplay between two basic factors, i.e., the attractive interactions in the systems and the repulsion in the film built of grafted chains. The bonded chains can act as a barrier for the solvent molecules that tend to the substrate. On the other hand, they can provide additional 'adsorption sites'. The adsorption process depends in a very complicated way on the properties of all components of the system: the solid surface, grafted chains and the solvents.

The aim of our study is to show how the selected parameters affect adsorption and the composition of the surface layer. We present here the results of the systematic model calculations carried out for different combinations of the parameters. We try to mimic a polar adsorbent (e.g., silica gel) modified with tethered alkyl-like chains in contact with the mixture of solvents.

As already mentioned, the system under study comprises several parameters. In order to reduce their number to a minimum we assume that diameters of all the segments, as well as of all the free molecules are the same, $\sigma^{(\mathrm{P})} = \sigma^{(1)} = \sigma^{(2)} = 1$. Moreover, the calculations have been carried out by fixing the following parameters: $\varepsilon^{(\mathrm{PP})} = \varepsilon^{(\mathrm{P}1)} = \varepsilon^{(11)} = \varepsilon^{(22)} = \varepsilon^{(12)} = 1$ and $\varepsilon_s^{(1)} = 1$. The total bulk density of the fluid is $\rho^{(\mathrm{F})} = \rho^{(1)} + \rho^{(2)} = 0.8$. Notice that $\chi^{(\mathrm{P}2)} = 0$.

In this work the component 1 mimics an organic solvent that has a high affinity to tethered chains but a relatively weak affinity to the polar surface. The component 2, however, can be treated either as water or as an organic solvent, depending on the value of the parameter $\varepsilon_s^{(2)}$. We assumed that interactions of 'polar' surface with the solvent 2 are considerably stronger than the interactions with the solvent 1

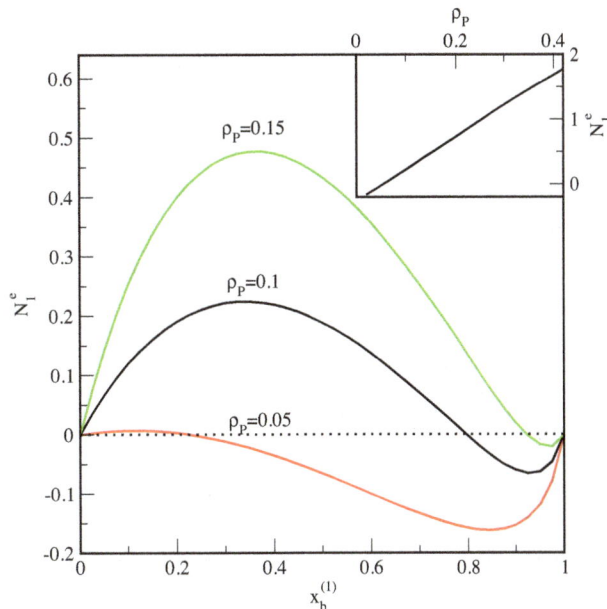

Figure 1. (Color online) Relative excess adsorption isotherms for different values of the grafting density: $\rho_P = 0.05, 0.1, 0.15$. The inset presents the influence of the grafting density on the relative adsorption, N_1^e, for the fixed composition of the bulk solution, $x_b^{(1)} = 0.35$. Parameters: $M = 8$, $\varepsilon^{(P1)} = 1.2$, $\varepsilon_s^{(2)} = 10$.

($\varepsilon_s^{(2)} > \varepsilon_s^{(1)}$). We focus our attention on the role of the modified adsorbent. To simplify the analysis of the results, we assume that the liquid mixture is ideal ($\chi^{(12)} = 0$).

We begin with the discussion of the influence of the grafting density on the adsorption from solutions on the chemically bonded phases. We consider here the surface layer built of grafted octamers ($M = 8$). The relative excess adsorption isotherms obtained for different grafting densities are shown in figure 1. The remaining parameters do not vary. In this case, the 1st solvent has only a slightly higher affinity to the grafted chains than the other component: $\varepsilon^{(P1)} = 1.2$ and $\varepsilon^{(P2)} = 1$. However, the interaction of the 1st component with the substrate is much weaker than the interactions of the 2nd solvent: $\varepsilon_s^{(1)} = 1$ but $\varepsilon_s^{(2)} = 10$. One sees that an increase of the grafting density causes a considerable increase of the relative excess adsorption N_1^e. In the inset, the relation N_1^e vs ρ_P is presented for the fixed composition ($x_b^{(1)} = 0.35$). The more grafted are the chains, the more profitable are the contacts between the molecules 1 and the chain segments. As a consequence, the sorption increases. Such a trend was observed in experimental data [21, 24, 25].

Adsorption azeotropy is observed in the considered systems. At the azeotropic point, $x_{b,az}^{(1)}$, the relative excess adsorption equals zero: $N_1^e(x_{b,az}^{(1)}) = 0$ for $0 < x_{b,az}^{(1)} < 1$. One can say that at the azeotropic point, the composition of the liquid in the surface layer is the same as the composition of the bulk solution. With an increasing grafting density, the azeotropic point shifts toward higher concentrations of the 1st component For $x^{(1)} > x_{b,az}^{(1)}$, the preferential adsorption of the 2nd solvent is found. The maximum value of the relative excess adsorption of component 1 increases with an increase of grafting density while an opposite relation is observed for the component 2.

Our conclusions agree with the analysis of experimental data [21, 24, 26]. Gritti et al. [21] measured the adsorption of acetonitrile, tetrahydrofuran and alcohols from water on end-capped silica. They have shown that an increase of the grafting density causes a rise of the relative excess adsorption of acetonitrile and tetrahydrofuran and shifts the adsorption azeotropic points toward the higher concentration of organic modifiers. In the case of the investigated alcohols, the same trend is observed for dilute solutions and the grafting densities below a certain threshold value. The analysis of the adsorption isotherms obtained for non-end-capped C_{18}-bonded phases leads to the same conclusions [24].

These results are qualitatively consistent with the predictions of the analytical theory of adsorption on energetically heterogeneous solid surfaces [50]. Adsorption azeotropy can be caused either by the

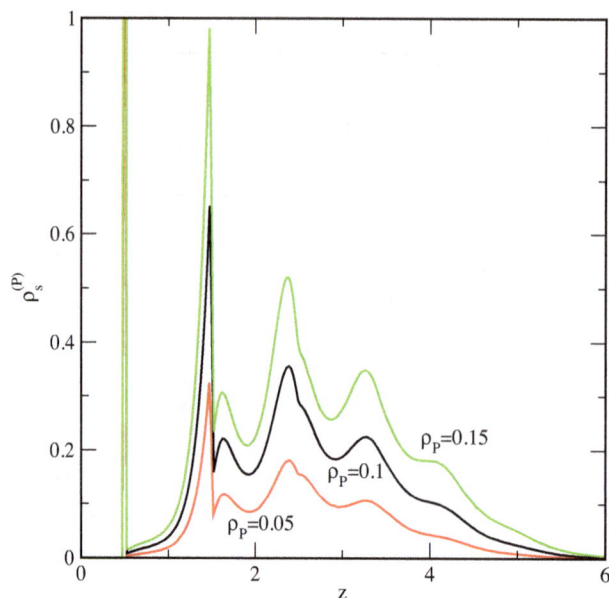

Figure 2. (Color online) Segment density profiles of grafted chains for different grafting densities, $\rho_P = 0.05, 0.1, 0.15$, at $x_b^{(1)} = 0.35$. The remaining parameters are the same as those in figure 1.

nonideality of the solution or by the heterogeneity of the surface. In the case of ideal solutions, as there increases the number of active sites exhibiting a stronger affinity to the molecules 1, the azeotropy point tends to unity [50]. Gritti and Gouichon [20] have shown that the chemically bonded phases behave as energetically heterogeneous adsorbents. They assumed that there are two kinds of adsorption sites on the surface: grafted chains and patches of bare substrate. Within the framework of their theory, with an increase of fraction of the surface covered with the chains, the adsorbent becomes more homogeneous and the azeotropic point tends to unity. In the Gritti-Gouichon [20] approach, the composition of the surface layer is assumed to be independent of the distance from the wall. There is no such a limitation in our treatment.

Figure 2 shows the segment density profiles of grafted chains for different grafting densities. The profiles have a typical liquid-like structure with peaks corresponding to successive layers of chain segments near the wall. In the outer region of the bonded phase, the chain density smoothly decreases to zero. The height of the chain layer considerably increases with an increase of the grafting density. The repulsive forces between the chains enforce them to stretch in the direction perpendicular to the wall.

The density profiles of the solution components presented in figure 3 provide data on the solvent distribution in the surface layer. The arrows show the brush edges for different grafting densities. The results have been obtained for the average concentration of the 1st component, $x_b^{(1)} = 0.35$. In the considered system, both solvents have relatively high affinity to the grafted chains, while the grafting density is rather low. Therefore, all fluid molecules penetrate the chain layer. There are several peaks at the local density profiles of both components. The positions of these peaks correspond to the maxima observed at the density profiles of the chains. The solvent molecules 'stick' to the grafted chains. Due to strong interactions with the solid surface, the molecules 2 accumulate on the wall and there is a significant peak near the substrate. The structure of the middle part of the surface layer considerably depends on the grafting density. With an increasing number of grafting chains, the density of the component 1, $\rho^{(1)}$, increases while the density of the other solvent decreases. In the outer region of the surface layer, the component densities gradually tends to their bulk values. Note that the molecules 1 accumulate not only within the brush but also atop the bonded phase.

It is instructive to analyze the local composition of the surface layer. The local mole fractions of liquid mixture are depicted in figure 4 for $x_b^{(1)} = 0.35$ (a) and $x_b^{(1)} = 0.9$ (b). For a given value of the grafting density, the mole fraction $x^{(1)}$ achieves a deep minimum near the surface and a maximum in the middle

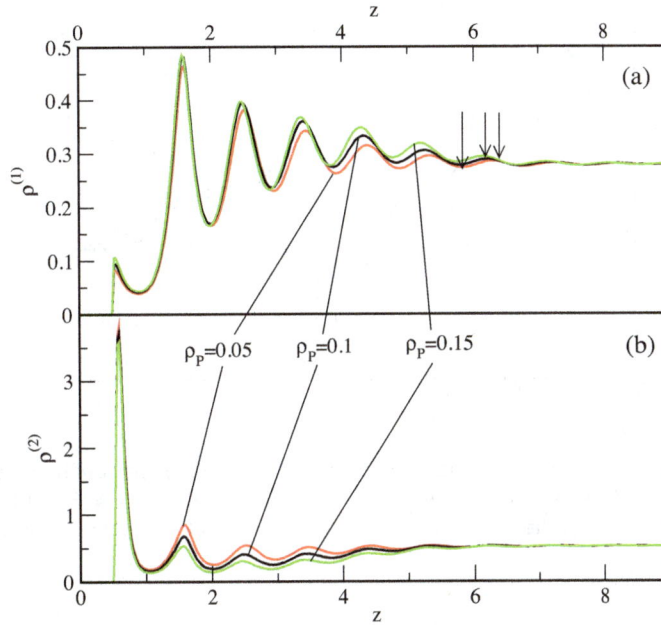

Figure 3. (Color online) Density profiles of the first solvent (panel a) and the second solvent (panel b) for different grafting densities, $\rho_P = 0.05, 0.1, 0.15$, at $x_b^{(1)} = 0.35$. The remaining parameters are the same as those in figure 1. Arrows indicate the chain layer edges.

part of the chain layer. Close to the wall, the mole fraction of the 1st (2nd) component is always considerably lower (greater) than in the bulk phase. In the region $1.5 < z < 6.5$, however, the inverse relation is found, the mole fraction of the 1st (2nd) component is greater (smaller) than its mole fraction in the bulk solution. There is a significant enrichment of the 'organic' solvent 1 within the bonded phase. With an increase of the grafting density, the mole fraction $x^{(1)}$ increases ($x^{(2)}$ decreases). These effects are much stronger for the lower mole fraction $x_b^{(1)} = 0.35$. We see that the liquid in the surface region is highly inhomogeneous. Our conclusions are in agreement with the results of molecular dynamic simulations reported by Rafferty et al. [10, 11].

Another way to moderate the adsorptive properties of the bonded phase is to change the length of the grafted chains. The examples of the relative excess adsorption isotherms calculated for different chain

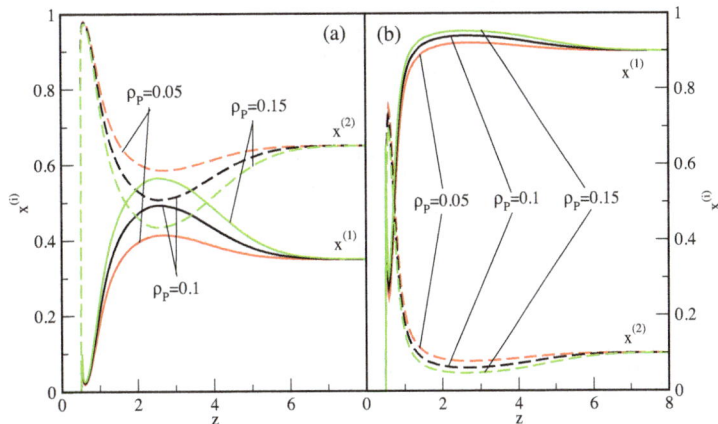

Figure 4. (Color online) Profiles of the solution composition for different values of the grafting density, $\rho_P = 0.05, 0.1, 0.15$, at fixed compositions of the bulk solution: $x_b^{(1)} = 0.35$ (panel a) and $x_b^{(1)} = 0.90$. Solid (dashed) lines correspond to the local mole fraction of the 1st (2nd) component. The remaining parameters are the same as those in figure 1.

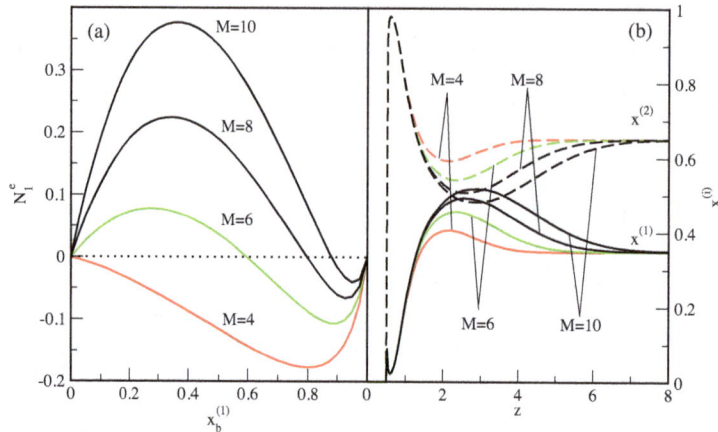

Figure 5. (Color online) Relative excess adsorption isotherms (panel a) and profiles of the solution com-position at $x_b^{(1)} = 0.35$ (panel b) for different lengths of grafted chains: $M = 4, 6, 8, 10$. In panel b, solid (dashed) lines correspond to the local mole fraction of the 1st (2nd) component. Parameters: $\rho_P = 0.1$, $\varepsilon^{(P1)} = 1.2$, $\varepsilon_s^{(2)} = 10$.

lengths are shown in figure 5 (a). The average value of grafting density was assumed to be $\rho_P = 0.1$. The remaining parameters are the same as those in figure 1. Under these conditions, the relative excess adsorption N_1^e is greater for longer grafted chains. On the surface covered by short chains ($M = 4$), the 2nd component is adsorbed preferentially in the whole concentration region ($N_1^e < 0$). For longer chains, adsorption of the 1st solvent is favored at lower mole fractions, $x_b^{(1)}$. An increase of the number of chain segments causes a rise of a profitable contact between the molecules 1 and the grafted chains, and thus the relative excess adsorption N_1^e increases. The same relation has been found for adsorption of organic solvents from aqueous solutions on the bonded phases with different chain lengths attached to silica [22, 23]. For denser brushes and very long polymers, an impact of chain lengths weakens [22].

In figure 5 (b), the composition profiles at $x_b^{(1)} = 0.35$ are presented. Near the substrate the composi-tion of the liquid is almost independent of the grafted chain length. However, one sees a significant effect of the chain length on the liquid composition in the remaining part of the surface layer. The mole fraction of the 1st component increases for longer grafted chains. Obviously, the opposite is observed for the mole fraction $x_b^{(2)}$. Analogous results have been obtained from simulation [14].

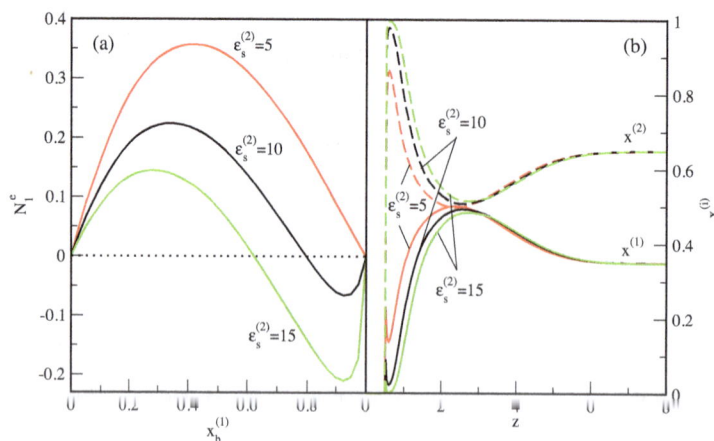

Figure 6. (Color online) Relative excess adsorption isotherms (panel a) and profiles of the solution compo-sition at $x_b^{(1)} = 0.35$ (panel b) for different values of the energy parameter $\varepsilon_s^{(2)} = 5, 10, 15$. In panel b, solid (dashed) lines correspond to the local mole fraction of the 1st (2nd) component. Parameters: $\rho_P = 0.1$, $M = 8$, $\varepsilon^{(P1)} = 1.2$.

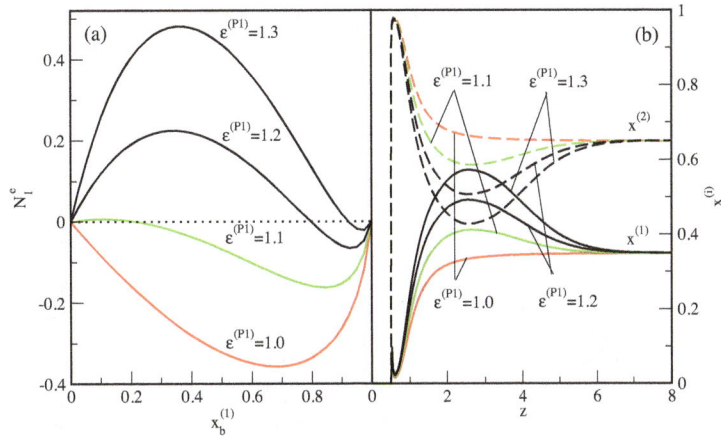

Figure 7. (Color online) Relative excess adsorption isotherms (panel a) and profiles of the solution composition at $x_{\text{b}}^{(1)} = 0.35$ (panel b) for different values of the energy parameter $\varepsilon^{(\text{P1})} = 1.0, 1.1, 1.2, 1.3$. In panel b, solid (dashed) lines correspond to the local mole fraction of the 1st (2nd) component. Parameters: $\rho_{\text{P}} = 0.1$, $M = 8$, $\varepsilon_{\text{s}}^{(2)} = 10$.

Now we turn to the effects of interactions with the whole adsorbing material, i.e., with the substrate and with the grafted chains. Let us consider the role of interactions with the solid surface. In figure 6 (a), the relative excess isotherms are plotted for a fixed value of energy parameter $\varepsilon_{\text{s}}^{(1)} = 1$ and for three values of the energy parameter $\varepsilon_{\text{s}}^{(2)} = 5, 10, 15$. We moderate the relative solvent affinity by changing the solvent 2. The relative excess adsorption isotherm reflects the competition in accumulation of components 1 and 2 within the surface layer. The relative affinity of the 1st component to the whole adsorbing material increases as the difference $\Delta = \varepsilon_{\text{s}}^{(1)} - \varepsilon_{\text{s}}^{(2)}$ increases. Indeed, one sees that the preferential adsorption of the component 1 is greater for lower values of the parameter $\varepsilon_{\text{s}}^{(2)}$. At the same time, the azeotropic point shifts to unity, namely $x_{\text{b,az}}^{(1)} = 0.62$ for $\varepsilon_{\text{s}}^{(2)} = 15$ and $x_{\text{b,az}}^{(1)} = 0.80$ for $\varepsilon_{\text{s}}^{(2)} = 10$. Such a behavior is predicted by the simple theories of adsorption form solutions [50]. These conclusions are in an agreement with the experimental observations [24].

The profiles of the mixture composition are plotted in figure 6 (b) for $x_{\text{b}}^{(2)} = 0.35$. The molecules 2 accumulate at the surface, and the mole fraction $x^{(2)}$ decreases with an increase of the parameter $\varepsilon_{\text{s}}^{(2)}$ (dashed lines). On the contrary, the mole fraction of the 1st component is considerably lower for higher values of $\varepsilon_{\text{s}}^{(2)}$. These effects are significant only in the immediate proximity to the wall.

Adsorption of the 1st component can be also altered by changing the parameter $\varepsilon^{(\text{P1})}$ [see figure 7 (a)]. The relative affinity of the components to the chains is well quantified by the difference $\Delta_{\text{P}} = \varepsilon^{(\text{P1})} - \varepsilon^{(\text{P2})}$. The energy parameter $\varepsilon^{(\text{P2})} = 1$. An increase of the parameter $\varepsilon^{(\text{P1})}$ leads to a considerable rise of the preferential adsorption of the 1st component. The attractive interactions of the molecules 1 with segments of the grafted chains markedly affect the composition of the middle and the outer parts of the surface region. This is clearly demonstrated in figure 7 (b).

Adsorptive properties of the bonded phases can be modelled by the use of the grafted chains with active groups. Such chains are copolymers containing a few segments which very strongly attract fluid molecules. We have carried out the calculations using the method described in reference [30]. We consider the grafted chains built of M_A segments A and M_B segments B ($M_A + M_B = M$). The segments A mimic methylene groups in alkyl chains, while the segments B correspond to the functional (active) group. Within the framework of the model, the parameters characterizing the interactions of the kth component with segments A and B are different ($\varepsilon^{(Ak)} \neq \varepsilon^{(Bk)}$). Here we show the results for the chains that contain $M_B = 2$ segments of the B-type. These segments can be placed at consecutive positions of the backbone: i_{s} and $i_{\text{s+1}}$. The given type of chains is labeled as $B i_{\text{s}} i_{\text{s+1}}$. We consider two positions of the functional group: in the close proximity to the wall (as the 3rd and 4th segments, $B34$) and at the chain end ($B78$). The first type of the grafted chains corresponds to polar embedded stationary phases which are often used in RPLC. The latter mimics stationary phases with functionalized terminal groups of the ligands [51]. We have focused only on the interactions of solvents with different segments. Therefore, we

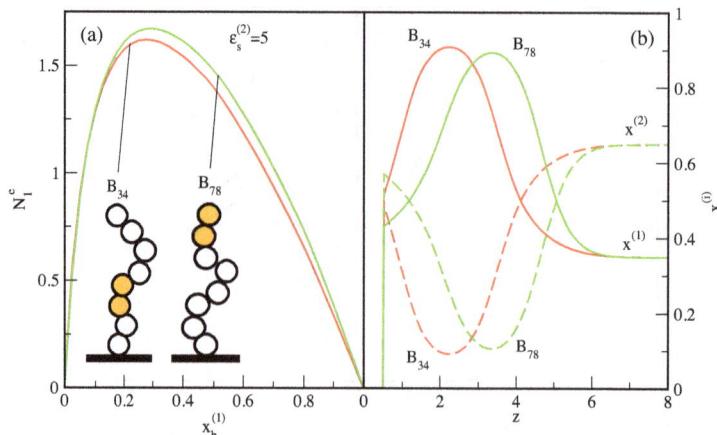

Figure 8. (Color online) Relative excess adsorption isotherms (panel a) and profiles of the solution com-position at $x_b^{(1)} = 0.35$ (panel b) for two isomers of grafted chains $B34$ and $B78$. In panel b solid (dashed) lines correspond to the local mole fraction of the 1st (2nd) component. Parameters: $\rho_P = 0.1$, $M = 8$, $\varepsilon^{(P2)} = 1$, $\varepsilon^{(A1)} = 1.2$ and $\varepsilon^{(B1)} = 3$.

have assumed that both components are inert with respect to the wall. In the considered case, affinity of the solvent 1 to the segments B is much larger than the affinity to the 'usual' segments A, $\varepsilon^{(A1)} = \varepsilon^{(P1)} = 1.2$ and $\varepsilon^{(B1)} = 3$. Moreover, $\varepsilon^{(AB)} = \varepsilon^{(PP)} = 1$ ($\chi^{(AB)} = 0$). Interactions of the 2nd solvent with all segments are weaker, $\varepsilon^{(A2)} = \varepsilon^{(B2)} = \varepsilon^{(P2)} = 1$.

Figure 8 illustrates the effect of the position of functional groups in grafted chains on the relative excess adsorption isotherms and the composition of the surface layer. The shapes of the isotherm are typical of a strong preferential adsorption of the 1st component. The position of the functional group affects the relative excess adsorption for $x_b^{(1)} > 0.1$. Adsorption is higher for terminal functional groups. In panel b, the local mole fractions of the components are shown for $x_b^{(1)} = 0.35$. There is a considerable enrichment of the component 1 near the active groups. The maximum in $x^{(1)}$-profile is shifted toward the outer part of the surface layer for $B78$.

It follows from the above discussion that the adsorption of liquid mixtures on chemically bonded phases and the composition of the surface layers depend on the relations between the parameters char-acterizing the system. The model is very sensitive to the choice of a particular set of the parameters. Changing these parameters one can simulate various systems.

Finally, we consider the problem of 'a range of the surface layer'. In chromatographic applications, a delimitation between the surface and bulk phases is necessary. The retention factor is proportional to the volume of the stationary (surface) phase. Moreover, the calculation of real adsorption (the number of molecules in the surface phase) requires a definition of the boundary of the adsorbed phase. In interface science, the Gibbs dividing surface (GDS) is introduced to define the volumes of both phases. The choice of the position of the Gibbs dividing surface is arbitrary [20]. When the density profiles are known, the GDS is usually defined via the standard equal area construction [52]. In analytical theories, the position of GDS is chosen using the procedure that ensures the thermodynamic consistency of the results [20]. Unfortunately, the problem is not trivial.

Various quantities can be treated as a measure of the thickness of the surface (stationary) phase. For example, one can use characteristics of the chain layer: (i) the effective height of the grafted layer, h_{eff} (the distance from the wall at which the segment density of the chains decreases to zero), and (ii) the average brush height calculated from the following equation[53, 54]

$$h = 2 \frac{\int dz z \rho_s^{(P)}(z)}{\int dz \rho_s^{(P)}(z)}. \tag{13}$$

The height of the polymer brush can quite well approximate the surface phase boundary since the partitioning is a dominant mechanism of the sorption.[5] On the other hand, when adsorption plays a

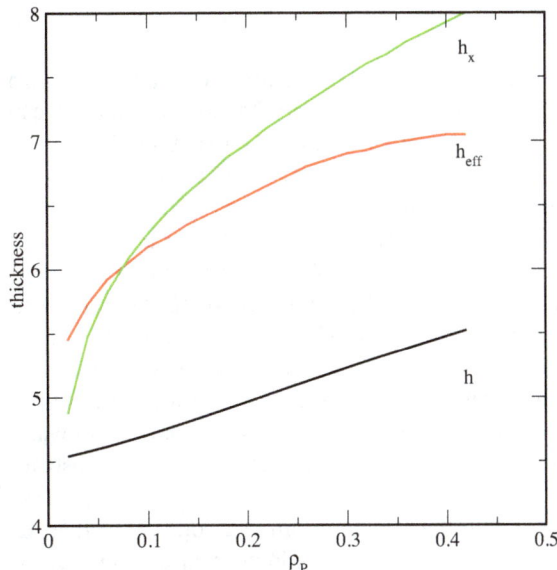

Figure 9. (Color online) The effect of the grafting density on the average height of the chain layer, h, calculated from equation (13), on the effective thickness of the chain layer (the distance from the wall at which the segment density of the chains decreases to zero), h_{eff}, and on the effective thickness of the surface layer (the maximum distance above which the liquid composition is the same as in the bulk solution), h_x, at $x_b^{(1)} = 0.35$. The remaining parameters are the same as those in figure 1.

significant role in the process, the fluid density profiles should be analyzed to estimate: (i) the location of the GDS [20] or (ii) the 'effective' surface phase boundary. The latter is the maximum distance, z, above which the liquid is considered to be identical to the bulk solution [8].

In all investigated systems, the bonded phases are highly inhomogeneous. One can divide the surface layer into three parts: (i) the interfacial region located near the wall, (ii) the interface between the brush and the bulk solution, and (iii) the middle ('bulk') part of the bonded phase. For relatively short grafted chains, the volume of the 'bulk' stationary phase is comparable with the volumes of the interfacial regions. There is no well-pronounced flat part in the segment density profiles (cf. figure 1) that could be treated as the 'bulk' bonded phase. Therefore, the standard method for the location of the GDS [52] cannot be sufficiently precise. We assume that the boundary is located at the distance at which the composition of the adsorbed liquid becomes identical with the bulk solution (with a precision of 1 percent of $x_b^{(1)}$).

Figure 9 shows how the grafting density affects the average height of the chain layer h [equation (13)], the 'effective' height of polymer brush, h_{eff}, and the 'effective' thickness of the surface layer, h_x. All the 'thicknesses' increase as the grafting density increases. The dependence of the average brush height on the grafting density, h vs ρ_P, has been the subject of numerous studies [54, 55]. The 'effective' thickness of the brush is considerably greater than the average height, $h_{eff} > h$ and $h_x > h$. As one can expected, the true adsorbed phase can be more expanded due to the secondary adsorption on the brush. Indeed, for $\rho_P > 0.08$, the 'effective' thickness of the surface phase h_x is considerably higher than h_{eff}. This just reflects a competitive adsorption in the region above the chain layer. However, at low surface coverages, the opposite relation is found, $h_x < h_{eff}$. In the outer region of the brush, the density of chain segments is too low to change the liquid composition. After penetration of the solution into the chain layer, the composition starts to vary. The problem of the delimitation between surface and bulk phases requires a further analysis.

The thickness of the surface layer varies with the change of the bulk solution. However, these effects are not very significant [55].

4. Conclusions

We have performed the density functional calculations to study the adsorption from binary solutions on the chemically bonded phases. We have assumed that solvent molecules are spherical. The segments and the solvent molecules interact via Lennard-Jones (12–6) potential. All chain segments but the bonding segment are inert with respect to the substrate. The solvent molecules interacts with the solid surface via Lennard-Jones (9–3) potential or by the hard-wall potential.

We have systematically analyzed the effect of the selected factors on the composition of the surface layer. In the model, both solvents are capable of penetrating the brush. Solvent molecules 'stick' to the chains. Therefore, the solvent density profiles reflect the structure of the brush. We have assumed that the solvent 1 exhibits high affinity to the grafted chains ($\varepsilon^{(P1)} > \varepsilon^{(P1)}$), while the solvent 2 very strongly interacts with the substrate ($\varepsilon_s^{(1)} < \varepsilon_s^{(2)}$). Molecules of different components compete for room inside the brush. As a consequence, the composition of the liquid mixture in the surface region changes with the distance from the wall. A considerable enrichment of the 1st component was observed in the middle and in the outer parts of the surface region. On the contrary, close to the substrate, high concentration of the 2nd component was found.

We have discussed the effect of the system parameters on the profiles of local mole fractions of the components. We have studied the impact of the following parameters: the grafting density, the grafted chain length, interactions of solvent molecules with grafted chains and the substrate and the presence of active groups in the chains. We have shown that the mole fraction of the 1st component increases as the grafting density increases. The same effect is observed as the parameter $\varepsilon^{(P1)}$ rises or the adsorption energy of the 2nd component, $\varepsilon_s^{(2)}$, decreases. The impact of the difference in interactions with the substrate is limited to its proximity while the effects of interactions with grafted chains are well pronounced almost in the whole surface layer. An increase of the local mole fraction $x^{(1)}$ is also found for longer grafted chains. The composition of the liquid mixture markedly changes near the active groups in the chains.

The local mole fractions of the liquid components inside the bonded phase cannot be experimentally measured. However, this is possible for the relative adsorption isotherms. Therefore, we have presented the relative adsorption isotherms for the model systems considered. We have qualitatively compared our results with experimental isotherms. The theory well predicts general trends observed in experiments [20–26].

We have also analyzed the problem of the delimitation between the surface phase and the bulk phase. The thickness of the surface phase is necessary for calculating the retention factors in chromatography. We have compared three methods for the estimation of the thickness of the surface phase. The quantities characterizing the brush, namely, the average brush height, h [53] or the effective brush thickness (the location of the brush edge), h_{eff}, can be effectively used since the solvent molecules deeply penetrate the chain layer and do not accumulate 'on the brush'. However, when the composition of the liquid mixture above the chain layer differs from that in the bulk solution, the effective surface layer thickness, h_x should be estimated.

Our results confirmed that the density functional theory provides a flexible and effective tool for modelling the adsorption from binary solutions on the surfaces modified with grafted chains. The study of adsorption from binary solutions on the chemically bonded phase can be a starting point for the theoretical prediction of the solute retention from mixed mobile phases. The extension of this approach is straightforward. Such a model system would contain at least four components: the bonded chains and three free components, i.e., two solvents and a solute. More realistic models would be used to investigate chromatographic systems, e.g., the models involving associative and electrostatic interactions between components [36, 38].

References

1. Mittal V., In: Polymer Brushes: Substrates, Technologies and Properties, Mittal V. (Ed.), CRC Press, Boca Raton, 2012, Chapter 1.
2. Dorsey J.G., Dill K.A., Chem. Rev., 1989, **89**, 331; doi:10.1021/cr00092a005.
3. Martire D.E., Boehm R.E., J. Phys. Chem., 1983, **87**, 1045; doi:10.1021/j100229a025.
4. Borówko M., Ościk-Mendyk B., Borówko P., J. Phys. Chem., 2005, **109**, 21056; doi:10.1021/jp053195b.

5. Bohmer M.R., Koopal L.K., Tijssen R., J. Phys. Chem., 1991, **95**, 6285; doi:10.1021/j100169a041.
6. Leermakers F.A.M., Philipsen H.J.A., Klumperman B., J. Chromatogr. A, 2002, **959**, 37; doi:10.1016/S0021-9673(02)00382-5.
7. Borówko M., Rżysko W., Sokołowski S., Staszewski T., J. Phys. Chem., 2009, **113**, 4763; doi:10.1021/jp811143n.
8. Borówko M., Sokołowski S., Staszewski T., J. Chromatogr. A, 2011, **1218**, 711; doi:10.1016/j.chroma.2010.12.029.
9. Klatte S.J., Beck T.L., J. Chem. Phys., 1995, **99**, 16024; doi:10.1021/j100043a049.
10. Rafferty J.L., Siepmann J.I., Schure M.R., J. Chromatogr. A, 2008, **1204**, 11; doi:10.1016/j.chroma.2008.07.037.
11. Rafferty J.L., Siepmann J.I., Schure M.R., J. Chromatogr. A, 2008, **1204**, 20; doi:10.1016/j.chroma.2008.07.038.
12. Rafferty J.L., Siepmann J.I., Schure M.R., Anal. Chem., 2008, **80**, 6214; doi:10.1021/ac8005473.
13. Rafferty J.L., Siepmann J.I., Schure M.R., J. Chromatogr. A, 2009, **1216**, 2320; doi:10.1016/j.chroma.2008.12.088.
14. Rafferty J.L., Siepmann J.I., Schure M.R., J. Chromatogr. A, 2012, **1223**, 24; doi:10.1016/j.chroma.2011.11.039.
15. Linsey R.K., Rafferty J.L., Eggimann B.L., Siepmann J.I., Schure M.R., J. Chromatogr. A, 2013, **1287**, 60; doi:10.1016/j.chroma.2013.02.040.
16. Fouqueau A., Meuwly M., Bermish R.J., J. Phys. Chem. B, 2007, **111**, 10208; doi:10.1021/jp071721o.
17. Braun J., Fouqueau A., Bermish R.J., Meuwly M., Phys. Chem. Chem. Phys., 2008, **10**, 4765; doi:10.1039/B807492E.
18. Dill K.A., J. Phys. Chem., 1987, **91**, 1980; doi:10.1021/j100291a060.
19. Jaroniec M., Martire D.E., Borówko M., Adv. Colloid Interface Sci., 1985, **22**, 177; doi:10.1016/0001-8686(85)80005-1.
20. Gritti F., Guiochon G., J. Chromatogr. A, 2007, **1155**, 85; doi:10.1016/j.chroma.2007.04.024.
21. Gritti F., Kazakevich Y.V., Guiochon G., J. Chromatogr. A, 2007, **1169**, 111; doi:10.1016/j.chroma.2007.08.071.
22. Kazakevich Y.V., LoBrutto R., Chan F., Patel T., J. Chromatogr. A, 2001, **913**, 75; doi:10.1016/S0021-9673(00)01239-5.
23. Bocian S., Felinger A., Buszewski B., Chromatographia, 2008, **68**, S19; doi:10.1365/s10337-008-0519-4.
24. Bocian S., Vajda P., Felinger A., Buszewski B., Anal. Chem., 2009, **81**, 6334; doi:10.1021/ac9005759.
25. Bocian S., Vajda P., Felinger A., Buszewski B., Chromatographia, 2010, **71**, S5; doi:10.1365/s10337-010-1522-0.
26. Buszewski B., Bocian S., Nowaczyk A., J. Sep. Sci., 2010, **33**, 2060; doi:10.1002/jssc.201000101.
27. Borówko M., Sokołowski S., Staszewski T., J. Phys. Chem. B, 2012, **116**, 3115; doi:10.1021/jp300114y.
28. Borówko M., Sokołowski S., Staszewski T., J. Phys. Chem. B, 2012, **116**, 12842; doi:10.1021/jp305624n.
29. Borówko M., Staszewski T., Condens. Matter Phys., 2012, **15**, 1; doi:10.5488/CMP.15.23603.
30. Borówko M., Sokołowski S., Staszewski T., Mol. Phys., 2015, **113**, 1014; doi:10.1080/00268976.2014.962636.
31. Yu Y.X., Wu J., J. Chem. Phys., 2002, **117**, 2368; doi:10.1063/1.1491240.
32. Yu Y.X., Wu J., J. Chem. Phys., 2002, **117**, 10156; doi:10.1063/1.1520530.
33. Yu Y.X., Wu J., J. Chem. Phys., 2003, **118**, 3835; doi:10.1063/1.1539840.
34. Bryk P., Sokołowski S., J. Chem. Phys., 2004, **120**, 8299; doi:10.1063/1.1695554.
35. Bryk P., Sokołowski S., J. Chem. Phys., 2004, **121**, 11314; doi:10.1063/1.1814075.
36. Bryk P., Pizio O., Sokołowski S., J. Chem. Phys., 2005, **122**, 174906; doi:10.1063/1.1888425.
37. Bryk P., Pizio O., Sokołowski S., J. Chem. Phys., 2005, **122**, 194904; doi:10.1063/1.1898484.
38. Tscheliessnig R., Billes W., Fischer J., Sokołowski S., Pizio O., J. Chem. Phys., 2006, **124**, 164703; doi:10.1063/1.1898484.
39. Bucior K., Fischer J., Patrykiejew A., Tscheliessnig R., Sokołowski S., J. Chem. Phys., 2007, **126**, 094704; doi:10.1063/1.2566372.
40. Borówko M., Rżysko W., Sokołowski S., Staszewski T., J. Chem. Phys., 2007, **126**, 214703; doi:10.1063/1.2743399.
41. Matusewicz M., Patrykiejew A., Sokołowski S., Pizio O., J. Chem. Phys., 2007, **127**, 174707; doi:10.1063/1.2780890.
42. Patrykiejew A., Sokołowski S., Tscheliessnig R., Fischer J., Pizio O., J. Phys. Chem. B, 2008, **112**, 4552; doi:10.1021/jp710978t.
43. Fundamentals of Inhomogeneous Fluids, Henderson D. (Ed.), Marcel Dekker, New York, 1992.
44. Rosenfeld J., Phys. Rev. Lett., 1989, **63**, 980; doi:10.1103/PhysRevLett.63.980.
45. Wertheim M.S., J. Chem. Phys., 1987, **87**, 7323; doi:10.1063/1.453326.
46. Borówko M., Sokołowski S., Staszewski T., J. Colloid Interface Sci., 2011, **356**, 267; doi:10.1016/j.jcis.2011.01.023.
47. Weeks J.D., Chandler D., Andersen H.C., J. Chem. Phys., 1971, **54**, 5237; doi:10.1063/1.1674820.
48. Everett D.H., Trans. Faraday Soc., 1965, **62**, 2478;
49. Roe R.J., J. Chem. Phys., 1974, **60**, 4192; doi:10.1063/1.1680888.
50. Borówko M., In: Adsorption, Theory, Modeling and Analysis, Surfactant Science Series Vol. 107, Tóth J. (Ed.), Marcel Dekker, New York, 2002, Chapter 3.
51. O'Sullivan G.P., Scully N.M., Glennon J.D., Anal. Lett., 2010, **43**, 1609; doi:10.1080/00032711003653973.
52. Zangwill A., Physics at Surfaces, Cambridge University Press, Cambridge, 1988.
53. Pastorino C., Binder K., Mueller M., Macromolecules, 2009, **42**, 401; doi:10.1021/ma8015757.
54. Binder K., Milchev A., J. Polym. Sci., Part B: Polym. Phys., 2012, **50**, 1515; doi:10.1002/polb.23168.
55. Borówko M., Sokołowski S., Staszewski T., J. Phys. Chem. B, 2013, **117**, 10293; doi:10.1021/jp4027546.

4

Phase transitions of fluids in heterogeneous pores

A. Malijevský[1,2]

[1] Department of Physical Chemistry, University of Chemistry and Technology Prague,
166 28 Praha 6, Czech Republic

[2] Laboratory of Aerosols Chemistry and Physics, Institute of Chemical Process Fundamentals,
Academy of Sciences, 16502 Prague 6, Czech Republic

We study phase behaviour of a model fluid confined between two unlike parallel walls in the presence of long range (dispersion) forces. Predictions obtained from macroscopic (geometric) and mesoscopic arguments are compared with numerical solutions of a non-local density functional theory. Two capillary models are considered. For a capillary comprising of two (differently) adsorbing walls we show that simple geometric arguments lead to the generalized Kelvin equation locating capillary condensation very accurately, provided both walls are only partially wet. If at least one of the walls is in complete wetting regime, the Kelvin equation should be modified by capturing the effect of thick wetting films by including Derjaguin's correction. Within the second model, we consider a capillary formed of two competing walls, so that one tends to be wet and the other dry. In this case, an interface localized-delocalized transition occurs at bulk two-phase coexistence and a temperature $T^*(L)$ depending on the pore width L. A mean-field analysis shows that for walls exhibiting first-order wetting transition at a temperature T_w, $T_s > T^*(L) > T_w$, where the spinodal temperature T_s can be associated with the prewetting critical point, which also determines a critical pore width below which the interface localized-delocalized transition does not occur. If the walls exhibit critical wetting, the transition is shifted below T_w and for a model with the binding potential $W(\ell) = A(T)\ell^{-2} + B(T)\ell^{-3} + \cdots$, where ℓ is the location of the liquid-gas interface, the transition can be characterized by a dimensionless parameter $\kappa = B/(AL)$, so that the fluid configuration with delocalized interface is stable in the interval between $\kappa = -2/3$ and $\kappa \approx -0.23$.

Key words: *capillary condensation, wetting, Kelvin equation, adsorption, density functional theory, fundamental measure theory*

1. Introduction

It is very well known that structure and phase behaviour of a confined fluid is quite distinct from that of its bulk counterpart. A familiar example of a confining geometry is a slit pore formed by two parallel, identical and infinite plates, a distance L apart. The combination of finite-size effects and fluid adsorption at the walls leads to a shift in the liquid-vapour phase boundary and in the critical point compared to a bulk fluid [1–5]. The location of this capillary condensation (or evaporation) transition is macroscopically described by Kelvin's equation which predicts that the chemical potential at which the vapour in the slit condenses into the liquid-like phase is shifted from its saturation value μ_{sat} by an amount

$$\delta\mu \equiv \mu_{sat} - \mu_{cc} = \frac{2\gamma\cos\theta}{(\rho_l - \rho_v)L}, \tag{1.1}$$

where ρ_l and ρ_g are the coexisting bulk liquid and gas densities, respectively, γ is the liquid-gas surface tension and θ is the contact angle of a macrosocpic liquid droplet sitting on isolated wall. The role of wetting layers adsorbed at the walls when $\theta = 0$, i.e., for $T > T_w$, where T_w is the wetting temperature corresponding to a semi-infinite system $L \to \infty$, has also been appreciated but this effect is mostly quantitative, such that the presence of the layers effectively reduces the pore width [6]. If the isolated walls exhibit first-order wetting transition at $T = T_w$ (and $\mu = \mu_{sat}$), a prewetting transition corresponding to a finite jump in the wetting layers thickness can also occur, although the transition is typically metastable with respect to the capillary condensation unless the pore width is fairly large [7].

The scenario of a liquid-vapour coexistence can, however, be very different in pores made of unlike walls. This was demonstrated by Parry and Evans [8, 9] who proposed a model, treated within the Landau theory in the language of magnets, with perfectly antisymmetric surface fields of the walls. They have shown that, for sufficiently large pores, the capillary condensation transition is replaced by the interface localization-delocalization phase transition which occurs at a two-phase bulk coexistence at a temperature $T^*(L)$ near the wetting temperature T_w of the wall with the affinity to "+" phase (which, owing to the symmetry of the system, is identical to the wetting temperature of the opposing wall which tends to be wetted by "−" phase). The transition separates a regime, present for $T < T^*(L)$, when the equilibrium density profile corresponds to a very thick "+" or to a very thick "−" phase with a "+−" interface pinned to either of the walls, from the high temperature regime at which the interface is unbounded from either of the walls. In the latter case, sometimes referred to as soft-mode phase, the interface finds a compromise between the antagonistic wetting preferences of the walls, such that it develops around the midpoint of the pore being a subject of large fluctuations causing the interface to wander along the pore. Compared with the former model of pores with identical walls, the case of antisymmetric walls makes much closer link with wetting properties of the walls. In the theory of wetting phenomena, the concept of a binding potential proved to be very useful (at least on a mean-field level) and can be used to describe the phase behaviour of the fluid confined between antisymmetric walls. If the walls, when isolated, exhibit critical wetting as assumed in references [8, 9], the binding potential at each wall acquires a single minimum whose location shifts continuously from the wall to infinity as the temperature increases towards T_w along the phase coexistence line [$\mu = \mu_{sat}(T)$]. It means that, when the pore is sufficiently wide, the binding potential has two minima that are of the same depth and are located symmetrically around the midpoint of the pore. There are thus two equally stable solutions for the density profile corresponding to large and small adsorptions. The adsorption difference, or equivalently the distance between the two minima, decreases with an increasing temperature and ultimately disappears at $T^*(L)$, which is a finite-size shift of T_w. Above $T^*(L)$, the binding potential landscape adopts a U-shape with a very shallow single minimum which enables the interface to drift around the centre with a very small free energy cost.

In contrast to pores with identical walls, the nature of wetting transition at the (isolated) walls becomes much more important when the walls are antisymmetric. If they exhibit first-order wetting, the characteristic feature of the binding potential is a competition between a minimum at a finite distance from the wall with the unbounded state corresponding to a minimum at infinity. Assuming the pore is sufficiently wide, the binding potential now possesses three local minima, such that two minima near the walls are the global minima for temperatures below the finite-size shifted wetting temperature $T^*(L)$, whilst the middle minimum becomes a global minimum for $T > T^*(L)$. The nature of the transition (at fixed L and T varying) thus reflects the nature of wetting at the walls: while the transition is continuous for a critical wetting, in which case two potential minima continuously merge in the middle of the pore, it becomes discontinuous for first-order wetting due to a jump in the location of the global minimum. Consequently, $T^*(L)$ is a critical point for critical wetting, whereas it is a triple point for first-order wetting. However, all these conclusions are only valid for sufficiently large pores. By decreasing the pore width, the space to accommodate all three minima of the binding potential is reduced and when the middle minimum disappears, the order of the transition becomes second order [10, 11].

These predictions have been verified by extensive Monte Carlo simulations by Binder et al. for Ising-like models [12–17]. More recently, the properties of the interface localization-delocalization transition were also studied for fluid models of soft matter systems [18–20]. Compared to magnets, however, the situation with fluids is a bit more intricate. Firstly, owing to unequal entropy of coexisting phases, they lack the perfect symmetry of Ising-like models which makes the concept of antisymmetric walls less clear. Secondly, ubiquity of dispersion forces in fluid systems is an important extra ingredient to be considered which, in fact, prevents complete drying. The latter problem can be avoided by considering binary (colloid-polymer) mixtures such as in references [18–20]. The long-range dispersion interaction was included in reference [21] by Stewart and Evans for a simple, one-component fluid; in this study, extensive density functional theory (DFT) calculations have been made to confirm scaling predictions for several thermodynamic quantities. The wall parameters have been set such that they ensure a complete wetting on one wall and a complete drying on the opposite wall and that, for the given fixed subcritical temperature, the corresponding Hamaker constants, i.e., the coefficients of the lowest order term in the respective binding potentials, are identical. However, even if neglecting the higher order contributions in the binding potential, the system does not exhibit the same "antisymmetry" as in the case of magnets. This is because the potential exerted by both walls must contain a repulsive part for the walls to be impenetrable, but only one of them (the "solvophilic" wall) also contains an attractive portion. This produces a binding potential well so that there exists a wetting temperature T_w below which the wall is only partially wet. By contrast, the other, "solvophobic" wall is purely repulsive and thus dried (i.e., wet by gas) at all temperatures. Consequently, the binding potential landscape for a fluid confined between these competing walls is clearly different from that described above for magnetic systems. Furthermore, since the Hamaker constants depend on a temperature (via coexisting densities), any change in the temperature would necessarily break the "antisymmetry" of the system, which is thus not a property of the system itself but also depends on thermodynamic parameters.

In this paper, we study the phase behaviour of a fluid confined between asymmetric walls that interact with the fluid via dispersion forces. There are two models that we use to represent such a heterogeneous pore. Within the first model, both walls exhibit the wetting transition but at different temperatures. We present simple geometric arguments that lead to the extended Kelvin equation which predicts the location of capillary condensation in the pore. For the case, when the temperature of the system is greater than the wetting temperature of at least one of the walls, we modify Kelvin's equation by incorporating the effect of the presence of the wetting layer(s). Both of these predictions are tested against numerical results obtained from a non-local density functional theory (DFT). The second model represents the case of antisymmetric walls, so that one of the walls is wetted by liquid whereas the other wall by gas. We present a mean-field analysis for the location of the interface localization-delocalization transition. The comparison with DFT shows that the analytic predictions are surprisingly accurate down to very narrow pores at least for the case, when one wall exhibits first-order wetting transition (as opposed to critical wetting) and the other wall is completely dried (wetted by gas) at all temperatures.

The remainder of the paper is organized as follows. In section 2 we present geometric arguments to determine the location of capillary condensation for pores with two unlike walls and also show how this extended Kelvin's equation can be modified to embrace Derjaguin's correction due to the presence of wetting layers. In this section we also consider the pore model consisting of antisymmetric walls for which we present a mean-field analysis to locate the interface localization-delocalization transition. In section 3 we set the molecular model and show the DFT results. Although the planar symmetry permits to treat the system as a one-dimensional problem, we employ a two-dimensional DFT to test the plausibility of geometric arguments. We then examine the properties of the interface localization-delocalization transition for antisymmetric walls by determining the binding potential and study its behaviour as the temperature and pore width vary. We summarise and discuss our results in the concluding section 4.

2. Heuristic arguments

We start by recalling macroscopic arguments leading to a condition for a liquid-vapour equilibrium in a homogeneous pore, i.e., a pore made of identical walls that exhibit a wetting transition (by liquid) at a temperature T_w. Within the purely macroscopic treatment, the distance between the walls L is taken to be large and it is assumed that a first-order transition between a state corresponding to a gas-like and a liquid-like phase occurs for any temperature T below the bulk critical temperature T_c. We then expect that the pressure p (or the chemical potential μ) at which the transition occurs in the pore is shifted below the saturation value p_{sat} (or μ_{sat}). Based on the surface thermodynamics, this value can be determined by a simple free energy balance for a low- and a high-density state which leads to the well known Kelvin's equation (1.1). Apart from this thermodynamic picture, Kelvin's equation has also a geometric interpretation as it is illustrated in figure 1. Based on this approach, one considers a single pore in which both phases coexist. Since the equilibrium occurs off bulk two-phase coexistence, the interface between the gas and vapour phases in the pore must be curved in the direction perpendicular to the walls with a Laplace radius $R = \gamma/\delta p$, where $\delta p = p_{sat} - p$. Assuming that δp is small, we can write

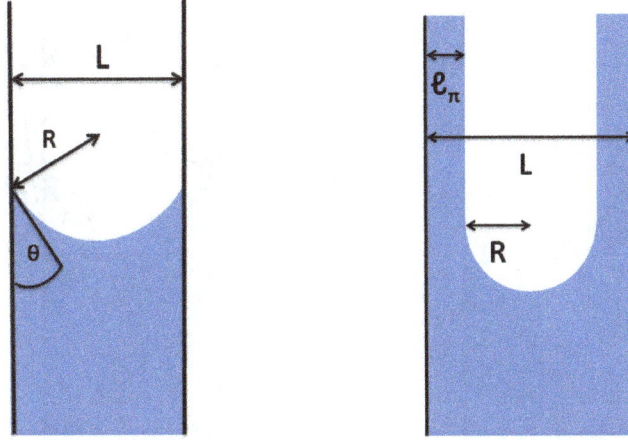

Figure 1. (Color online) A schematic picture illustrating a coexistence of liquid-like and gas-like phases in a homogeneous pore of a macroscopic width L. In the left panel, a contact angle at the walls is assumed to be $\theta > 0$ meaning that the temperature of the system is below the wetting temperature. The radius R of the cylindrical meniscus separating the phases is given by the Laplace pressure $R = \gamma/\delta\mu(\rho_1 - \rho_v)$. In the left panel, the contact angle is zero, so that the meniscus meets tangentially the wetting layers of thickness ℓ_π.

$\delta p = \delta\mu(\rho_1 - \rho_v)$ according to the Gibbs-Duhem relation, where ρ_1 and ρ_v are the particle densities of coexisting liquid and vapour phases, respectively. Substituting from $R = L/(2\cos\theta)$ yields equation (1.1).

Kelvin's equation has proven to be fairly accurate down to surprisingly small values of L for temperatures $T < T_w$. Above the wetting temperature, when liquid layers adsorb at the walls, the appropriate geometric picture of the phase coexistence in the pore is shown in the right-hand panel of figure 1. In this case, the cylindrically-shaped meniscus tangentially connects the wetting films that are of thickness ℓ_π. For sufficiently wide pores, ℓ_π can be considered as a thickness of a complete wetting layer adsorbed on an isolated wall. This geometric interpretation suggests that the pore width L appearing in the denominator of equation (1.1) should be replaced by its effective value $L_{\mathrm{eff}} = L - 2\ell_\pi$. Explicit calculations based on a free energy balance including the effective interaction between interfaces show that L_{eff} depends on the range of the molecular forces and that $L_{\mathrm{eff}} = L - 3\ell_\pi$ when long range nonretarded dispersion forces are involved.

We now turn our attention to a heterogeneous pore, where, within the macroscopic treatment, the walls are characterised by contact angles θ_1 and θ_2. For the rest of the paper we shall assume without any loss of generality that the contact angle at the left-hand wall θ_1 is not larger than θ_2 and that $\theta_1 \leqslant \pi/2$ [if $\theta_1 > \pi/2$, the problem is either reversed spatially (if $\theta_2 < \pi/2$) or thermodynamically (if $\theta_2 > \pi/2$) in which case the gas and liquid phases interchange their roles]. We can again construct a geometric picture of the liquid-vapour coexistence in the pore, as is shown in figure 2. Now, the center of the meniscus is no more in the centre of the pore but is shifted towards the right-hand wall or even beyond in case of $\theta_2 > \pi/2$. A simple geometry then leads to the following generalization of Kelvin's equation for heterogeneous pores [8]:

$$\delta\mu \equiv \mu_{\mathrm{sat}} - \mu_{\mathrm{cc}} = \frac{\gamma(\cos\theta_1 + \cos\theta_2)}{(\rho_1 - \rho_v)L}. \tag{2.1}$$

Usually, Kelvin's equation is interpreted such that it tells us what is the chemical potential at the capillary condensation for a given pore width L. This view can also be reversed and we can ask what is the equilibrium distance L when we fix $\delta\mu$ and T. From the geometric construction shown in figure 2 it is clear that the "more heterogenous" the pore is, the shorter the distance L must be for thermodynamic criteria to be met.

Next, we wish to adopt a more microscopic view for the case when either $\theta_1 = 0$ or $\theta_2 = 0$ (or both). We then expect that the wetting layer of a thickness $\ell_\pi^{(i)}$ which is formed at the wall i will somewhat modify the purely macroscopic prediction of equation (2.1). When a wetting layer of thickness $\ell_\pi^{(i)}$ intrudes between a single wall i and the bulk gas, the corresponding surface free energy of the wall-gas interface

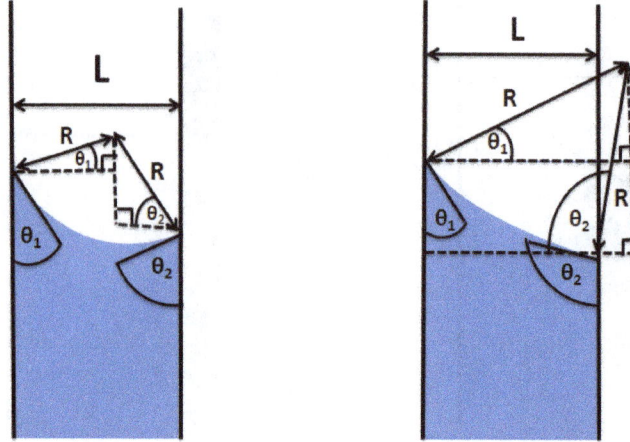

Figure 2. (Color obline) Sketch of a vapour-liquid coexistence in a heterogeneous pore of a width L. In this picture, the left-hand wall is partially wet and the right-hand wall is either partially wet but with a larger contact angle (left-hand panel) or partially dried ($\theta_2 > \pi/2$, as shown in the right-hand panel). Accordingly, the centre of the meniscus of a radius $R = \gamma/\delta\mu(\rho_l - \rho_v)$ is now closer to the right-hand wall.

is as follows:

$$\gamma_{w_i g} = \gamma_{w_i l} + \gamma + W_i\left(\ell_\pi^{(i)}\right), \tag{2.2}$$

where $\gamma_{w_i l}$ is the surface tension between the wall i and the liquid, and

$$W_i(\ell) = \delta p \ell + \frac{A_i}{\ell^2} + \frac{B_i}{\ell^3} + \cdots \tag{2.3}$$

is the effective potential between the wall-liquid and liquid-gas interfaces due to the wetting layer of a thickness ℓ. The coefficient A_i is called the Hamaker constant which must be positive for $T > T_w$. This form of the binding potential assumes that the wall-fluid or fluid-fluid interactions are dominated by nonretarded dispersion forces at large distances. The minimum of W determines the equilibrium thickness of the wetting layer

$$\ell_\pi^{(i)} \approx \left(\frac{2A_i}{\delta p}\right)^{1/3}, \qquad T \gtrsim T_w^{(i)}. \tag{2.4}$$

If only a microscopic wetting layer forms at the wall, i.e., if $T < T_w^{(i)}$, then the global minimum of $W_i(\ell)$ is negative and the liquid-vapour interface is pinned to the wall to a microscopic distance $\ell_\pi^{(i)}$ even for $\delta p = 0$, in which case the comparison of (2.2) with Young's equation reveals that

$$W_i(\ell_\pi^{(i)}) = \gamma(\cos\theta_i - 1), \qquad T < T_w^{(i)}. \tag{2.5}$$

Now, in the pore of a large width L, we assume that the thickness of the wetting layer adsorbed at either wall is the same as the one on a single wall. The free-energy difference per unit area between a low-density state (with the grand-potential per unit area $\omega_g = -pL + \gamma_{w_1 g} + \gamma_{w_2 g}$) and a high-density state (with the grand-potential per unit area $\omega_g = -p_1^+ L + \gamma_{w_1 l} + \gamma_{w_2 l}$, where p_1^+ denotes a pressure of the metastable liquid at a given $\mu < \mu_{sat}$ and T) is then given by

$$\Delta\omega = \omega_g - \omega_l = (p_1^+ - p)L + 2\gamma + W_1(\ell_\pi^{(1)}) + W_2(\ell_\pi^{(2)}), \tag{2.6}$$

where we have neglected effective interactions other than between the liquid-gas interface and the nearest wall.

In particular, when $\theta_1 = 0$ and $\theta_2 > 0$, a substitution of equations (2.3) and (2.5) into (2.6) gives

$$\Delta\omega = (p_1^+ - p)L + \delta p \ell_\pi^{(1)} + \frac{A_1(T)}{\left(\ell_\pi^{(1)}\right)^2} + \gamma(1 + \cos\theta_2). \tag{2.7}$$

Identifying $\delta p \approx p - p_1^+ \approx \delta\mu(\rho_1 - \rho_g)$ to first order in $\delta\mu$ and using equation (2.4), we obtain

$$\delta\mu = \frac{\gamma(1 + \cos\theta_2)}{\left(L - \frac{3}{2}\ell_\pi^{(1)}\right)(\rho_1 - \rho_g)}. \tag{2.8}$$

This result can be generalised as follows:

$$\delta\mu = \frac{\gamma(\cos\theta_1 + \cos\theta_2)}{\left[L - \frac{3}{2}\left(\ell_\pi^{(1)} + \ell_\pi^{(2)}\right)\right](\rho_1 - \rho_g)}, \tag{2.9}$$

where $\ell_\pi^{(i)} = 0$ if $\cos\theta_i > 0$.

The case of "antisymmetric" walls when $\cos\theta_1 = -\cos\theta_2$ deserves special attention. The modified Kelvin equation implies that the only phase transition can occur at $\mu = \mu_{\text{sat}}$. In this case, apart from the gas-like and liquid-like states, a configuration consisting of a thick film of liquid of width ℓ and a film of gas of thickness $L - \ell$ with the liquid-gas delocalized interface parallel to the walls should be also taken into account. The appropriate grand potential per unit area of this configuration is:

$$\omega_{\text{deloc}}(\ell) = \gamma_{\text{w}_1 l} + \gamma + \gamma_{\text{w}_2 g} + \frac{A_1}{\ell^2} + \frac{A_2}{(L-\ell)^2} + \frac{B_1}{\ell^3} + \frac{B_2}{(L-\ell)^3}, \tag{2.10}$$

where the higher-order terms were neglected.

Assuming that the walls are perfectly opposite, $A_1 = A_2 \equiv A$ and $B_1 = B_2 \equiv B$ we separately consider the cases $T < T_{\text{w}_1} = T_{\text{w}_2} \equiv T_{\text{w}}$ and $T > T_{\text{w}}$. The respective binding potentials are shown in figure 3 for the walls that undergo critical (upper panels) and first-order (lower panels) wetting transitions, when L is macroscopic. From the inspection of figure 3 we can conclude that for L macroscopic T_{w} represents a triple point for first-order wetting in which case three minima of $W(\ell)$ corresponding to low-density, high-density and delocalized states are of equal depths. For critical wetting, there are only two minima in the binding potential below T_{w}. Upon increasing the temperature, the two minima are getting closer to each other continuously and finally merge at the midpoint of the pore at T_{w} which thus represents a critical temperature in this case.

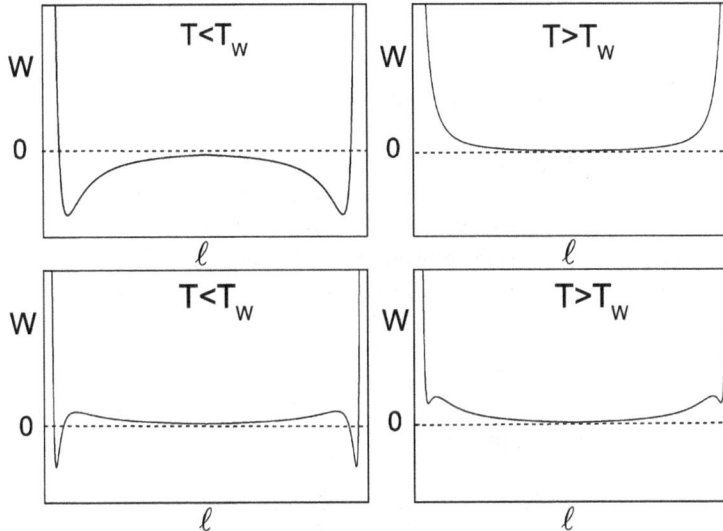

Figure 3. Schematic behaviour of a binding potential in a macroscopically wide pore with perfectly anti-symmetric walls. Below the wetting temperature T_{w}, the binding potential has two minima located near the walls that correspond to a high-density and a low-density state in which case the pore is predominantly filled by one phase with only a microscopic layer of the other fluid phase adsorbed on one of the walls. For $T > T_{\text{w}}$, there is a single minimum of the biding potential in the middle of the pore, so that thick liquid-like and gas-like films form at respective walls. The upper panel describes the case when the walls exhibit critical wetting and the lower panel is appropriate for first-order wetting.

We now wish to include the finite-size effects due to finite values of L, still considering the binding potential of the form (2.3). The grand potential of the delocalized state with the liquid-gas interface in the middle of the pore is as follows:

$$\omega_{\text{deloc}} = \gamma_{\text{w}_1\text{l}} + \gamma + \gamma_{\text{w}_2\text{g}} + \frac{8A}{L^2} + \frac{16B}{L^3},\tag{2.11}$$

whereas the grand potentials of the low-density configuration

$$\omega_{\text{g}} = \gamma_{\text{w}_1\text{g}} + \gamma_{\text{w}_2\text{g}}\tag{2.12}$$

and the high-density configuration

$$\omega_{\text{l}} = \gamma_{\text{w}_1\text{l}} + \gamma_{\text{w}_2\text{l}}\tag{2.13}$$

are identical by symmetry. For $T < T_{\text{w}}$, the comparison between (2.12) (say) and (2.11), yields, upon using Young's equation,

$$\gamma \cos\theta_1 = \gamma + \frac{8A}{L^2},\tag{2.14}$$

where we have neglected the contributions of $\mathcal{O}(L^{-3})$. For $T \lesssim T_{\text{w}}$ it follows that

$$-\theta_1^2 = \frac{8A}{\gamma L^2},\tag{2.15}$$

which can only be satisfied when the Hamaker constant is negative, i.e., only when the walls exhibit critical wetting (in which case the extreme of the binding potential at the centre of the pore is a local maximum). In this case, the critical temperature T^* at which the system breaks the symmetry is given implicitly by

$$\theta_1(T^*) = \frac{1}{L}\sqrt{\frac{8|A(T^*)|}{\gamma(T^*)}}.\tag{2.16}$$

Using an abbreviation $t \equiv (T_{\text{w}} - T)$ and noting that $\theta(T) \sim t^{3/2}$ and $A \sim t$ for critical wetting and small t, we obtain an asymptotic behaviour of $T^*(L)$:

$$T^* = T_{\text{w}} - \frac{f(T^*)}{L}, \qquad L \to \infty,\tag{2.17}$$

where $f(T^*) > 0$, in agreement with the result obtained on the basis of finite-size scaling arguments [8] $T^* = T_{\text{w}} - L^{-1/\beta_{\text{s}}}$, where β_{s} is the surface critical exponent for critical wetting, for algebraically decaying binding potential (2.3) in three bulk dimensions.

In a more microscopic manner, the phase behaviour of a fluid confined between antisymmetric walls exhibiting critical wetting can be analyzed for $T < T_{\text{w}}$ by comparing the grand potential of the delocalized state (2.11) with the localized (to wall 1) one. The latter is obtained by a substitution of $\ell = -3B/2A$ (recall that $A < 0$ and $B > 0$ for critical wetting below T_{w}), as given by a minimization of (2.3), back to (2.3) with $\delta p = 0$:

$$\omega_{\text{loc}} = \gamma_{\text{w}_1\text{l}} + \gamma + \gamma_{\text{w}_2\text{g}} + \frac{4}{27}\frac{A^3}{B^2} + \frac{A}{\left(L + \frac{3}{2}\frac{B}{A}\right)^2} + \frac{B}{\left(L + \frac{3}{2}\frac{B}{A}\right)^3}.\tag{2.18}$$

Balancing ω_{loc} and ω_{deloc} leads to the equation

$$\frac{4}{27}\frac{1}{\kappa^2} + \frac{1}{\left(1 + \frac{3}{2}\kappa\right)^2} + \frac{\kappa}{\left(1 + \frac{3}{2}\kappa\right)^3} - 16\kappa = 8,\tag{2.19}$$

where we have introduced a dimensionless parameter $\kappa \equiv B/AL$. Equation (2.19) can be solved graphically as is shown in figure 4 revealing four solutions with three of them, corresponding to negative κ, being relevant. This analysis suggests that within the interval between $\kappa = -2/3$, where the LHS of equation (2.19) has a pole and $\kappa \approx -0.23$, the delocalized state is the stable solution, while the localized state is the more stable solution otherwise. There is another point $\kappa = -1/3$ at which the localized and delocalized states are equally stable; however, this solution seems to be just an artifact of the current analysis

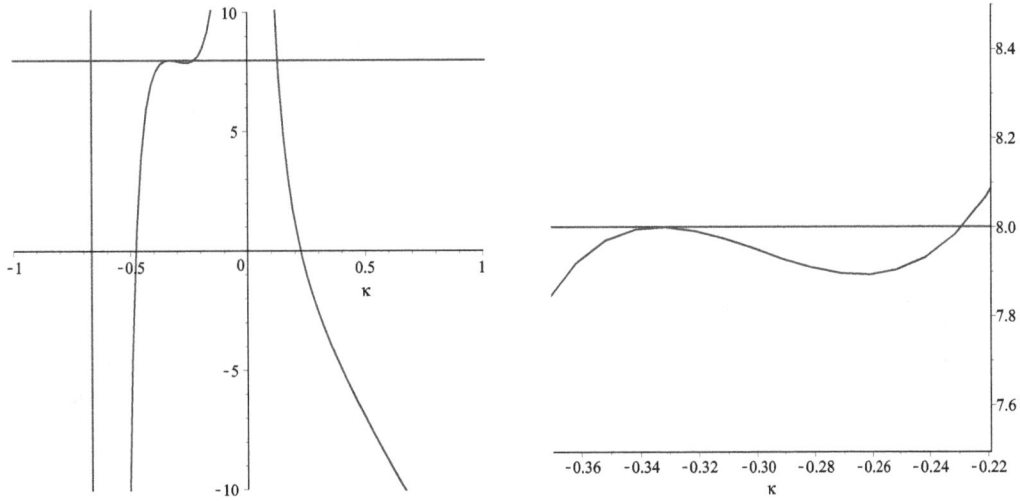

Figure 4. A graphical solution of equation (2.19). The curve (blue line) corresponds to the left-hand side of equation (2.19) as a function of $\kappa = B/AL$, the horizontal line (red) belongs to the right-hand side of equation (2.19). The plot in the right-hand panel magnifies the graph in the interval $\kappa = -0.37$ to $\kappa = -0.22$

which presumably disappears when higher order terms of $\mathcal{O}(L^{-4})$ in the binding potential are taken into account. We note that this result is not inconsistent with the asymptotic result given by equation (2.17), since in the latter case L is large and A is small, so that $|\kappa|$ is not necessarily small in this limit.

For $T > T_w$, there is only a delocalized state present for the walls undergoing critical wetting. For the walls exhibiting first-order wetting transition, the localized states become metastable with respect to the delocalized one in the macroscopic limit of the separation of the walls. For L finite, the binding potential is non-vanishing in the middle of the pore which makes the delocalized state less stable. The grand potential per unit area for the localized state is given by

$$\omega_{\text{loc}} = \gamma_{\text{w}_1\text{l}} + \gamma + \gamma_{\text{w}_2\text{g}} + W_1(\ell_{\text{loc}}) + W_2(L - \ell_{\text{loc}}), \tag{2.20}$$

where ℓ_{loc} is the minimum of the binding potential near the (left-hand) wall. In order to describe the binding potential for the case of first-order wetting (see figure 3), we should consider one more term in (2.3)

$$W_i(\ell) = \frac{A_i}{\ell^2} + \frac{B_i}{\ell^3} + \frac{C_i}{\ell^4} + \cdots, \tag{2.21}$$

where A_i, $C_i > 0$ and $B_i < 0$ and where we set $\delta p = 0$. Minimization of $W_i(\ell)$ gives $\ell_{\text{loc}} = -3B/4A - \sqrt{9B^2/A^2 - 32C/A}/4$, where we again assume that the walls are perfectly antisymmetric ($A_1 = A_2$, $B_1 = B_2$ and $C_1 = C_2$ and thus $W_1 = W_2 \equiv W$). The grand potential per unit area for the delocalized state is as follows:

$$\omega_{\text{deloc}} = \gamma_{\text{w}_1\text{l}} + \gamma + \gamma_{\text{w}_2\text{g}} + 2W(L/2). \tag{2.22}$$

The energy barrier of $W(\ell)$ disappears when its local minimum and maximum coincide which leads to the condition for the spinodal temperature T_s:

$$\frac{B^2}{AC} = \frac{32}{9}, \qquad T = T_s. \tag{2.23}$$

Above T_s, only the delocalized state is present. For $T_w < T < T_s$, there exists a wall separation $L^*(T)$ for which $\omega_{\text{loc}} = \omega_{\text{deloc}}$, which determines the location of the interface delocalization transition. At this wall separation, the presence of the walls rises (relative to the case of L macroscopic) the central minimum of the binding potential (corresponding to the delocalized state) up to the level of the minima near the walls (corresponding to the localized states). For $T \approx T_s$ we approximately obtain

$$L_c \equiv L^*(T_s) \approx \frac{3}{\sqrt{2}} \frac{|B|}{A}, \tag{2.24}$$

which is the critical value of L below which the interface localized-delocalized transition never occurs. In the opposite limit, $L \to \infty$, the transition occurs right at T_w in line with the macroscopic arguments. Below T_w, only the states with the interface pinned to either walls are stable for any value of L.

Finally, we note that in a more general case of competing walls when $A_1 \neq A_2$, the position of the delocalized interface is shifted towards the wall with a weaker surface field, such that

$$\ell_{eq} \approx \frac{L}{\left(1 + \sqrt[3]{A_2/A_1}\right)}. \tag{2.25}$$

3. Density functional theory

In order o describe microscopic properties of a simple fluid in heterogeneous pores we adopt a classical density function theory [22]. Within DFT, the equilibrium density profile is found by minimizing the grand potential functional

$$\Omega[\rho] = \mathscr{F}[\rho] + \int d\mathbf{r} \left[V(\mathbf{r}) - \mu \right] \rho(\mathbf{r}), \tag{3.1}$$

where $V(\mathbf{r})$ is the external field due to the walls and μ is the chemical potential. The intrinsic free energy functional $\mathscr{F}$ can be split into the contribution from the ideal gas and the remaining excess part

$$\mathscr{F}[\rho] = \mathscr{F}_{id}[\rho] + \mathscr{F}_{ex}[\rho], \tag{3.2}$$

where $\mathscr{F}_{id}[\rho] = k_B T \int d\mathbf{r}\rho(\mathbf{r}) \left[\ln(\Lambda^3 \rho(\mathbf{r})) - 1 \right]$ with Λ being the thermal de Broglie wavelength that can be set to unity.

The fluid model is characterised by a pair potential consisting of a hard-sphere repulsion of the range of σ, and a Lennard-Jones-like attractive portion given by

$$u(r) = \begin{cases} 0, & r < \sigma, \\ -4\varepsilon \left(\frac{\sigma}{r}\right)^6, & \sigma < r < r_c, \\ 0, & r \geq r_c, \end{cases} \tag{3.3}$$

where the cut-off is set to $r_c = 2.5\sigma$. The repulsive contribution to the excess free energy functional is approximated using Rosenfeld's fundamental measure theory [23] and the attractive part is treated within the mean-field approximation:

$$\mathscr{F}_{ex}[\rho] = k_B T \int d\mathbf{r}\,\Phi(\{n_\alpha\}) + \frac{1}{2} \int d\mathbf{r}\rho(\mathbf{r}) \int d\mathbf{r}'\rho(\mathbf{r}')u(|\mathbf{r} - \mathbf{r}'|), \tag{3.4}$$

The free energy density Φ is a function of a set of three independent weighted densities $\{n_\alpha\}$ for which we use the original Rosenfeld prescription.

The external field due to the parallel impenetrable walls a distance L apart is given by $V(\mathbf{r}) = V(z) = V_1(z) + V_2(L - z)$, where

$$V_i(z) = \begin{cases} \infty, & z < \sigma, \\ -\frac{2}{3} \frac{\pi \varepsilon_{w_i} \rho_{w_i} \sigma^6}{z^3}, & z \geq \sigma. \end{cases} \tag{3.5}$$

The parameter ρ_{w_i} is a density of uniformly distributed atoms forming the wall i, such that each atom exerts a potential $u(r)$ according to expression (3.3) with a parameter ε_{w_i} replacing ε and without a cut-off, i.e., for $r_c = \infty$. The potential V_i of a single wall i is given by integrating all contributions of atoms over the entire wall. The diameter σ of the wall atoms has been chosen equal to that of fluid atoms. Since the parameter ρ_{w_i} is always associated with ε_{w_i}, we can characterise the wall strength by tuning just a single parameter ε_{w_i} by setting $\rho_{w_1} = \rho_{w_2} = \sigma^{-3}$.

In order to make a link with the more phenomenological model introduced in the previous section, the Hamaker constant defined by equation (2.3) can be expressed in terms of the microscopic parameters using the sharp-kink approximation [24]. For the interaction between wall-liquid and liquid-gas interfaces we have

$$A_i = \frac{1}{3}\pi\varepsilon_{w_i}\rho_{w_i}(\rho_l - \rho_g)\sigma^6, \tag{3.6}$$

$$B_i = -\frac{2}{3}\pi\varepsilon_{w_i}\rho_{w_i}(\rho_l - \rho_g)\sigma^7, \tag{3.7}$$

where it should be noted that A_i is always positive in our model owing to the fact that the fluid-fluid interaction (3.3) is only short ranged and thus does not contribute to A_i. For the interaction between wall-gas
and gas-liquid interfaces, the expressions in equations (3.6) and (3.7) have opposite signs. However, for
"antisymmetric" walls, $\varepsilon_{w_1} = -\varepsilon_{w_2}$, in which case $A_1 = A_2$ and $B_1 = B_2$. This "antisymmetry" is decep-
tive, however, even if $\varepsilon_{w_1} = -\varepsilon_{w_2}$ one does not expect B_1 and B_2 to be identical beyond the sharp-kink
approximation, since these second-order contributions are due to excluded volume effects at the walls
and it is reasonable to expect that those of the wall-liquid interface are much stronger than those of the
wall-gas interface. Indeed, one does not expect the existence of a drying temperature at all for the purely
repulsive wall, so that a link between a microscopic and a mesoscopic model at the level of the sharp-kink
approximation is fully justified only up to the leading order with a coefficient A_i.

The minimization of (3.1) is carried out numerically on a 2D grid with spacing $dx = dz = 0.05\sigma$. Although the symmetry of the external field permits to treat the problem as one-dimensional $\rho(\mathbf{r}) = \rho(z)$, we
also wish to test the geometric arguments for microscopic values of L by constructing equilibrium density
profiles of a single pore filled with liquid- and gas-like coexisting phases, in which case $\rho(\mathbf{r}) = \rho(x, z)$. For
a given set of model and thermodynamic parameters, the 1D DFT results are used as boundary condi-
tions for the 2D DFT, such that we set $\rho_{2D}(x, z_{max}) = \rho_{1D}^{low}(x)$ and $\rho_{2D}(x, z_{min}) = \rho_{1D}^{high}(x)$. Here, z_{max} and
z_{min} are respectively the maximal and minimal values of the vertical coordinate z which are considered
within the 2D calculations and are set to $z_{min} = 0$ and $z_{max} = 40\sigma$. The functions $\rho_{1D}^{low}(x)$ and $\rho_{1D}^{high}(x)$ are
1D density profiles corresponding to the low and high density states, respectively. The low density state
is a configuration in which the system is filled primarily with a gas (the liquid-gas interface is pinned to
the left-hand wall), whereas the high density state is a configuration in which the system is either filled
with liquid (the interface is pinned to the right-hand wall) or a delocalized state in which case the liquid-
gas interface is around the centre of the pore. Further details and particularly the implementation of
Rosenfeld's functional within the 2D-DFT treatment can be found in reference [25].

We start the discussion of our numerical DFT results by examining the wetting properties of a single
wall. In figure 5 we display the temperature dependence of the contact angle for two walls with surface
fields $\varepsilon_{w_1} = 1.2\varepsilon$ (wall 1) and $\varepsilon_{w_2} = \varepsilon$ (wall 2). The walls exhibit first-order wetting transition at temperatures $T_{w_1} = 0.83 T_c$ and $T_{w_2} = 0.93 T_c$, with $k_B T_c/\varepsilon = 1.414$ corresponding to the bulk critical temperature
[26]. We now consider a pore formed by a wall 1 and a wall 2 at a temperature $T = 0.81 T_c$ ($k_B T/\varepsilon = 1.15$)
at which both walls, when separated, are only partially wet. In figure 6 we display the equilibrium 2D
density profiles corresponding to the liquid-gas coexistence in pores of widths $L = 20\sigma$, $L = 10\sigma$, and
$L = 5\sigma$. In this case, the walls of the pore in the low-density phase are only microscopically wet and the
liquid-gas meniscus meets both walls at angles that appear in a good agreement with the predicted values of the respective contact angles (cf. figure 5), including, somewhat surprisingly, even the narrow pore

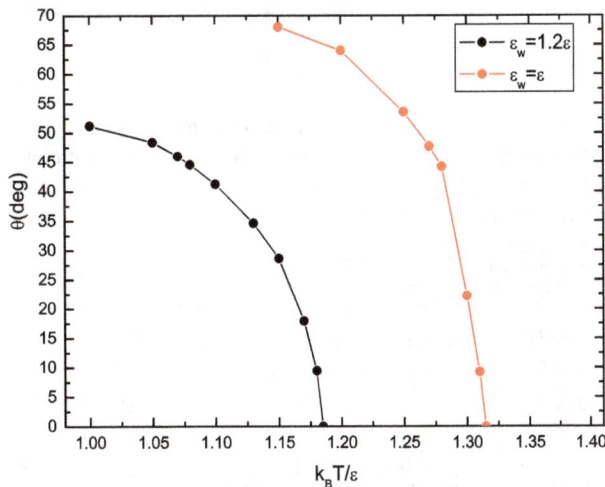

Figure 5. (Color online) Variation of the contact angle with temperature for walls with strengths $\varepsilon_{w_1} =$
1.2ε and $\varepsilon_{w_2} = \varepsilon$.

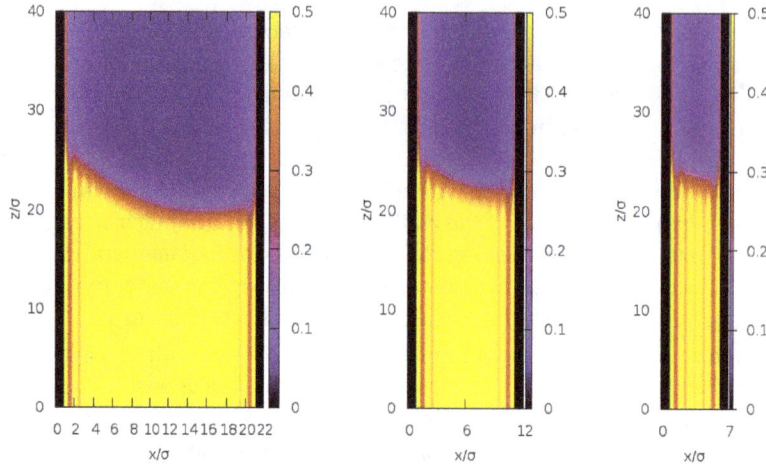

Figure 6. (Color online) Two-dimensional equilibrium density profiles of a fluid confined between walls with surface fields $\varepsilon_{w_1} = 1.2\varepsilon$ and $\varepsilon_{w_2} = \varepsilon$ at a temperature $T = 0.81\,T_c$ ($k_B T/\varepsilon = 1.15$) which is below both T_{w_1} and T_{w_2}. The contact angles at the single walls are $\theta_1 = 29°$ and $\theta_2 = 68°$, as given by the results displayed in figure 5. The width of the pores accessible for fluid particles is (from left to right) $L = 20\sigma$, $L = 10\sigma$, and $L = 5\sigma$. All configurations correspond to the liquid-gas coexistence in the respective pores, i.e., the capillary condensation which occurs at the chemical potentials that are shifted below $\mu_{sat} = -4.0309\,\varepsilon$ by the values $\delta\mu(L = 20\sigma) = 0.0275\,\varepsilon$, $\delta\mu(L = 10\sigma) = 0.0547\,\varepsilon$ and $\delta\mu(L = 5\sigma) = 0.1934\,\varepsilon$.

with $L = 5\sigma$. We have further determined the location of capillary condensation for each of the pores, and the results are in perfect agreement with the modified Kelvin's equation (2.1), cf. table 1.

We next consider a temperature $T = 0.92\,T_c$ ($k_B T/\varepsilon = 1.3$), in which case the wall 2 is still partially wet but the wall 1, when separated and at bulk two-phase coexistence, would be completely wet. The equilibrium density profiles at two-phase coexistence are shown in figure 7. Now, the comparison of the DFT results with the macroscopic Kelvin equation (2.1) is less satisfactory than in the previous case, where both walls are non-wet, especially for narrow pores, see table 1. Nevertheless, if we employ the modified Kelvin equation as given by (2.9), we obtain a prediction in a very good agreement with DFT, provided the capillaries are sufficiently wide. For instance, for $L = 50\sigma$, the relative difference in $\delta\mu$ between DFT and the modified Kelvin equation is only 4% (compared to 18% for the macroscopic Kelvin equation). Only for those large capillaries, the assumptions leading to equation (2.9) are justified. For the pores widths considered here, one can still use equation (2.9) but with ℓ_π replaced by the real thickness of the wetting layer. This can be read off from the 1D density profiles by determining e.g., the Gibbs dividing surface of the liquid-gas interface. The inclusion of this correction substantially improves the agreement with the

Table 1. The predicted values of a location of capillary condensation, as given by Kelvin's equation (2.1) and its modified version taking into account the effect of wetting layers [equation (2.9)], are compared with DFT results for three pore widths and temperatures $T < T_{w_1} < T_{w_2}$, $T_{w_1} < T < T_{w_2}$ and $T_{w_1} < T_{w_2} < T$. The comparison is made for the heterogeneous pore with $\varepsilon_{w_1} = 1.2\varepsilon$ and $\varepsilon_{w_2} = \varepsilon$.

Temperature	Pore width	DFT	equation (2.1)	equation (2.9)
$k_B T/\varepsilon = 1.15$	$L = 20\sigma$	$\delta\mu = 0.027502\,\varepsilon$	$\delta\mu = 0.026998\,\varepsilon$	–
	$L = 10\sigma$	$\delta\mu = 0.054670\,\varepsilon$	$\delta\mu = 0.053996\,\varepsilon$	–
	$L = 5\sigma$	$\delta\mu = 0.101500\,\varepsilon$	$\delta\mu = 0.107001\,\varepsilon$	
$k_B T/\varepsilon = 1.3$	$L = 20\sigma$	$\delta\mu = 0.027415\,\varepsilon$	$\delta\mu = 0.020330\,\varepsilon$	$\delta\mu = 0.026229\,\varepsilon$
	$L = 10\sigma$	$\delta\mu = 0.059021\,\varepsilon$	$\delta\mu = 0.040635\,\varepsilon$	$\delta\mu = 0.058079\,\varepsilon$
	$L = 5\sigma$	$\delta\mu = 0.121618\,\varepsilon$	$\delta\mu = 0.081302\,\varepsilon$	$\delta\mu = 0.116158\,\varepsilon$
$k_B T/\varepsilon = 1.35$	$L = 20\sigma$	$\delta\mu = 0.0234047\,\varepsilon$	$\delta\mu = 0.014348\,\varepsilon$	$\delta\mu = 0.026087\,\varepsilon$

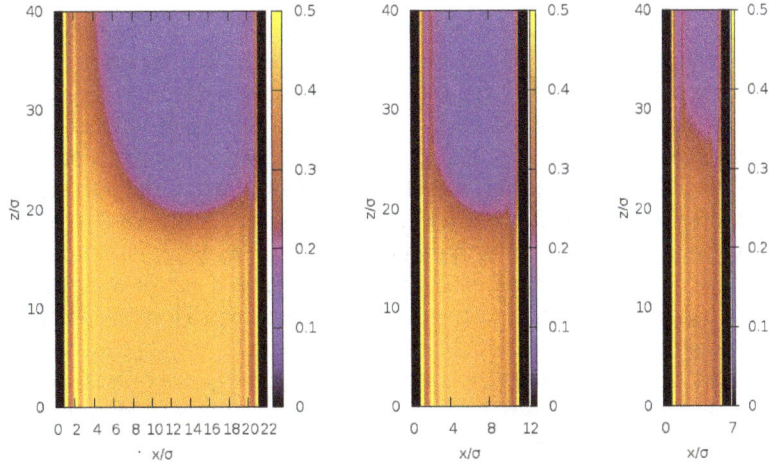

Figure 7. Two-dimensional equilibrium density profiles of a fluid confined between walls with surface fields $\varepsilon_{w_1} = 1.2\varepsilon$ and $\varepsilon_{w_2} = \varepsilon$ at a temperature $T = 0.92\,T_c$ ($k_B T/\varepsilon = 1.3$) which is below T_{w_2} but above T_{w_1}. The contact angle of the partially wet wall is $\theta_1 = 22°$, as given by the results displayed in figure 5. The width of the pores accessible for fluid particles is (from left to right) $L = 20\sigma$, $L = 10\sigma$, and $L = 5\sigma$. All configurations correspond to the liquid-gas coexistence in the respective pores, i.e., the capillary condensation which occurs at the chemical potentials that are shifted below $\mu_{sat} = -3.9651\,\varepsilon$ by the values $\delta\mu(L = 20\sigma) = 0.0275\,\varepsilon$, $\delta\mu(L = 10\sigma) = 0.0590\,\varepsilon$ and $\delta\mu(L = 5\sigma) = 0.1216\,\varepsilon$.

DFT results, as is shown in table 1.

Finally, if we increase the temperature above T_{w_2}, then both walls are covered with wetting films but of different thickness, as can be seen from the equilibrium density profile for $L = 20\sigma$ and $\mu = \mu_{cc}$ in figure 8. As expected, the presence of the wetting layers deteriorates the quality of the prediction given by the macroscopic Kelvin equation [equation (2.1)] but the modified Kelvin equation [equation (2.9)] is still fairly accurate, when again both $\ell_\pi^{(i)}$ are replaced by the respective film thicknesses, see table 1. At this temperature ($T = 0.95\,T_c$), there is no capillary condensation for the intermediate ($L = 10\sigma$) and narrow ($L = 5\sigma$) pores. While the $\mu-\Omega$ dependence (not shown here) is smooth for the narrowest pore. It exhibits a kink for the intermediate pore suggesting that the critical pore width at this temperature is $L_c \approx 10\sigma$.

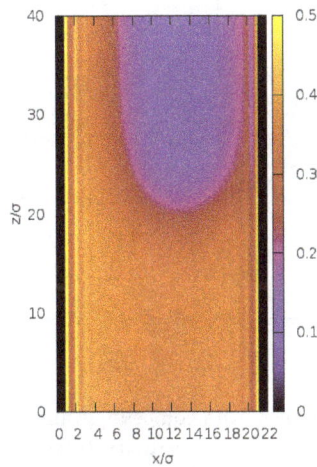

Figure 8. Two-dimensional equilibrium density profile of a fluid confined between walls with surface fields $\varepsilon_{w_1} = 1.2\varepsilon$ and $\varepsilon_{w_2} = \varepsilon$ at a temperature $T = 0.95\,T_c$ ($k_B T/\varepsilon = 1.35$) which is above both T_{w_1} and T_{w_2}. The width of the pores accessible for fluid particles is $L = 20\sigma$ and the configuration corresponds to the liquid-gas coexistence and occurs at the chemical potentials that is shifted below $\mu_{sat} = -3.9569\,\varepsilon$ by the value $\delta\mu(L = 20\sigma) = 0.0234\,\varepsilon$.

At last, we turn our attention to the case of asymmetric walls, such that one wall tends to be wet and the other dry. To this end, we consider a capillary in which the wall 1 has the same strength as before, i.e., $\varepsilon_{w_1} = 1.2\varepsilon$ but the opposite wall is purely repulsive. The repulsive wall exerts the potential according to (3.5) with a *negative* surface field $\varepsilon_{w_2} = -1.2\varepsilon$. Although the Hamaker constants and thus the asymptotic behaviour of binding potentials for both walls are assumed to be the same, it is only the attractive wall (wall 1) which exhibits (first order) wetting transition at temperature T_w. The purely repulsive wall (wall 2), does not induce a drying transition, i.e., the wall is always completely wet by gas. Consequently, the binding potential of wall 2 is purely repulsive (monotonously decaying) in contrast with the binding potential of wall 1 which, at least for sufficiently low temperatures, has two competing minima that are of equal depth at $T = T_w$. In figure 9 we display numerically constructed binding potentials for this model pore using a constrained minimization of (3.1) at bulk two-phase coexistence for several representative temperatures and a relatively wide pore ($L = 50\,\sigma$). For $T < T_w$, the global minimum in the total binding potential lies near wall 1, which means that the liquid-gas interface is pinned to the adsorbing wall and thus the system is in a low-density state for $\mu \leqslant \mu_{sat}$. On the other hand, if the two-phase coexistence is approached from the liquid state, the pore exhibits capillary emptying for some $\mu > \mu_{sat}$, according to the Kelvin equation (2.1). For $T \approx T_w$, the binding potential has two equally deep minima: one near the wall 1 and the other at the middle of the pore. For a perfectly antisymmetric system (such as for magnets), this temperature would be identified with a triple point, where the states with a liquid-gas interface near both walls and the state with the liquid-gas interface in the centre of the pore are all equally stable. In this model, however, the configuration corresponding to the pore filled with liquid is missing (at $\mu = \mu_{sat}$), so that the transition at a temperature $T^* \approx T_w$ can be interpreted as a thin-to-thick transition in some analogy with prewetting transition on a single wall. However, at a single wall, the prewetting transition is induced by the appearance of a non-zero term $\delta p \ell$ in the binding potential (2.3), i.e., because the system is away from bulk coexistence, which shifts a local minimum of $W(\ell)$ at $\ell = \infty$ (at saturation, $\delta p = 0$) to a finite distance from the wall (for $\delta p > 0$). For our system, the shift of the minimum is due to the presence of the second wall. For a temperature $T > T_w$, the minimum of $W(\ell)$ at $\ell = L/2$ becomes a global minimum corresponding to a delocalized interface. Above the spinodal temperature T_s, the energy barrier disappears and the configuration with the bounded interface ceases to be even metastable.

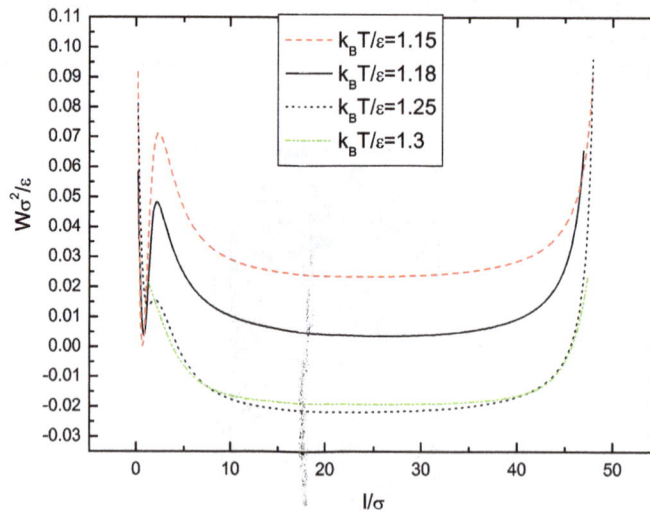

Figure 9. Binding potentials in a capillary with competing walls ($\varepsilon_{w_1} = -\varepsilon_{w_2} = 1.2\varepsilon$) and a width $L = 50\,\sigma$. For all temperatures below the spinodal temperature T_s the binding potential exhibits two minima corresponding to the liquid-gas interface being pinned at wall 1 and a delocalized interface with a mean location in the pore midpoint. The competition between the two minima determines the equilibrium configuration of the fluid. For this wide pore, the crossover temperature T^* between the states, at which the interface is bounded to the wall ($T < T^*$) and delocalized ($T > T^*$) is $T^* \approx T_w$ ($k_B T_w/\varepsilon = 1.18$ for wall 1). The highest temperature ($k_B T = 1.3\varepsilon$) is already above T_s and thus the binding potential has only one minimum in the middle of the pore. Also note a very small energetic barrier for $k_B T = 1.25\varepsilon$.

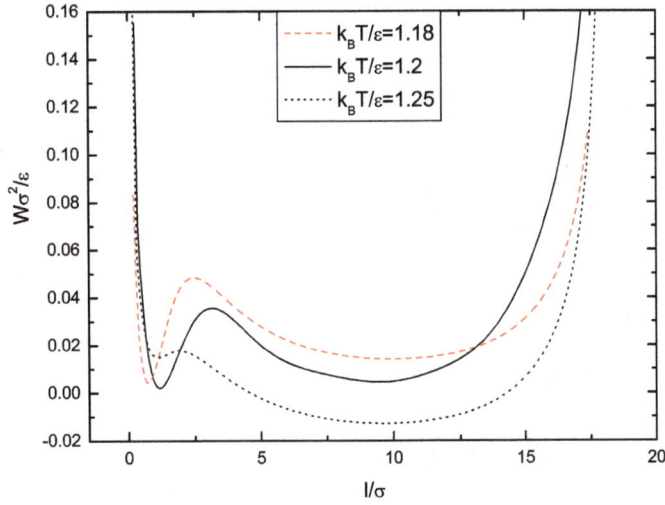

Figure 10. Binding potentials in a capillary with competing walls ($\varepsilon_{w_1} = -\varepsilon_{w_2} = 1.2\,\varepsilon$) and a width $L = 20\,\sigma$. In this case, in contrast to the case of a wide pore shown if figure 9, the temperature T^* at which two different fluid configurations coexist is above T_w and corresponds to $k_B\,T \approx 1.2\,\varepsilon$.

Thus far, the DFT results are in line with the predictions obtained from macroscopic considerations. However, if the pore width is reduced, the midpoint minimum of the binding potential becomes more affected by the (repulsive) interaction between the liquid-gas interface with both walls. Consequently, the midpoint minimum is pushed upwards which brings about a two-phase coexistence at $T^* > T_w$, as illustrated in figure 10 for the pore width of $L = 20\,\sigma$. In figure 11 we also show a density profile for a temperature $k_B\,T = 1.2\,\varepsilon$ at which the states corresponding to the localized and delocalized interface coexist. As we reduce the pore width even more, $T^*(L)$ still increases and terminates at T_s which determines the critical pore width $L_c = L(T^* = T_s)$ below which no phase transition occurs at $\mu = \mu_{\text{sat}}$. In figure 12 we display binding potentials for a near critical pore width ($L = 10\,\sigma$); note that the two-phase coexistence occurs now at a temperature $k_B\,T/\varepsilon = 1.25$ for which the binding potential for a single wall (or a wide pore) has only a very weak minimum near the wall (cf. figure 9). Also shown here is the binding potential for $T > T_s$ which does not permit any phase transition at $\mu = \mu_{\text{sat}}$ for any pore width.

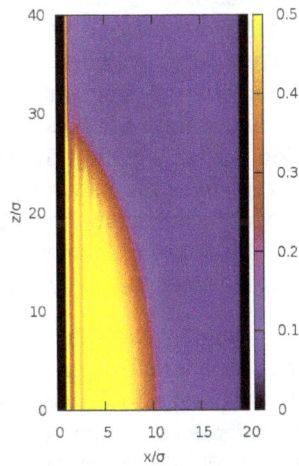

Figure 11. Two-dimensional density profile of a fluid confined between parallel walls with strengths $\varepsilon_w = 1.2\,\varepsilon$ (left-hand wall) and $\varepsilon_w = -1.2\,\varepsilon$ (right-hand wall), separated by a distance $L = 20\,\sigma$. The temperature of the system $k_B\,T = 1.2\,\varepsilon$ corresponds to the interface localized-delocalized transition ($\mu = \mu_{\text{sat}}$).

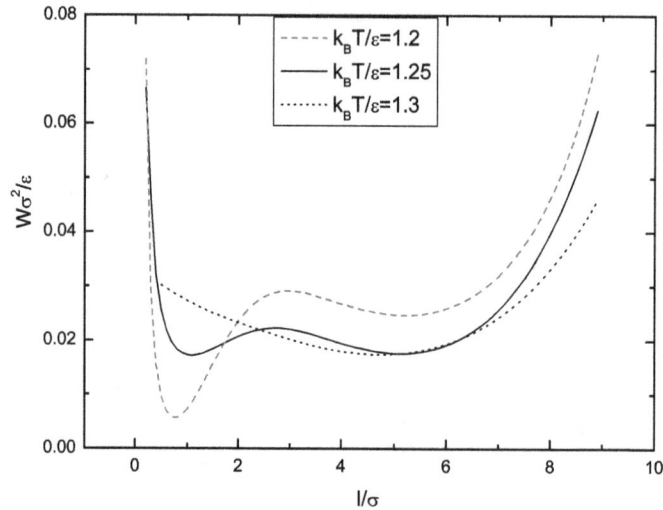

Figure 12. Binding potentials in a capillary with competing walls ($\varepsilon_{w_1} = -\varepsilon_{w_2} = 1.2\varepsilon$) and a width $L = 10\sigma$. The temperature T^* at which two different fluid configurations coexist is now $k_B T \approx 1.25\varepsilon$, which is only slightly below the spinodal temperature (cf. figure 9). For $T > T_s$, the binding potential has only one minimum in the pore midpoint.

4. Summary and concluding remarks

In this work we have studied the phase behaviour of a simple fluid confined between two parallel walls with different surface fields in the presence of long-range dispersion forces. We have shown that simple geometric arguments lead to the generalized macroscopic Kelvin equation which, when at least one of the walls is in complete wetting regime, should be modified by capturing the effect of thick wetting films by including Derjaguin's correction. We have also shown that simple mean-field arguments can be invoked to describe the fluid phase behaviour in a capillary at which one wall tends to be wet while the other tends to be dry and which exhibits interface localized-delocalized transition. Some of these predictions have been tested using a microscopic non-local density functional theory. The main results of this work can be summarised as follows:

1. Simple geometric arguments lead to a macroscopic prediction for a location (a shift in the chemical potential relative to that of bulk saturated fluid) of the capillary condensation/evaporation in a pore made of different walls. This leads to the (generalized) Kelvin equation (2.1) which has been tested against DFT calculations and the comparison revealed a very good agreement between these two approaches even for very narrow pores provided both walls are in a partial wetting/drying regime. The geometric interpretation of Kelvin's equation was also supported by constructing 2D density profiles. However, if the contact angle of at least one wall is zero (or π), the predictions of the macroscopic Kelvin equation worsen significantly, especially as the pore width decreases. Nevertheless, a substantial improvement of the Kelvin equation can be achieved by taking into account the effect due to the presence of wetting (or drying) films. The film thicknesses that appear in this corrected Kelvin equation (2.9) can be determined from equation (2.4) provided the pore is wide enough ($L \gtrsim 50\sigma$, with σ being the molecular diameter).

2. We have then revisited a model of a capillary slit with competing walls. Here, the Kelvin equation as well as symmetry considerations dictate that any phase coexistence must occur at $\mu = \mu_{sat}$. In this case, the liquid-vapour interface is parallel to the walls and can be pinned to either of the walls (in which case the system is in a condensed or evaporated state) or can be unbounded from both walls around a halfway between the walls; the latter state is referred to as a soft mode phase or a delocalized interface. Nature and the location of a transition when the interface unbinds (interface localized-delocalized transition) are determined by an interplay between the wetting properties of the walls and the finite-size effects due to finite separation L of the walls:

(a) According to macroscopic arguments ($L \gg \sigma$), the transition occurs at $T^* = T_w = T_d$, assuming the walls are perfectly antisymmetric. If the walls exhibit continuous wetting/drying transition at T_w, then T^* is a critical temperature, below which either of the localized state is equally stable. As $T \to T_w^-$, the liquid-gas unbinds continuously from a given wall and moves towards the pore midpoint. Above T_w only a single state corresponding to a delocalized interface is present. If the walls undergo first-order wetting transition, then $T^* = T_w$ (and $\mu = \mu_{sat}$) is a triple point at which all three configurations coexist. These macroscopic predictions were verified by DFT calculations for pores of widths $L \gtrsim 50\sigma$.

(b) As the pore width is reduced, T^* does not coincide with T_w anymore but becomes a function of L. The behaviour of $T^*(L)$ then strongly depends on the nature of the wetting transition of the walls and can be described by mean-field mesoscopic arguments that capture the effect of the effective interaction between the liquid-gas interface and the attractive tails of the wall potentials. For the walls exhibiting first-order wetting transition, the analysis shows that there are three temperature regimes. For $T < T_w$, the only stable solutions correspond to a bounded interface for any values of L. For $T_w < T < T_s$, there is an interface localized-delocalized transition for the wall separation $L^*(T)$ which decreases with an increasing T. Above T_s, there is no transition and there is a single configuration corresponding to the delocalized interface. The spinodal temperature T_s is characterised by a disappearance of the energy barrier in the binding potential and can be associated with the surface critical point of prewetting transition. The spinodal temperature determines the critical width $L_c = L^*(T_s)$, which is a minimal value of L allowing the interface localization-delocalization transition. Our DFT results suggest that L_c is slightly below 10σ, in a reasonable agreement with the mesoscopic prediction as given by equation (2.24) with the parameters A and B determined from equations (3.6) and (3.7), which yields $L_c \approx 6\sigma$. Conversely, these results can also be interpreted such that for each $L_c < L < \infty$ there exists a temperature T^* below which the fluid undergoes capillary filling (as μ approaches μ_{sat} from below) or emptying (as μ approaches μ_{sat} from above) and above which no transition is present and the interface is delocalized.

(c) If the walls exhibit critical wetting at T_w then, in contrast to the case of first-order wetting, the finite-size effects favour interface delocalization, so that the interface localized-delocalized transition is shifted below T_w. For very large L, this shift is proportional to L^{-1} with a (non-universal) temperature dependent amplitude that can be in principle determined from a given molecular model and this result is consistent with the one obtained on the basis of finite-size scaling [8]. For moderately large values of L and the model with the binding potential $W(\ell) = A(T)\ell^{-2} + B(T)\ell^{-3} + \cdots$, the mesoscopic analysis suggests that the phase behaviour can be characterized by a dimensionless parameter $\kappa = B/AL$, such that the localized-delocalized transition occurs for $\kappa = -2/3$ and $\kappa \approx -0.24$ with the delocalized interface being a stable solution within this interval.

3. Within the microscopic DFT model that has been considered in this work, the walls exert long range (decaying as z^{-3} for large z) potential while the fluid-fluid interaction was taken to be only short-ranged. This particular choice of the interaction model requires some comments:

(a) In terms of the study of interface localized-delocalized transition, this model has one advantage and one disadvantage. The great advantage is that the competing walls can be made antisymmetric by simply choosing $\varepsilon_{w_1} = -\varepsilon_{w_2}$ and this antisymmetry is maintained regardless of the temperature. This is because the only temperature dependent factor in both Hamaker constants (corresponding to the situation when the adsorbed film at the wall is a liquid or a gas phase) is $(\rho_l - \rho_g)$. Therefore, the Hamaker constants certainly vary with temperature but in identical manner for both walls which allowed us to study the phase behaviour of a fluid between antisymmetric walls for various temperatures. This is in contrast with the model used in reference [21], where the fluid-fluid interaction is also long-ranged which produces an extra temperature dependent factor in the respective Hamaker constants $A_{w1} \propto (\rho_l - \rho_w)$ and $A_{w2} \propto (\rho_g - \rho_w)$ that are identical at only one particular temperature. On the other hand, a disadvantage of our model is that since the Hamaker constant is always positive, a possibility that the walls undergo critical wetting is excluded, hence we could only test the cases when the walls exhibit first-order wetting.

(b) Setting $\varepsilon_{w_1} = -\varepsilon_{w_2}$ makes the wall 2 purely repulsive, so that even though $A_1 = A_2$ the walls cannot be viewed as perfectly antisymmetric in the sense $T_w = T_d$ (T_d is the drying temperature of the wall 2), because the wall 2 is completely dried at all temperatures for which vapour and liquid may coexist. This broken (anti)symmetry is reflected in the binding potential for the capillary with competing walls which exhibits at most two minima, missing a local minimum near wall 2. In contrast with the fully antisymmetric model (considered within the macro- and mesoscopic pictures), the temperature T^* for first-order wetting is not a triple temperature but a temperature appropriate to an ordinary first-order transition at which the liquid-gas interface jumps from the proximity of wall 1 to the centre of the pore in some analogy to prewetting transition (in contrast to the latter, the transition takes place at $\mu = \mu_{sat}$ and the jump is determined by L). Also note that below T_w, $\cos\theta_1 + \cos\theta_2$ appearing in Kelvin's equation becomes negative, meaning that for $T < T_w$ a two phase coexistence (capillary evaporation) occurs at $\mu > \mu_{sat}$. Finally, if the competing walls have different Hamaker constants, the interface localized-delocalized transition is still possible but with the position of the delocalized interface near $\ell = L/[1 + (A_2/A_1)^{1/3}]$ rather than $\ell = L/2$.

Acknowledgements

The financial support from the Czech Science Foundation, project number 13-09914S is acknowledged.

References

1. Fisher M.E., Nakanishi H., J. Chem. Phys., 1981, **75**, 5857; doi:10.1063/1.442035.
2. Nakanishi H., Fisher M.E., J. Chem. Phys., 1983, **78**, 3279; doi:10.1063/1.445087.
3. Evans R., Tarazona P., Phys. Rev. Lett., 1984, **52**, 557; doi:10.1103/PhysRevLett.52.557.
4. Evans R., J. Phys.: Condens. Matter, 1990, **2**, 8989; doi:10.1088/0953-8984/2/46/001.
5. Gelb L.D., Gubbins K.E., Radhakrishnan R., Sliwinska-Bartkoviak M., Rep. Prog. Phys., 1999, **62**, 1573; doi:10.1088/0034-4885/62/12/201.
6. Evans R., Marini Bettolo Marconi U., Chem. Phys. Lett., 1985, **114**, 415; doi:10.1016/0009-2614(85)85111-3.
7. Evans R., Marini Bettolo Marconi U., Phys. Rev. A, 1985, **32**, 3817; doi:10.1103/PhysRevA.32.3817.
8. Parry A.O., Evans R., Phys. Rev. Lett., 1990, **64**, 439; doi:10.1103/PhysRevLett.64.439.
9. Parry A.O., Evans R., Physica A, 1992, **181**, 250; doi:10.1016/0378-4371(92)90089-9.
10. Swift M.R., Owczarek A.L., Indekeu J.O., Europhys. Lett., 1991, **14**, 475; doi:10.1209/0295-5075/14/5/015.
11. Ferrenberg A.M., Landau D.P., Binder K., Phys. Rev. E, 1998, **58**, 3353; doi:10.1103/PhysRevE.58.3353.
12. Binder K., Landau D.P., Ferrenberg A.M., Phys. Rev. Lett., 1995, **74**, 298; doi:10.1103/PhysRevLett.74.298.
13. Binder K., Landau D.P., Ferrenberg A.M., Phys. Rev. E, 1995, **51**, 2823; doi:10.1103/PhysRevE.51.2823.
14. Binder K., Evans R., Landau D.P., Ferrenberg A.M., Phys. Rev. E, 1996, **53**, 5023; doi:10.1103/PhysRevE.53.5023.
15. Binder K., Landau D.P., Müller M., J. Stat. Phys., 2003, **110**, 1411; doi:10.1023/A:1022173600263.
16. Schulz B.J., Binder K., Müller M., Phys. Rev. E, 2005, **71**, 046705; doi:10.1103/PhysRevE.71.046705.
17. Albano E.V., Binder K., J. Stat. Phys., 2009, **135**, 991; doi:10.1007/s10955-009-9710-8.
18. De Virgiliis A., Vink R.L.C., Horbach J., Binder K., Europhys. Lett., 2007, 77, 60002; doi:10.1209/0295-5075/77/60002.
19. Binder K., Horbach J., Vink R.L.C., De Virgiliis A., Soft Matter, 2008, **4**, 1555; doi:10.1039/b802207k.
20. De Virgiliis A., Vink R.L.C., Horbach J., Binder K., Phys. Rev. E, 2008, **78**, 041604; doi:10.1103/PhysRevE.78.041604.
21. Stewart M.C., Evans R., Phys. Rev. E, 2012, 86, 031601, doi:10.1103/PhysRevE.86.031601.
22. Evans R., Adv. Phys., 1979, **28**, 143; doi:10.1080/00018737900101365.
23. Rosenfeld Y., Phys. Rev. Lett., 1989, **63**, 980; doi:10.1103/PhysRevLett.63.980.
24. Dietrich S., In: Phase Transitions and Critical Phenomena, Vol. 12, Domb C., Lebowitz J.L. (Eds.), Academic, New York, 1988.
25. Malijevský A., J. Chem. Phys., 2012, **137**, 214704; doi:10.1063/1.4769257.
26. Malijevský A., Parry A.O., Phys. Rev. Lett., 2013, **110**, 166101; doi:10.1103/PhysRevLett.110.166101.

5

Equilibrium clusters in suspensions of colloids interacting via potentials with a local minimum

A. Baumketner[1]* W. Cai[2]

[1] Institute for Condensed Matter Physics, National Academy of Sciences of Ukraine,
   1 Svientsistskii St., 79011 Lviv, Ukraine
[2] Beijing Computational Science Research Center, Beijing 100094, China

In simple colloidal suspensions, clusters are various multimers that result from colloid self-association and exist in equilibrium with monomers. There are two types of potentials that are known to produce clusters: a) potentials that result from the competition between short-range attraction and long-range repulsion and are characterized by a *global* minimum and a repulsive tail and b) purely repulsive potentials which have a soft shoulder. Using computer simulations, we demonstrate in this work that potentials with a *local* minimum and a repulsive tail, not belonging to either of the known types, are also capable of generating clusters. A detailed comparative analysis shows that the new type of cluster-forming potential serves as a bridge between the other two. The new clusters are expanded in shape and their assembly is driven by entropy, like in the purely repulsive systems but only at low density. At high density, clusters are collapsed and stabilized by energy, in common with the systems with competing attractive and repulsive interactions.

Key words: *colloids, clusters, local minimum, repulsive potential, computer simulations*

1. Introduction

The term "clusters" refers to a large variety of objects that range in size from small multimers to mesoscopic domains [1] and arise as a consequence of monomer self-association in a large variety of soft materials [2]. Most often, clusters are discussed in reference to colloidal suspensions [3], where they exist in equilibrium with monomers, but they were also reported for proteins [4], synthetic clays [5] and metal nanoparticles [6]. Equilibrium clusters become the dominant species in the solution at appropriate thermodynamic conditions. They may also arise transiently, as a consequence of arrested phase transition [7].

As a particular case of the self-assembly process, cluster formation is of key interest to basic research, in particular condensed matter physics. Additionally, it also has appreciable practical applications, for instance as a drug-delivery vehicle [8]. Clusters are capable of significantly altering the mechanical properties of aqueous solutions in which they assemble. This is the case, for instance, of solutions containing monoclonal antibodies, a known biopharmaceutical, which experience a considerable viscosity increase if clusters are present [9]. It is imperative, therefore, to develop a basic understanding of the principles underlying the formation of clusters, in order to use these systems successfully for therapeutic purposes.

Historically, it seems, the possibility of clusters emerging spontaneously in a homogeneous fluid was first raised within the concept of competing interactions. When discussing phase transitions in systems interacting via attractive potentials, Lebowitz and Penrose [10] asked about the effect of an additional repulsive part in the potential, whose range is longer than that of the attractive part. Their conclusion was that the liquid formed as a result of the normal first-order gas-liquid transition due to the attraction among particles would break into finite-size droplets whose size is large compared to the size of the colloid but small compared to the range of the repulsion. Over the past forty years, this scenario was seen to

---

* E-mail: andrij@icmp.lviv.ua

play out in a variety of systems, interacting by a variety of potentials. Kendrick et al. [11] considered a system of colloids interacting via repulsive Coulomb and attractive van der Waals forces and concluded that as a result of the competition between these two interactions, the critical point between the low density liquid phase and the high-density liquid phase is 'preempted by a finite wave vector critical point' [11]. As a result of the new phase transition, macrophases with inhomogeneous density distributions are formed. The same idea of competition between short-range attractive and long-range repulsive forces (SALR) in simple fluids was later researched by Sear et al. [12, 13] in the context of colloids lying on the air-water interface. Analytic mean-field theory [13] and computer simulations [12] indicated that, indeed, instead of the liquid-liquid phase separation (LLPS), colloids form various patterned (modulated) phases, including finite-size clusters. In many respects, these were similar to the patterns created by the competition between repulsion and attraction and described previously for complex fluids [14, 15]. Further progress was made with the work of Groenewold and Kegel [16] who showed how an elaborated model describing the amount of charge on a colloidal surface may lead to the stabilization of clusters of large size. Since that publication, a large amount of work on SALR clusters has followed (see references [17, 18] and references therein) focusing mainly on how the LLPS line is broken/modified by the presence of the repulsion in the potential. A sophisticated and accurate self-consistent Ornstein-Zernike approximation (SCOZA) within the integral-equation theory of the liquid state was used by the group of Reatto [19, 20] to delineate the line in the phase diagram separating the homogeneous fluid phase from the cluster fluid phase, or the so-called λ-line. The SCOZA predictions were later compared to the results of a density-functional theory [21] and simulation [22]. A partial list of other topics that have been covered includes: a) the effect of the attraction range [23, 24], b) the height of the barrier [25], and c) the role of the tail in the potential [26–30].

The second class of systems for which clusters have been reported are colloids interacting via repulsive potentials with a soft shoulder [2]. Stell and Hemmer pointed out in a seminal work [31] that due to the additional length scale, one set by the size of the colloid and the other by the size of the shoulder, such potentials possess very unusual properties. In addition to the usual gas-liquid transition ending in a critical point, for instance, potentials with shoulders may exhibit a liquid-liquid transition, which also ends in a critical point. There have been numerous follow-up studies that focused on the specific details of the new phase transitions as well as on whether or not this model is capable of explaining the thermodynamic anomalies of liquid water [32–36]. It was not until the work of Klein et al. [37] that clusters were made the specific subject of such studies. Using analytical theory and computer simulations, colloids interacting via a hard-core and soft shoulder (HCSS) potentials were seen in this work to form spherical clusters (or clumps) in the fluid phase. At low temperature, clusters were observed to freeze into cluster crystals, cluster glasses and a number of polycrystalline materials [37]. In a later study for the same potential but in 2D, Norizoe and Kawakatsu [38, 39] also reported clusters that are string-like or extended in shape. Extended, chain-like clusters were also observed by Camp [40] for a continuous repulsive potential with a shoulder (different from that of HCSS); the shoulder was determined to be responsible for the emergence of the clusters. In addition to clusters, HCSS are capable of forming a variety of other ordered phases at low temperature. Malescio and Pellicane [41] reported a regular striped phase while Glaser et al. [42] discovered cluster crystals and cluster fluids in MC simulations. Cluster fluids were also reported by Mladek et al. [43], lanes by Fornleitner and Kahl [44], and lamella, hexagonal-columnar and body-center cubic phases, as well as the associated inverse structures by Shin et al. [45]. Dotera et al. [46] describe various quasicrystalline structures that are formed in 2D at low temperature.

It should be noted that many of the properties displayed by the cluster-forming colloids with repulsive interactions actually do not require the presence of the hard core. Likos et al. [47] considered a number of potentials bound at the origin, including the flat portion of the HCSS lacking the hard core, and introduced a criterion for determining if a system with such potential should experience a micro-phase separation into a lattice state where multiple colloids occupy the same lattice site. This criterion was later successfully tested in computer simulations [48], which indeed uncovered cluster crystals. Repulsive [3, 7] as well as attractive [49] colloids were seen to form clusters in laboratory experiments.

In this work we demonstrate by computer simulations that potentials that do not belong to either of the classes mentioned above are also capable of forming equilibrium clusters. Specifically, we consider the potentials that, instead of the global minimum as in the SALR scheme, display a local minimum that has a positive energy and is separated from the longer-distance, zero-energy states by a finite barrier.

Unlike in the SALR scheme, however, potentials with local attractive minima and long-range repulsion (LALR) lack the energy incentive for the particles to self-associate. LALR model shares with the HCSS potential the global repulsive character and soft shoulder. Experimentally, such potentials may arise as a result of incomplete cancellation between the attractive and repulsive terms. Theoretically, potentials with local minima are little studied. Batten et al. [50] showed that such potentials have unique ground-state configurations, including kagome and honeycomb crystals, and stripes. Liu et al. [51] also examined ground-state configurations of a LALR system confined to a plane. Neither of these papers focused on clusters. To the best of our knowledge, this subject has never been studied.

A comparative analysis of clusters formed by systems of all three schemes, LALR, SALR and HCSS, indicates that the LALR serves as a bridge between the other two. Like in the HCSS model, LALR potentials induce entropy-driven expanded clusters but only at low overall system density and for small cluster sizes. With an increase in the density, the LALR system, like its SALR counterpart, leads to energy driven clusters of large size. The dual identity of the LALR model is also revealed in its temperature behavior. Small clusters of this system are stabilized by temperature, like in the HCSS, but larger clusters are destabilized, in common with the SALR.

2. Methods

The shape of the model potential considered in this work is inspired by the inter-molecular potential of aqueous solutions of lysozyme. Using structural functions available from scattering experiments for this protein, effective potentials were derived in our prior work (data not published) for a number of pH values. For pH 2, the potential was seen to possess a local minimum at short distances followed by a repulsive term decaying to zero at longer distances. To reflect these features, the following analytical function was considered:

$$
u(r) = \begin{cases}
+\infty, & r < \sigma_H, \\
-ar + b + \epsilon_w, & \sigma_H < r < \dfrac{\sigma_H + \sigma_A}{2}, \quad a = \dfrac{2\delta\epsilon}{\sigma_A - \sigma_H}, \\
ar - b + \epsilon_w, & \dfrac{\sigma_H + \sigma_A}{2} < r < \sigma_A, \quad b = \delta\epsilon\dfrac{\sigma_A + \sigma_H}{\sigma_A - \sigma_H}, \\
\epsilon_b \sigma_A \dfrac{e^{-\kappa(r - \sigma_A)}}{r}, & \sigma_A < r.
\end{cases}
\tag{2.1}
$$

The shape of the potential is determined by a set of six basic parameters, σ_H, σ_A, ϵ_w, ϵ_b, $\delta\epsilon$ and κ. A hard-core wall is placed at σ_H. The short-range (SA) part of the potential contains a minimum (either local or global) located between σ_H and σ_A. The minimum is triangular in shape. Its energy is defined by ϵ_w while the depth is given by $\delta\epsilon$. The long-range part (LR), defined by distances greater than σ_A is either missing or given by the Yukawa potential characterized by the inverse decay length κ. The SA and LR parts meet at $r = \sigma_A$ where the energy is ϵ_b. The schematic of the model potential is shown in figure 1 together with its experimental prototype. Depending on the choice of parameters, the potential may represent a number of soft-matter systems. Throughout this article, we use σ_A as the unit of distance and ϵ_b as the unit of energy. The hard-core diameter is set at $\sigma_H = \sigma_A/3$. The remaining three parameters are varied to achieve the following five representative shapes: 1) Hard-core soft-shoulder (HCSS) potential studied extensively previously in 2- and 3-D [38, 39, 41, 42, 52], 2) Hard-core long-range repulsion potential (HCLR), which, to the best of our knowledge, has not been studied yet in this specific form, but similar potentials were considered in the past [40, 53], 3) short-range attraction and long-range repulsion potential (SALR), which was also studied extensively but in a different form [2, 53, 54] , 4) local attraction soft-shoulder (LASS) potential, which has a minimum but no long-range tail and 5) local attractive minimum and long-range repulsion (LALR) potential. The last two potentials are considered here for the first time. The summary of all the employed parameters is shown in table 1.

All systems were studied using the standard Metropolis Monte Carlo (MC) method [55]. For a quick scan in the phase space, simulation boxes with 216 particles were considered at 6 values of the total density. The densities ranged from 0.17 to 1.69 in reduced units. A number of temperatures were investigated, depending on the needs of each system, as discussed in the main text. Temperature is reported

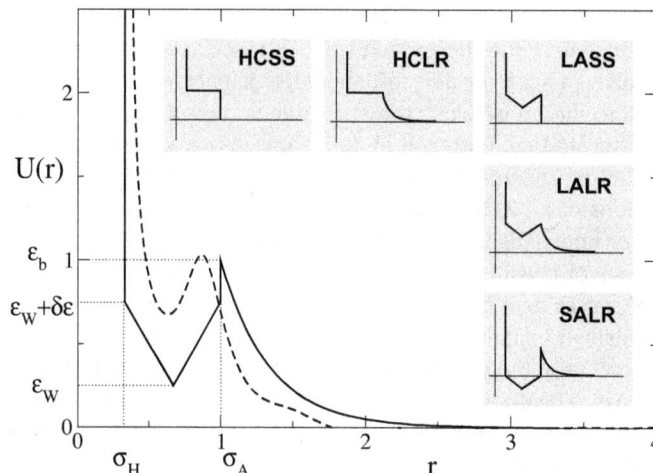

Figure 1. Schematic representation of the potential considered in this work. Depending on the choice of parameters, different types of soft-matter systems can be studied. Five specific shapes investigated in detail are shown as icons. Broken line shows potential, appropriately scaled, obtained for lysozyme solutions at acidic pH.

in units of ϵ_b/k_B, where k_B is the Boltzmann constant. In potentials with tails, HCLR, LALR and SALR, the truncation distance $r_c = 2\sigma_A$ was used. Tests were performed with longer r_c to make sure that the results are not affected by this parameter. Additionally, simulations with larger boxes containing 5832 particles were conducted for select thermodynamic points in order to extract cluster statistics. The maximum displacement of particles in the MC moves was adjusted to achieve 30% success rate. All simulations were run for more than 1×10^6 MC steps and tests were performed to make sure that the reported results are converged.

Table 1. Parameters of five model potentials studied in this work. All energies are measured in units of ϵ_b and all distances in units of σ_A. Abbreviations are as in the main text.

Model	ϵ_w	$\delta\epsilon$	κ
HCSS	1	0	
HCLR	1	0	4.05
LASS	0.525	0.475	
LALR	0.525	0.475	4.05
SALR	−0.475	0.475	4.05

To quantify the process of self-association, all the recorded conformations were clustered. The standard clustering algorithm was used which assigns a particle to the given cluster if its separation from any particle in the cluster is less than a cut-off distance R_c. For the cut-off, the diameter of the soft-shoulder σ_A was used.

In order to characterize the progress of clusterization, a ratio x of the number of particles in monomeric state to the total number of particles was analyzed. Configurations saved in simulations were used to construct a distribution $P(x)$ as a function of density and other thermodynamic parameters. The distribution function yields the free energy profile $\Delta F(x) = -k_B T \log P(x)$, where k_B is the Boltzmann constant and T is the temperature, which has a minimum at the most likely value of x. Function $\Delta F(x)$ measures the free energy difference between conformations with two different values of parameter x, for instance 0 and 1, in which case it reports the full cost/benefit of converting the system at the given thermodynamic point into clusters. The free energy can be decomposed into internal energy and entropy contributions using the standard thermodynamic relationship $\Delta F(x) = \Delta U(x) - T\Delta S(x)$. The unknown functions $\Delta U(x)$ and $-T\Delta S(x)$ can be determined if the temperature dependence of $\Delta F(x)$ is available.

First, numerical differentiation of $\Delta F(x)$ can yield $-T\Delta S(x)$ (ignoring the entropy dependence on temperature in the numerical algorithm) and then entropy and free energy can be combined to produce internal energy. This method is commonly used in literature to monitor the progress of various biochemical reactions (see for instance reference [56] and references therein). In the analysis of the free energy profiles, the small size of the simulation box was not seen to affect the conclusions of the work.

3. Results

3.1. All of the studied models form equilibrium clusters

Several models were studied by computer simulations, as described in detail in the 'Methods' section. All of the studied systems were seen to form equilibrium clusters at appropriate densities and tempera-

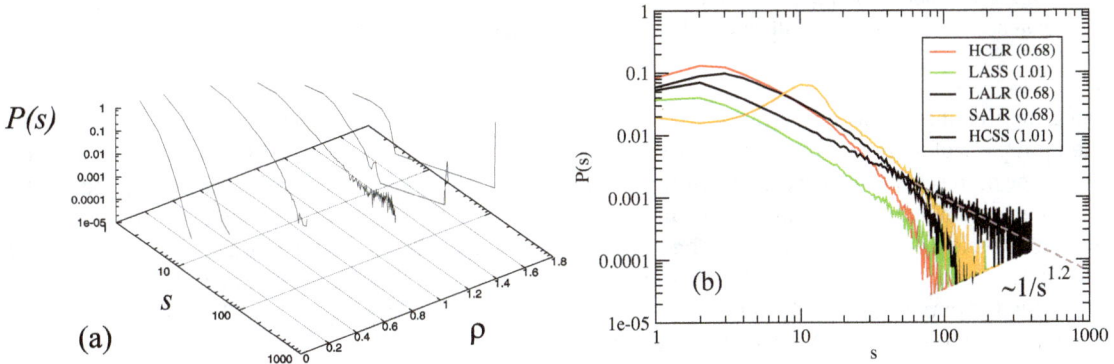

Figure 2. (Color online) The number of particles observed in clusters of specific type relative to the total number of particles in the simulation box. In (a) data for LALR model are shown for select densities. In (b) data for all models are shown, each model at a specific density, indicated between parentheses. All data were collected at a reduced temperature $T = 0.62$.

ture. As an illustration, figure 2 (a) shows $P(s)$, the ratio of particles engaged in clusters of size s to the total number of particles in the simulation box observed for the HCSS model at a reduced temperature $T = 0.62$. At low density $\rho = 0.17$ (shown in units of σ_A^{-3}), the majority of all particles are monomers. As the density is increased, the proportion of monomers drops at the benefit of dimers, trimers and other multimers. At $0.34 < \rho < 0.6$, the population of monomers drops below 0.5, at which point the system becomes dominated by clusters. Clusterization continues for higher densities, and at $\rho = 1.01$, the number of monomers drops below the number of particles involved in dimers. At still higher densities, $\rho > 1.35$, the system undergoes a percolation transition, above which only a single cluster encompassing the entire simulation box survives. Figure 2 (b) shows $P(s)$ recorded for different systems at the density just below the percolation threshold. While all curves demonstrate a majority cluster population, two specific features stand out. First, the SALR simulations show a strong preference for cluster size of ~10 monomers. All other systems populate a wide spectrum of sizes with only a small population maximum for dimers and trimers. Second, in the limit of large s, the data of HCSS simulations demonstrate a clear power-law dependence, $P(s) \sim 1/s^n$. The observed exponent of $n = 1.2$ is in excellent agreement with the prediction of the random percolation limit [39] [note that the usual cluster-size distribution function $n(s)$ is related to the distribution function used here as $n(s) = P(s)/s$]. In all other systems, $P(s)$ decays faster.

3.2. Cluster statistics critically depends on the type of the potential

All the studied systems populate the clusters that differ widely in size and shape. Radius of gyration $R_g(s)$ is used to characterize how the size of a cluster depends on the number of particles it contains. The most interesting dependence is observed for the SALR potential. Figure 3 shows $R_g(s)$ for this model recorded for several values of the total density of the solution ρ. At low densities, the radius of gyration is a monotonous function of the number of particles. Starting at $\rho = 0.68$, it begins to develop a plateau,

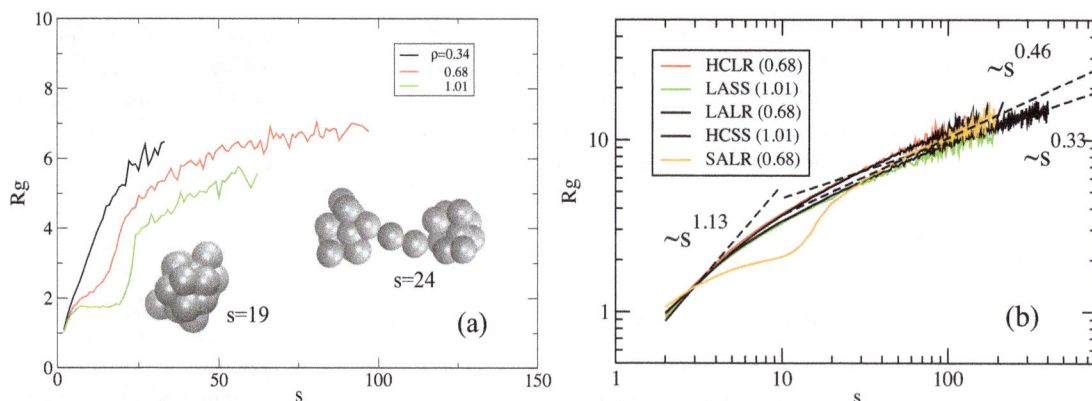

Figure 3. (Color online) Radius of gyration R_g as a function of the number of particles s. Data for SALR potential are shown in (a). All potentials with appropriate densities are shown in (b).

signaling the onset of a configurational change. At $\rho = 1.01$, R_g remains unchanged for $3 < s < 19$ and then jumps twofold for $s > 23$. It is easy to figure out what happens at the transition point by examining the shape of the clusters. Representative clusters for $s = 19$, which is just before the transition, and $s = 24$, which is immediately after it, are shown in figure 3 (a). The smaller, or primary, clusters form dense and almost spherical clumps of particles. The larger clusters are made of two (or more) such clumps joined together. Thus, cluster formation is driven by hierarchical supramolecular assembly: small clusters are assembled at initial stages and later on serve as building blocks for larger clusters. This scenario explains the sudden increase in $R_g(s)$ at a certain transition value s_c.

The unusual cluster statistics can be rationalized from the standpoint of equilibrium thermodynamics. In common with other SALR potentials [53, 57], the ground-state (GS) configuration of the current model is a 1D Bernal spiral. The energy of the spiral is a monotonously decreasing function of s that continues to decline until all available monomers are absorbed onto the cluster. At zero and sufficiently low temperature, the spiral is the only observable conformation. At finite temperatures, however, this picture changes dramatically. The entropy that is lost by the monomers during the assembly process becomes important. The lowest free energy is achieved through a combination of both internal energy and entropy. The balance between energy and entropy contributions dictates the size of the primary cluster s_c. The loss of energy due to the fracturing of the ground state is compensated for by the gain in the entropy arising from the translational freedom of the resulting clusters. As the density of the solution increases (volume is lowered), the entropic gain decreases. So will do the associated loss of energy. As a consequence, the GS conformation will break into a smaller number of pieces. This means that the size of the primary cluster should go up with density. This is exactly what one sees in figure 3 (a) after comparing $R_g(s)$ for $\rho = 0.68$ and 1.01.

The statistics of all other systems are similar to one another and can be described as those of a polydisperse mixture of clusters of different sizes and shapes. As in the case of SALR, the size of the clusters also decreases with the increase of density but this effect is much less pronounced. No non-trivial behavior is seen. A detailed comparison of $R_g(s)$ obtained for different systems [considered at the same densities as in figure 2 (b)] is presented in figure 3 (b). The plateau in the SALR curve is seen again. Moreover, it is possible to discern the scaling statistics for large clusters. In that limit, the radius of gyration has a power-law dependence $R_g(s) \sim s^n$ with $n = 0.46$, which is independent of the density. This exponent is close to $n = 0.5$ observed for ideal, or Gaussian, polymer chains which lack the excluded volume interactions. The Gaussian model mechanism may be applied to the assembly of large colloidal clusters. Although the primary clusters do have the excluded volume, they are allowed to inter-penetrate via the exchange of particles. Thus, from the statistical point of view, they act as zero-size particles. For the same reason, SALR clusters do not experience compression from the environment, as do HCSS clusters, and remain swollen compared to maximally compact states.

The statistics of other systems in the large-cluster limit are governed by $n = 0.33$, which corresponds to a maximally compact object. This observation is illustrated in figure 3 (b) for HCSS potential. For the same system in the limit of small s, the radius of gyration scales as $R_g \sim s^m$, where $m = 1.13$. Note that for the linear arrangement of the smallest clusters, dimer and trimer, the exponent $m = 1.2$. Therefore, HCSS behaves as a linear chain for small s. This is in agreement with the prior work on this system [39], which also shows that the linear scaling persists to large clusters for a specific choice of potential parameters (different from those of the present work). HCLR, LASS and LALR potentials also exhibit almost a linear scaling but with the exponent that is system-specific. In common with the HCSS model, clusters formed by these potentials experience a cross-over from a linear regime for small clusters to a collapsed chain for large clusters. The transition is gradual and takes place over two orders of magnitude. This behavior is in sharp contrast to the SALR model and can be explained by the specific character of the HCSS interaction. Although the formation of clusters in this model is driven by entropy (see below), the specific shape of different clusters of the same size is also influenced by energy. Since the interaction is purely repulsive, particles within a cluster will choose to reside as far from their neighbors as possible. This results in linear conformations for small clusters, which minimize the overlap between constituent particles. For larger clusters, the minimal overlap can be achieved in string-like but bended conformations, similarly to those observed in the self-avoiding polymers. Like in the polymers, colloidal clusters should experience an entropic collapse with the pertinent exponent of the radius of gyration of $m = 0.591$, or approximately 3/5 as predicted by Flory [58]. The fact that the observed exponent 0.33 is lower indicates that large clusters experience additional compression from the environment, which may only result from the repulsive interactions of clusters with other species in the solution, both clusters and monomers alike. These theoretical arguments lead us to the conclusion that the specific details of the cross-over will depend on a) the extent of the soft shoulder in the potential, which governs how much linear chains gain in energy compared to the collapsed ones and b) the strength of the repulsive tail, which determines how strongly various colloidal species repel each other, thus inducing cluster collapse.

4. Discussion

4.1. Two classes of cluster-forming systems use different assembly mechanisms

Our simulations clearly demonstrate that all of the studied systems form equilibrium clusters. In view of this finding, the relevant question to ask is: 'Why does that happen? What are the pertinent mechanisms?'. Perhaps the simplest to explain is the mechanism of the SALR model. By design, this potential favors association of particles by providing a negative potential energy to configurations with close separation among particles. Clusters of different topologies represent a ground-state configuration of SALR potentials [2, 57], including the one studied here. Thus, for these systems, cluster assembly is expected to be driven by internal energy. To quantify this assessment, it makes sense to analyze the free energy profile of the clusterization reaction as a function of a certain order parameter. Since we are interested in the transition into the cluster state in general, not into a specific type of cluster, it is most convenient to monitor the number of particles in the monomeric state as a progress variable. To make a comparison between systems of different sizes easier, the number of monomers will be normalized by the total number of particles, producing a ratio of monomers $0 < x < 1$. Conformations with no clusters have $x = 1$ while those lacking monomers (complete transformation into clusters) are characterized by $x = 0$. Internal energy, $\Delta U(x)$, and entropy, $-T\Delta S(x)$, can be computed for each x (see 'Methods' section) to elucidate the role of these functions in the cluster formation. Figure 4 (a) shows $\Delta U(x)$ and $-T\Delta S(x)$ obtained for the SALR model at $\rho = 0.68$. As predicted above, the cluster formation is driven by energy and opposed by entropy. The two curves combined yield a minimum of free energy at about $x = 0.015$.

The same analysis was repeated for the HCSS potential. Note that this model in 2D, given the appropriate choice of parameters, forms ordered configurations in the ground and low-energy states [59] as well as quasi-crystals [46] . Thus, one may expect the remnants of ordered structures to survive at finite temperature, giving rise to clusters stabilized by energy [42]. Figure 4 (b), depicting $\Delta U(x)$ and $-T\Delta S(x)$ demonstrates that this expectation is not justified. Clusters are stabilized entirely by entropy and destabilized by energy. The two systems, SALR and HCSS, differ dramatically in their cluster formation scenarios. While clusters in SALR tend to stay intact due to the mutual attraction, clusters in the HCSS model use a different mechanism. Namely, particles at distances $\sigma_H < r < \sigma_A$ experience no force and move freely

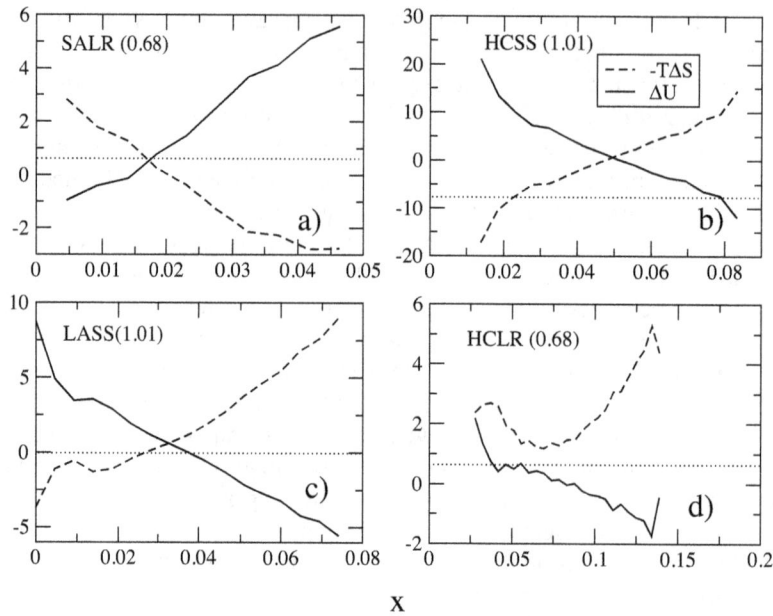

Figure 4. Internal energy, $\Delta U(x)$, and entropy, $-T\Delta S(x)$, as a function of the fraction of monomeric particles x generated in this work for different cluster-forming potentials. All models except SALR exhibit the clusters stabilized by entropy. Here, and elsewhere in the article, energy is measured in units of ϵ_b.

about each other. Particles in the monomeric state experience no force either, since the potential is zero for $r > \sigma_A$. Therefore, clusters stay together because the volume created by joining the soft-shoulder regions of all constituent particles is greater than the volume available to them in the monomeric state (at a distance $r > \sigma_A$ from any other monomer or cluster). This point is illustrated in figure 5. For a monomeric particle, the moment of joining an existing cluster is accompanied by an increase in energy as well as by an increase in entropy. As long as the entropy wins, particles spend the majority of their time in cluster configurations. Note that these clusters will have a very short residence time of the order of the duration of binary collisions. They will also rapidly disintegrate in the event of an increase in the available volume.

Cluster formation in HCRL and LASS models is driven by the same mechanism as in the HCSS model. Adding a repulsive tail to the potential, as in HCLR, changes the appearance of $-T\Delta S(x)$: at small x it

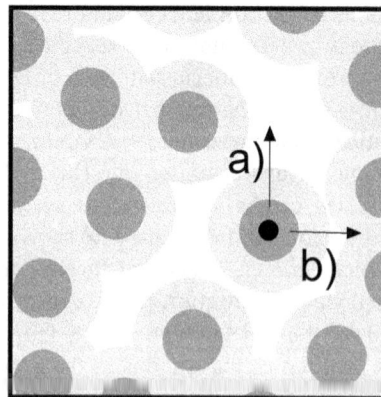

Figure 5. Illustration of how entropy stabilizes the clusters in the HCSS model. The central monomeric particle, denoted by black circle, has the choice of either to: a) remain monomeric, in which case it is allowed to explore white-colored space, or b) join a cluster, in which case it is allowed to explore light-gray space. Dark gray space indicates hard-core areas which are not accessible to the particle. Clusters become stable when the volume of the light-gray area exceeds that of the white area.

begins to disfavor clusters, see figure 4 (d). However, the destabilizing effect of energy is unaffected. An additional minimum in the soft shoulder, as in LASS, decreases the destabilizing effect of energy, compare figure 4 (c) and (b), but not to the extent of reversing its role. Cluster formation of both systems is driven by entropy.

4.2. The model with local minimum exhibits a dual identity

The model with the local minimum and repulsive tail stands out from the rest of the studied systems. Figure 6 shows its energy and entropy evaluated at two different densities, $\rho = 0.34$ and $\rho = 0.68$. At a lower density, the mechanism is as in the HCSS system with the entropy stabilizing the clusters. For a higher density, the mechanism changes and becomes similar to that of the SALR potential, where the internal energy drives the assembly of clusters. The transformation is not complete as entropy still favors the clusters at high density instead of disfavoring them, as in the SALR potential. Nevertheless, the reversal of the energy role is quite remarkable. This is not seen in any other potential and indicates that the process of cluster formation in the systems interacting via potentials with local minima can be extraordinarily complex. It takes the presence of both monomers and small clusters in such systems for the clusterization to be driven by energy. Any theory that aims at predicting the cluster distribution quantitatively should be capable of capturing that effect. That means a proper description of monomer/cluster mixture, including the association/dissociation balance, across a wide range of densities. It is not clear which of the currently available liquid-state theories would be capable of accomplishing this task.

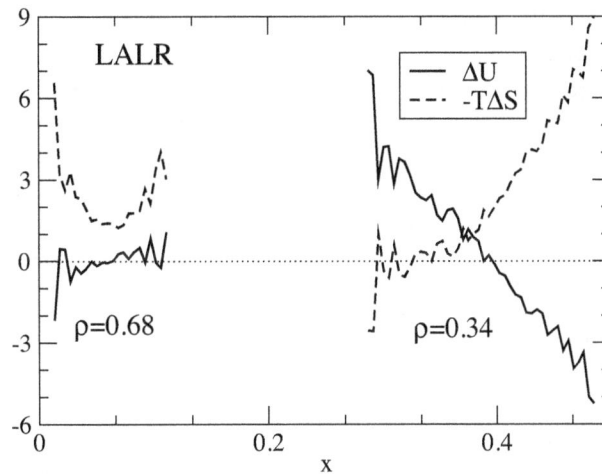

Figure 6. The same as figure 4, but for the potential with local minimum and repulsive tail, LALR. Both entropy- and energy driven mechanisms of cluster formation are present, depending on the total density of the solution.

The split identity of the LALR model is also evident in its temperature behavior. Since the assembly of clusters in the HCSS potential is entropic in nature, it should be enhanced by temperature. The opposite is true for the SALR model, where clusters should become less stable at higher temperature. Figure 7, showing the fraction of monomers as a function of temperature and density for different models, confirms this conjecture. Different symbols in this figure represent different systems, broken lines correspond to a high temperature of 0.62 while solid lines stand for the low temperature of 0.21. For the HCSS model, the population of monomers at the lower temperature is higher than the population at a higher temperature, the solid line is above the broken line, at all densities, indicating that cluster formation is enhanced by temperature. For the SALR model, the broken line is above the solid line, which is the evidence that temperature hinders the formation of clusters. For the LALR potential, the broken line is above the solid line for $\rho < 0.68$ and vice versa for higher densities. Thus, this model exhibits the features characteristic of both classes of cluster-forming colloids. The entropy-driven mechanism dominates at low densities while the energy-driven mechanism — at high densities.

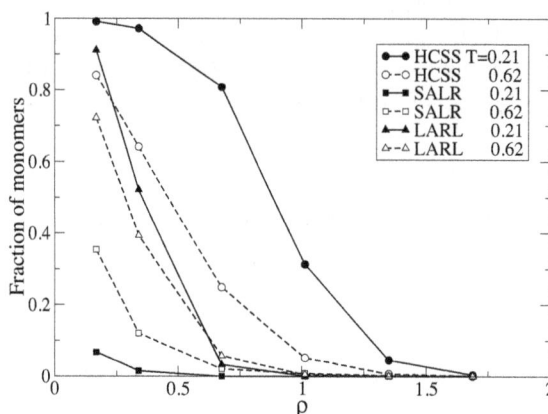

Figure 7. Fraction of monomers in the conformational ensemble obtained for different models, temperatures and densities. HCSS clusters are destabilized by temperature. SALR clusters are stabilized by temperature while LALR clusters exhibit both behaviors depending on density.

5. Conclusions

There are two known types of cluster-forming colloids: a) those that interact via purely repulsive potential with a soft shoulder and, possibly, a tail, and b) those that are characterized by a potential with the global attractive well at short distances followed by a repulsive tail at long distances. By adding an attractive term to the otherwise repulsive potential, it is easy to transform the potentials of the first type into the potentials of the second type. Thus, it may seem obvious that systems of the two types exhibit similar properties, in particular the capability of forming clusters. However, there is one caveat with this argument. Extrapolating properties of one system based on the results of another system makes sense only if the two are sufficiently similar so that perturbation theory can be used. This is clearly not the case for the cluster-forming colloids. Repulsive and attractive potentials exhibit completely different structural, dynamic and phase behaviors. Take, for instance, the gas-liquid transition in attractive colloids that is missing in the repulsive colloids. The differences also extend to the clusters of these systems. While attractive potentials form compact clusters of specific size, the repulsive potentials lead to extended, polydisperse multimers. There is only a limited amount of information that can be learned about one system by extrapolating the properties of the other. Simulations remain the most reliable and accurate tool in the studies of a new or unknown potential. Simulations played a key role in this work, helping us to reveal that a repulsive potential with a local minimum shares certain properties with both other types of cluster-forming potentials. The clusters it populates are similar to those of purely repulsive potentials, while the energy-driven mechanism is similar to the attractive potentials.

Our results were obtained for a specific shape of the potential with a local minimum. How general they are with respect to other systems remains to be seen. Of particular interest here would be to consider square-well potentials which have been studied extensively in the literature in regard to phase transitions. Also, of interest would be to examine the behavior of continuous potentials, like those that can be constructed with the help of two Yukawa functions.

References

1. Vcldlov N.O., Rev. Chem. Eng., 2011, 27, No. 1 0, 1, doi:10.1010/revco.2011.000.
2. Dinsmore A.D., Dubin P.L., Grason G.M., J. Phys. Chem. B, 2011, **115**, No. 22, 7173; doi:10.1021/jp202724b.
3. Osterman N., Babic D., Poberaj I., Dobnikar J., Ziherl P., Phys. Rev. Lett., 2007, **99**, No. 24, 248301; doi:10.1103/PhysRevLett.99.248301.
4. Stradner A., Sedgwick H., Cardinaux F., Poon W.C.K., Egelhaaf S.U., Schurtenberger P., Nature, 2004, **432**, No. 7016, 492; doi:10.1038/nature03109.
5. Atmuri A.K., Bhatia S.R., Langmuir, 2013, **29**, No. 10, 3179; doi:10.1021/la304062r.

6. Tam J.M., Murthy A.K., Ingram D.R., Nguyen R., Sokolov K.V., Johnston K.P., Langmuir, 2010, **26**, No. 11, 8988; doi:10.1021/la904793t.
7. Janai E., Schofield A.B., Sloutskin E., Soft Matter, 2012, **8**, No. 10, 2924; doi:10.1039/C2SM06808G.
8. Xu S., Sun C., Guo J., Xu K., Wang C., J. Mater. Chem., 2012, **22**, No. 36, 19067; doi:10.1039/C2JM34877B.
9. Yearley E.J., Godfrin P.D., Perevozchikova T., Zhang H.L., Falus P., Porcar L., Nagao M., Curtis J.E., Gawande P., Taing R., Zarraga I.E., Wagner N.J., Liu Y., Biophys. J., 2014, **106**, No. 8, 1763; doi:10.1016/j.bpj.2014.02.036.
10. Lebowitz J.L., Penrose O., J. Math. Phys., 1966, 7, No. 1, 98; doi:10.1063/1.1704821.
11. Kendrick G.F., Sluckin T.J., Grimson M.J., Europhys. Lett., 1988, **6**, No. 6, 567; doi:10.1209/0295-5075/6/6/016.
12. Sear R.P., Chung S.W., Markovich G., Gelbart W.M., Heath J.R., Phys. Rev. E, 1999, **59**, No. 6, R6255; doi:10.1103/PhysRevE.59.R6255.
13. Sear R.P., Gelbart W.M., J. Chem. Phys., 1999, **110**, No. 9, 4582; doi:10.1063/1.478338.
14. Wu D., Chandler D., Smit B., J. Phys. Chem., 1992, **96**, No. 10, 4077; doi:10.1021/j100189a030.
15. Seul M., Andelman D., Science, 1995, **267**, No. 5197, 476; doi:10.1126/science.267.5197.476.
16. Groenewold J., Kegel W.K., J. Phys. Chem. B, 2001, **105**, No. 47, 11702; doi:10.1021/jp011646w.
17. Zhang T.H., Kuipers B.W.M., Tian W.D., Groenewold J., Kegel W.K., Soft Matter, 2015, **11**, No. 2, 297; doi:10.1039/C4SM02273D.
18. Sweatman M.B., Fartaria R., Lue L., J. Chem. Phys., 2014, **140**, No. 12, 124508; doi:10.1063/1.4869109.
19. Pini D., Ge J.L., Parola A., Reatto L., Chem. Phys. Lett., 2000, **327**, No. 3–4, 209; doi:10.1016/S0009-2614(00)00763-6.
20. Pini D., Parola A., Reatto L., J. Phys.: Condens. Matter, 2006, **18**, No. 36, S2305; doi:10.1088/0953-8984/18/36/S06.
21. Archer A.J., Pini D., Evans R., Reatto L., J. Chem. Phys., 2007, **126**, No. 1, 014104; doi:10.1063/1.2405355.
22. Archer A.J., Wilding N.B., Phys. Rev. E, 2007, **76**, No. 3, 031501; doi:10.1103/PhysRevE.76.031501.
23. Charbonneau P., Reichman D.R., Phys. Rev. E, 2007, **75**, No. 1, 011507; doi:10.1103/PhysRevE.75.011507.
24. Tarzia M., Coniglio A., Phys. Rev. E, 2007, **75**, No. 1, 011410; doi:10.1103/PhysRevE.75.011410.
25. Costa D., Caccamo C., Bomont J.M., Bretonnet J.L., Mol. Phys., 2011, **109**, No. 23–24, 2845; doi:10.1080/00268976.2011.611480.
26. Groenewold J., Kegel W.K., J. Phys.: Condens. Matter, 2004, **16**, No. 42, S4877; doi:10.1088/0953-8984/16/42/006.
27. Imperio A., Reatto L., J. Phys.: Condens. Matter, 2004, **16**, No. 38, S3769; doi:10.1088/0953-8984/16/38/001.
28. Hutchens S.B., Wang Z.G., J. Chem. Phys., 2007, **127**, No. 8, 084912; doi:10.1063/1.2761891.
29. Liu Y., Chen W.R., Chen S.H., J. Chem. Phys., 2005, **122**, No. 4, 044507; doi:10.1063/1.1830433.
30. Schwanzer D.F., Kahl G., J. Phys.: Condens. Matter, 2010, **22**, No. 41, 415103; doi:10.1088/0953-8984/22/41/415103.
31. Hemmer P.C., Stell G., Phys. Rev. Lett., 1970, **24**, No. 23, 1284; doi:10.1103/PhysRevLett.24.1284.
32. Stell G., Hemmer P.C., J. Chem. Phys., 1972, **56**, No. 9, 4274; doi:10.1063/1.1677857.
33. Young D.A., Alder B.J., Phys. Rev. Lett., 1977, **38**, No. 21, 1213; doi:10.1103/PhysRevLett.38.1213.
34. Jagla E.A., J. Chem. Phys., 1999, **111**, No. 19, 8980; doi:10.1063/1.480241.
35. Rzysko W., Pizio O., Patrykiejew A., Sokołowski S., J. Chem. Phys., 2008, **129**, No. 12, 124502; doi:10.1063/1.2970884.
36. Buldyrev S.V., Malescio G., Angell C.A., Giovambattista N., Prestipino S., Saija F., Stanley H.E., Xu L., J. Phys.: Condens. Matter, 2009, **21**, No. 50, 504106; doi:10.1088/0953-8984/21/50/504106.
37. Klein W., Gould H., Ramos R.A., Clejan I., Melcuk A.I., Physica A, 1994, **205**, No. 4, 738; doi:10.1016/0378-4371(94)90233-X.
38. Norizoe Y., Kawakatsu T., Europhys. Lett., 2005, **72**, No. 4, 583; doi:10.1209/epl/i2005-10288-6.
39. Norizoe Y., Kawakatsu T., J. Chem. Phys., 2012, **137**, No. 2, 024904; doi:10.1063/1.4733462.
40. Camp P.J., Phys. Rev. E, 2003, **68**, No. 6, 061506; doi:10.1103/PhysRevE.68.061506.
41. Malescio G., Pellicane G., Phys. Rev. E, 2004, **70**, No. 2, 021202; doi:10.1103/PhysRevE.70.021202.
42. Glaser M.A., Grason G.M., Kamien R.D., Kosmrlj A., Santangelo C.D., Ziherl P., EPL, 2007, **78**, No. 4, 46004; doi:10.1209/0295-5075/78/46004.
43. Mladek B.M., Gottwald D., Kahl G., Neumann M., Likos C.N., J. Phys. Chem. B, 2007, **111**, No. 44, 12799; doi:10.1021/jp074652m.
44. Fornleitner J., Kahl G., EPL, 2008, **82**, No. 1, 18001; doi:10.1209/0295-5075/82/18001.
45. Shin H.M., Grason G.M., Santangelo C.D., Soft Matter, 2009, **5**, No. 19, 3629; doi:10.1039/B904103F.
46. Dotera T., Oshiro T., Ziherl P., Nature, 2014, **506**, No. 7487, 208; doi:10.1038/nature12938.
47. Likos C.N., Lang A., Watzlawek M., Lowen H., Phys. Rev. E, 2001, **63**, No. 3, 031206; doi:10.1103/PhysRevE.63.031206.
48. Mladek B.M., Gottwald D., Kahl G., Neumann M., Likos C.N., Phys. Rev. Lett., 2006, **96**, No. 4, 045701; doi:10.1103/PhysRevLett.96.045701.
49. Klix C.L., Royall C.P., Tanaka H., Phys. Rev. Lett., 2010, **104**, No. 16, 165702; doi:10.1103/PhysRevLett.104.165702.
50. Batten R.D., Huse D.A., Stillinger F.H., Torquato S., Soft Matter, 2011, 7, No. 13, 6194; doi:10.1039/c0sm01380c.
51. Liu Y.H., Chew L.Y., Yu M.Y., Phys. Rev. E, 2008, **78**, No. 6, 066405; doi:10.1103/PhysRevE.78.066405.

52. Malescio G., Pellicane G., Nat. Mater., 2003, **2**, No. 2, 97; doi:10.1038/nmat820.

53. Sciortino F., Tartaglia P., Zaccarelli E., J. Phys. Chem. B, 2005, **109**, No. 46, 21942; doi:10.1021/jp052683g.

54. Zaccarelli E., J. Phys.: Condens. Matter, 2007, **19**, No. 32, 323101; doi:10.1088/0953-8984/19/32/323101.

55. Allen M.P., Tildesley D.J., Computer simulations of liquids, Oxford University Press, Oxford, 1987.

56. Baumketner A., J. Phys. Chem. B, 2014, **118**, No. 50, 14578; doi:10.1021/jp509213f.

57. Mossa S., Sciortino F., Tartaglia P., Zaccarelli E., Langmuir, 2004, **20**, No. 24, 10756; doi:10.1021/la048554t.

58. De Gennes P.G., Scaling Concepts in Polymer Physics, Cornell University Press, New York, 1979.

59. Dobnikar J., Fornleitner J., Kahl G., J. Phys.: Condens. Matter, 2008, **20**, No. 49, 494220; doi:10.1088/0953-8984/20/49/494220.

6

Self-sorting in two-dimensional assemblies of simple chiral molecules

A. Woszczyk, P. Szabelski*

Department of Theoretical Chemistry, Maria-Curie Sklodowska University,
Pl. M.C. Sklodowskiej 3, 20-031 Lublin, Poland

Structural modification of adsorbed overlayers by means of external factors is an important objective in the fabrication of stimuli-responsive materials with adjustable physicochemical properties. In this contribution we present a coarse-grained Monte Carlo model of the confinement-induced chiral self-sorting of hockey stick-shaped enantiomers adsorbed on a triangular lattice. It is assumed that the adsorbed overlayer consists of "normal" molecules that are capable of adopting any of the six planar orientations imposed by the symmetry of the lattice and molecular directors having only one permanent orientation, that reflect the coupling of these species with an external directional field. Our investigations focus on the influence of the amount fraction of the molecular directors, temperature and surface coverage on the extent of the chiral segregation. The simulated results demonstrate that the molecular directors can have a significant effect on the ordering in enantiopure overlayers, while for the corresponding racemates their role is largely diminished. These findings can be helpful in designing strategies to improve methods of fabrication of homochiral surfaces and enantioselective adsorbents.

Key words: *self-assembly, chiral molecules, Monte Carlo simulations, chiral resolution*

1. Introduction

Achieving control over the structure of molecular systems, and tuning their properties using external inputs has been one the most intensively studied topics in material science during the recent years. A considerable interest in such adjustable molecular structures results mainly from the possibility of changing their physico-chemical characteristics in a reversible manner by means of noninvasive external factors including, for example, electric or magnetic fields. These directional inputs have been found to be capable of triggering the ordering in bulk systems comprising molecules and particles, which results in substantial changes in optical, magnetic or rheological properties [1–4].

Another class of systems in which the structure formation can be guided by an external bias are the adsorbed overlayers comprising different, often functionalized, organic molecules. In this case, corrugation of the molecule-substrate interaction potential that is provided by atomic steps or grooves of metallic crystals has been often used to direct the self-assembly of molecular building blocks [5, 6]. Moreover, this method, when combined with covalent linkage of the adsorbed molecules, has been effectively used to produce persistent macromolecular structures including wires, ribbons and networks [7, 8].

Regarding the directional external fields acting on 2D molecular assemblies, it has been demonstrated recently by Berg et al. that the structure of the adsorbed overlayer comprising prochiral functionalized biphenyl molecules can be imposed through the chiral imprinting process occurring at the liquid-solid interface [3, 4]. In this approach, the prochiral molecules of the bulk liquid crystal (LC) phase were in direct contact with the graphite surface. The uniaxial ordering of the LC molecules that was induced by the

---

*E-mail: szabla@vega.umcs.lublin.pl

directional magnetic field was the source of orientational confinement of the adsorbing molecules forming mirror-image homochiral domains on graphite. Systematic changes in the orientation of the applied field resulted in the variation of the relative population of the homochiral domains making it possible to create extended molecular assemblies with one handedness. The experiments by Berg and coworkers have clearly demonstrated that steering the chiral organization of adsorbed molecules can be achieved using entirely achiral inputs such as directional magnetic fields. This idea seems particularly interesting from the perspective of chiral separations and fabrication of homochiral surfaces. For example, the field-induced orientational confinement of adsorbed enantiomers can be used to trigger their mixing or chiral resolution. These processes are relevant to the on-surface separation of enantiomers as well as to the creation of 2D molecular overlayers with adjustable adsorptive (enantioselective), optical and catalytic properties.

In order to explore the structure formation in adsorbed systems comprising chiral molecules, different theoretical approaches have been proposed including computer simulations and statistical mechanical calculations. In those studies, simplified molecular shapes, such as hard bent needles [9, 10], tetrominoes [11], tripods [12, 13] or patchy disks [14] have been usually employed, and the influence of anisotropic intermolecular interactions and molecular shape on the miscibility of the model enantiomers has been explored. Recently, more advanced approaches have been also proposed in which the adsorbed chiral molecules were presented in a more detailed way, accounting for atomic charges, Lennard-Jones parameters of the composite atoms or functional groups as well as for corrugation of the substrate [15–17].

Even though the influence of intrinsic properties of chiral molecules on their performance in the 2D self-sorting process has been studied systematically, the effect of external anisotropic inputs has not yet been thoroughly explored. This refers especially to the role of the aforementioned orientational confinement of adsorbed enantiomers, which can be induced, for example, by external fields. Recently, using MC simulations, we have demonstrated the confinement-induced 2D chiral segregation of simple molecular tectons whose composite segments were allowed to occupy vertices of a square lattice [18–20]. These theoretical predictions suggest that the unidirectional positioning of surface enantiomers can be used to trigger the chiral resolution also on the surfaces having different symmetry. In this contribution we extend our MC investigations and explore the possibility of steering the chiral segregation that occurs on a triangular lattice. To that end, we consider simple hockey-stick-shaped molecules [10] which are chiral, that is, they are capable of adopting mirror-image conformations when adsorbed. Moreover, in this work we study adsorbed overlayers in which only a part of the molecules is orientationally confined, playing the role of directors (sergeants) for the remaining molecules (soldiers). The main objective of our simulations is to determine the effect of such parameters as the amount fraction of the director molecules, surface coverage and temperature on the extent of the ordering and chiral resolution in the systems investigated.

2. Simulation

The chiral molecules of our model were assumed to consist of four interconnected segments arranged in a hockey-stick shape shown schematically in figure 1. These molecules were allowed to adopt mirror-image configurations (R and S enantiomers) when adsorbed. Each molecular segment of R and S occupied one site on a triangular lattice. The molecules were neither allowed to overlap nor to change chirality. Depending on the considered version of the model, the molecules could take any of the six planar orientations shown in the figure (isotropic) or the orientation of some of them was fixed to mimic the coupling of these species with the directional external field (see the molecules in black, unidirectional). The molecules with fixed orientation are further referred to as directors. In the simplified approach presented here, the director molecules have special properties which make them couple strongly with a directional external field while the rest of the molecules remain inert. In real situations, all of the molecules can be energetically coupled to the field and have orientationally dependent energies. In this case, the chiral segregation would depend on the individual energetic parameters which characterize each type of molecules. To minimize the number of parameters in our model, we considered an extreme case, in which the coupling energy of the director molecules is always large enough to orient these molecules parallel to the field. On the other hand, we also assumed that the coupling of the rest of the molecules is weak enough to

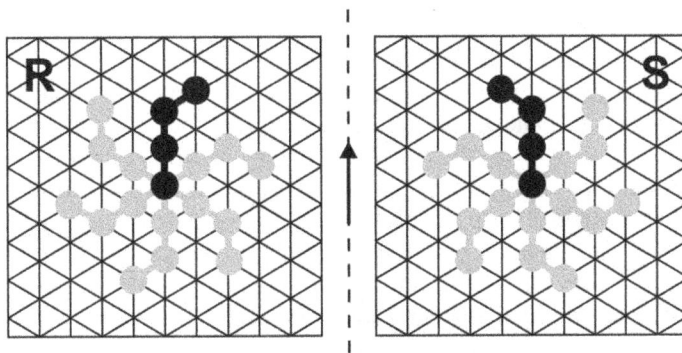

Figure 1. Schematic structure of the four-membered enantiomers R and S adsorbed on a triangular lattice. The preferred orientation assumed for the director molecules is marked in black while the orientations accessible to the remaining molecules are shown in grey. The dashed line represents the mirror symmetry plane. The black arrow indicates the preferred orientation (long molecular arm) assumed in the model.

be neglected. This approach permits to study the potential guiding role of the director molecules having a definite shape, without going into details of chemical composition of these species. Accordingly, the number of intermolecular interaction parameters of the model can be minimized. It was assumed that the interaction between adsorbed enantiomers can be described by an attractive short-range segment-segment interaction potential whose range was limited to the nearest neighbors on the lattice. The energy of elementary segment-segment interaction was characterized by the dimensionless parameter $\epsilon = -1$.

The MC simulations were performed on a 100 by 100 rhombic fragment of a triangular lattice imposing periodic boundary conditions in both planar directions. The calculations were carried out using the conventional MC method with Metropolis sampling [18–22], and they were additionally coupled with the cooling procedure in which the overlayer was equilibrated in a sequence of decreasing temperatures. The simulation protocol was organized as follows. In the first step N molecules in total were randomly distributed on the surface. Depending on the assumed composition of the overlayer, homochiral (R) and racemic (R+S, 1:1) assemblies were modelled. Moreover, in each of these cases, the fraction of director molecules, f, was systematically changed, and for the racemate, the directors contributed equally to R and S. At the beginning of the MC run, the director molecules were randomly distributed on the surface together with the remaining molecules and their orientation was selected at random. This corresponds to a situation in which the director molecules are not initially orientationally confined. Such a confinement, meaning a strong tendency to orient in one direction, can result from, for example, the coupling of the director molecules with external directional field parallel to the surface. During the simulation, when the confinement was imposed, the director molecules were allowed to translate but their orientation was changed (in case it was different from the original one) to the selected one shown in black in figure 1. To equilibrate the system, a series of random displacement moves was performed. In this procedure, the orientation of the director molecules was always fixed and the orientation of the remaining molecules was randomly changed by a multiple of 60 degrees. To accept or reject the move, the energies in the new U_n and old U_o positions were calculated for the selected molecule by summing up the segment-segment interactions with the neighboring molecules. Next, a random number $r \in (0, 1)$ was generated and compared with the probability $p = \min[1, \exp(-\Delta U / k_B T)]$ where $\Delta U = U_n - U_o$ and k_B and T are the Boltzmann constant and temperature, respectively. If $r < p$, the move was accepted; otherwise it was rejected. In the simulations, we used dimensionless units, i.e., the energies are expressed in units of ϵ and temperature in units of $|\epsilon|/k$. In these calculations we used $2 \times 10^8 \times N$ MC steps with one MC being a single attempt to move (and rotate if allowed) an adsorbed molecule. In such a MC run, the cooling procedure was realized within 200 temperature decrements of equal length; from $T = 4.2$ down to $T = 0.2$. Specifically, at each temperature, the system was equilibrated during $10^6 \times N$ steps of which the last 10 % were used for averaging. The adsorbed configuration equilibrated at one temperature was used for the subsequent lower temperature run. The cooling sequence was also used to calculate the corresponding specific heat capacities based on the fluctuations of the internal energy of the adsorbed overlayer [23]. All the results discussed in the manuscript are the averages over ten independent system replicas.

3. Results and discussion

To examine the effect of orientational confinement on the structure formation in our systems, we first performed simulations for the enantiopure overlayers. Top part of figure 2 shows exemplary snapshots obtained for 1600 molecules R at $T = 0.2$. As it can be easily noticed in panel (a), the rotational freedom of the adsorbed molecules is responsible for the formation of a compact disordered domain with sparse void defects. Although the obtained structure lacks a long-range order, some clustering of uniformly aligned molecules, shown in the inset, can be observed. Similar to puzzle tiles, the R clusters having different orientations are capable of packing densely and creating diversified local motifs. The tendency of R to form the uniformly aligned local molecular structures becomes evident when the orientational confinement is switched on. In this case [panel (b)], we can observe an extended lamellar pattern characterized by a rectangular $(1 \times 2\sqrt{3})$ unit cell.

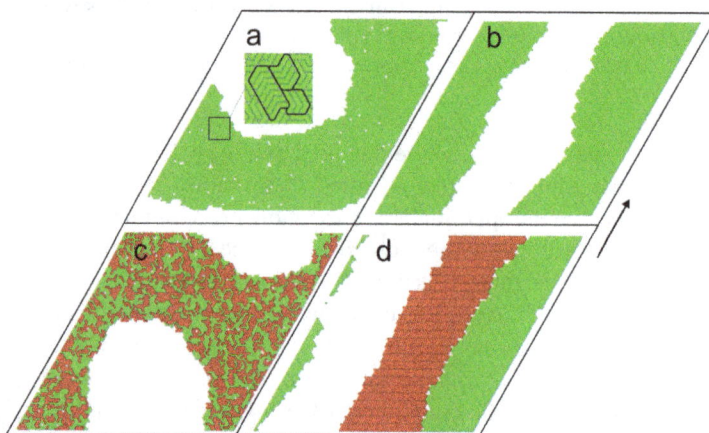

Figure 2. (Color online) Snapshots of the overlayers comprising 1600 molecules adsorbed on a 100 by 100 triangular lattice; $T = 0.2$. Panels (a) and (b) correspond to the enantiopure (R) systems while panels (c) and (d) show the results obtained for the corresponding racemates (800 R + 800 S). The effect of orientational confinement, indicated by the black arrow, on the structure formation is shown in the right-hand panels (b) and (d). The inset in panel (a) shows a magnified fragment of the overlayer in which clusters of uniformly aligned molecules can be noticed (encircled with black lines).

Regarding the racemic mixtures, in the case of unrestricted model, we can observe the formation of randomly mixed overlayers in which the enantiomers R and S exhibit some tendency to form homochiral clusters with uniform orientation [panel (c)]. However, this trend is much less pronounced compared to the enantiopure overlayer and it results mainly from the increased number of local tightly-packed motifs which can be formed by R and S adsorbed together. The increased structural diversity of molecular packings in the racemic assembly prevents an efficient propagation of the enantiopure lamellar domains. These domains, as clearly seen in panel (d), are formed when the orientational confinement is imposed on the enantiomers, leading to a complete demixing of R and S. The main source of the observed chiral resolution is the shape incompatibility of unlike enantiomers which are not capable of creating a dense mixed overlayer when uniformly oriented. The demixing is energetically favorable because it allows each molecule (R or S) to reach a maximum coordination (except for the phase boundary regions) characterized by the interaction energy equal to 9ϵ.

To quantify the differences between phase transformations occurring in the homochiral (a), (b) and racemic (c), (d) overlayers from figure 2, we calculated the mean potential energy of the adsorbed phase $\langle U \rangle$ and the associated specific heat, C_v as functions of temperature. These results are shown in figure 3. As it is seen in the left-hand panels of figure 3, for the enantiopure systems we can observe a considerable shift in the position of the $\langle U \rangle$ and C_v curves when the orientational confinement is switched on (dashed lines). Specifically, both $\langle U \rangle$ and C_v move towards higher temperatures, from about 1.38 to 2.19, as estimated from the corresponding peak maxima. In this case, a strong tendency for the formation of an ordered overlayer that is inherent to the unidirectional model enhances the propagation of

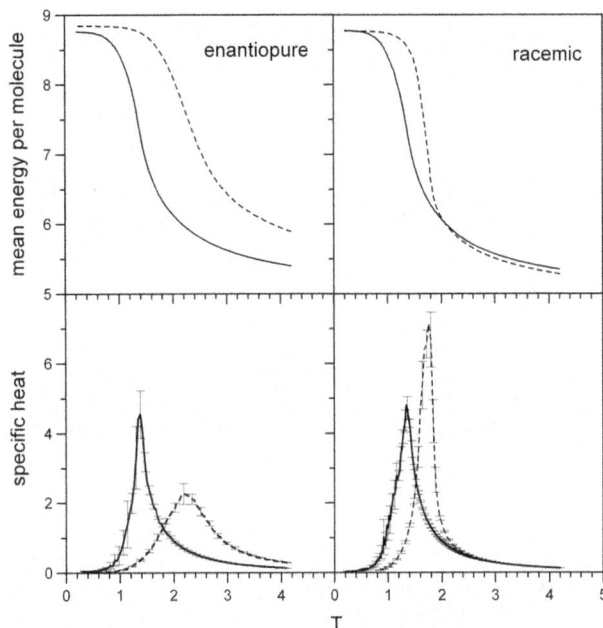

Figure 3. Temperature dependence of the average potential energy of an adsorbed molecule (top) and specific heat capacity calculated for the isotropic (solid) and unidirectional (dashed) model, for the enantiopure and racemic overlayers comprising 1600 molecules. These results are the averages over ten independent system replicas.

uniaxially aligned molecular pattern. Consequently, the transition is capable of occuring yet at higher temperatures. On the contrary, for the isotropic model, the increased orientational freedom of the adsorbed molecules lowers the temperature needed to reach the self-assembly. The results obtained for the enantiopure overlayers are qualitatively similar to those simulated for the corresponding racemates. However, for the racemic mixtures (right-hand panels), the analogous shifts in the position of the curves from the left-hand panels are much smaller. Here, the peak maximum in C_v shifts from about 1.36 to 1.74 and, interestingly, the position of the transition corresponding to the isotropic model (solid line) is very close to that observed for the enantiopure system. The proximity of the transition points observed in C_v for both types of isotropic models is likely to originate from the similar structure of the assemblies shown in the left-hand part of figure 2 (a), (c). For both these structures, we are dealing with randomly oriented, densely packed multimolecular clusters which fit in a way that a minimal number of void defects is created. In that sense, the presence of the other enantiomer (S) does not significantly change the mechanism of self-assembly. For the unidirectional model (dashed lines), however, the other enantiomer is responsible for an increased number of mixed molecular configurations (R-S) which can prevent somewhat the formation of separated homochiral domains. For this reason, the ordering transition shifts to lower temperatures.

To trace the extent of chiral segregation in the racemic systems, we introduced the parameter h, which is equal to the average number of heterogeneous R-S intermolecular interactions per adsorbed enantiomer. In other words, this parameter provides the number of bonds that can be drawn between the composite segments of an R(S) molecule, and the neighboring segments belonging to S(R) molecules. The parameter h characterizes the R-S interface length and, as it can be easily predicted, should be close to zero when the demixing occurs. On the other hand, for complete mixing of the enantiomers, h should take the values close to 9, since in a compact racemic R-S domain, statistically half of 18 bonds of a molecule, say R (full coordination), are formed with S-segments. To visualize these dependencies in figure 4 we plotted the parameter h as a function of temperature, for both versions of the model, for different surface coverages. From the results plotted in figure 4, it clearly follows that the phase transformations in the systems from figures 2 (c) and 2 (d) are manifested in the totally different shapes of the corresponding $h(T)$ curves. Namely, for the unidirectional model, we can observe a gradual decrease of h with a decreasing

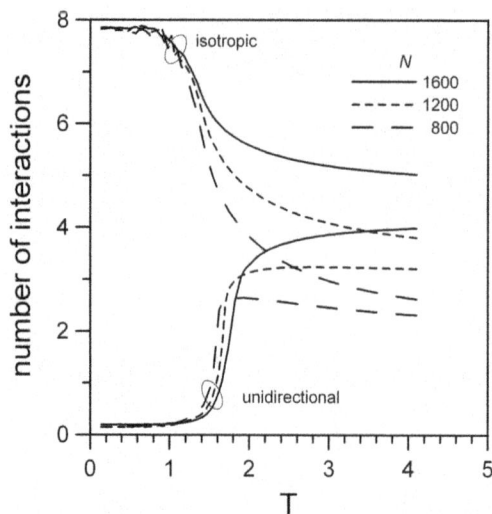

Figure 4. The average number of heterogeneous interactions per molecule, h, as a function of temperature, calculated for the isotropic and unidirectional model, for a different number of adsorbed enantiomers (racemic overlayers). These results are the averages over ten independent system replicas.

temperature down to about zero, regardless of the surface coverage. This behavior is a clear indication of the chiral segregation and the residual value of $h \approx 0.2$ comes from the inter-domain R-S contacts in the demixed racemic systems. On the other hand, for the unrestricted model, h increases with a decreasing temperature reaching a plateau at about 8. This value is consistent with the result of the simple mean field considerations made previously, meaning that the overlayer is randomly mixed with statistically half of the bonds of a molecule being of the heterogeneous (R-S) type. Obviously, the obtained value $h \approx 8$ is lower than the ideal one (9) due to the clustering tendency of like molecules and due to the presence of peripheral molecules with fractional coordination. The relative position of the curves, in which the one corresponding to the mixing is always placed above the other curve, is a natural consequence of the increased number of bimolecular R-S configurations allowed in the unrestricted model. This can be seen even at high temperatures, close to 4.

To explore the possibility of triggering the ordering transition using the director molecules we performed additional simulations in which the amount fraction, f, of these species was systematically chan-

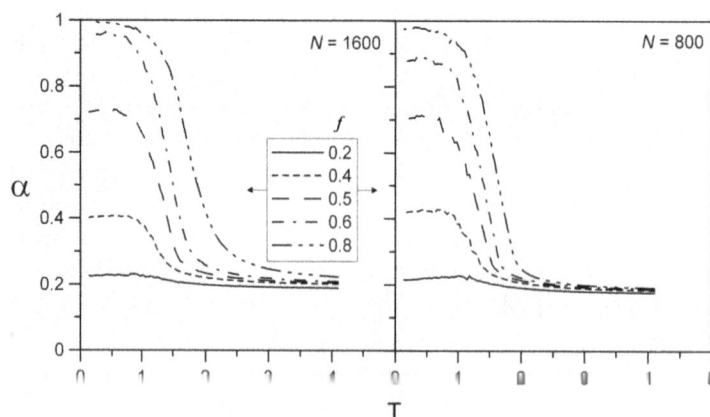

Figure 5. Relative gain in the number of vertically oriented molecules, α, as a function of temperature, calculated for different amount fractions of the director molecules. The results of the figure correspond to enantiopure overlayers with a different number of adsorbed molecules indicated in each panel. These results are the averages over ten independent system replicas.

ged. Figure 5 presents the effect of the active dopant on the parameter α, calculated for the enantiopure overlayers. This parameter is defined as follows:

$$\alpha = \frac{\hat{f} - f}{1 - f} \qquad (3.1)$$

in which $\hat{f}$ is the actual fraction of molecules with an upward orientation (the same orientation as the one assumed for the directors, see figure 1). The parameter α measures the relative gain in the number of vertically oriented molecules that is induced by injecting the director molecules at fraction f. From figure 5, it follows that in the enantiopure overlayers, below $T \approx 2$, the increase in the parameter α becomes noticeably large (> 0.5) when the fraction of the director molecules exceeds about one half. For example, for $f = 0.6$, the fraction of vertically oriented molecules among all the remaining molecules (which are not the director ones) reaches over 90 % highlighting quite effective guided ordering mechanism. This also refers to the overlayers in which the coverage is lower (see the right-hand panel), and for which the obtained dependencies are very similar to those shown in the left-hand panel. To compare these results with the corresponding dependencies calculated for the racemates, in figure 6 we plotted two additional sets of data. These, apart from the parameter α, also include the average number of heterogeneous interactions per molecule, h. The results of the left-hand panel of figure 6 clearly demonstrate a diversified

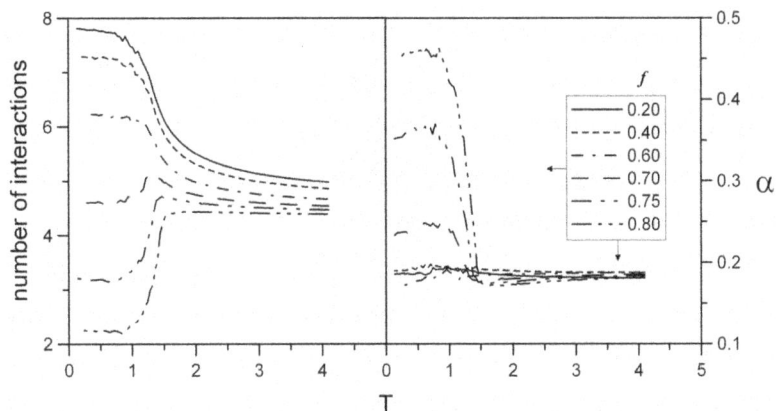

Figure 6. The average number of heterogeneous interactions per molecule, h, and relative change in the number of molecules with upward orientation, α, as functions of temperature calculated for the recamic overlayers for different amount fractions of the director molecules, f; $N = 1600$. These results are averages over ten independent system replicas.

phase behaviour of the racemate upon adding the director molecules at a different fraction. Specifically, for f lower than 0.6, we observe a distinct increasing trend in the average number of R-S heterogeneous interactions when the temperature drops. Here, we are dealing with an enhanced tendency to form a mixed R-S pattern. However, an increase of f above 0.6 does not lead to a significant drop of h (i.e., down to zero, see figure 4) at low temperatures where, for example, at $f = 0.7$, h is equal to about 4.7 which means an incomplete resolution of the enantiomers. To show this incomplete demixing occurring even at relatively high amounts of the director molecules, let us analyse the curves α from the right-hand panel of figure 6. In this case, we can observe that the relative increase in the fraction of vertically aligned molecules remains lower than 0.2 even for $f = 0.6$. A further increase of f, however, does not lead to a significant growth (up to 1) of the parameter α. On the contrary, even at a significant amount of the director molecules ($f = 0.8$), the relative content of the new molecules with upward orientation is still lower than 50 %. To illustrate how the director molecules contribute to the uniaxial ordering in figure 7, we showed an exemplary initial and final adsorbed configuration obtained for the system comprising 1600 molecules of which 60 % were the director molecules (dark grey). The results obtained for the racemic mixtures show that the role of the director molecules in the ordering in the corresponding 2D assemblies is much smaller than for the enantiopure systems. In other words, a much larger amount of the director molecules should be added to the racemate in order to induce its complete uniaxial ordering and

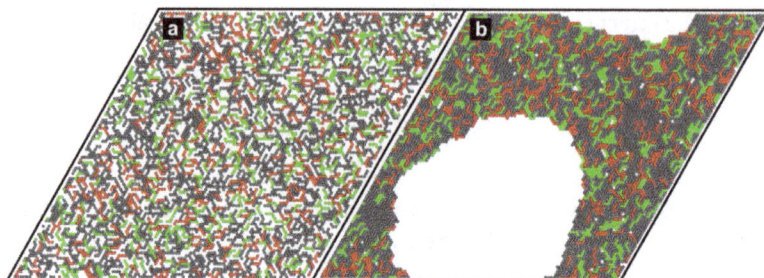

Figure 7. (Color online) The initial (a) and final (b) configuration obtained for the racemic mixture comprising 1600 molecules (800 R + 800 S) of which 60 % were the director molecules (dark grey, $f = 0.6$).

the resulting chiral resolution. As mentioned previously, this effect originates mainly from the increased number of tightly packed molecular motifs which can be formed in the quasi-two component R-S system.

In the next figure 8, we presented the effect of the surface coverage (the number of adsorbed molecules) on the extent of chiral resolution in the racemic overlayers. As it follows from the left-hand panel, at a high director content, the curves calculated for the two lower coverages ($N = 1200$ and $N = 800$) exhibit a noticeable increase at moderate temperatures followed by steep drops after passing through the transition temperature (about 1.5). In these cases, the number of R-S interactions is naturally low at higher temperatures where we deal with 2D gas phase. A further cooling of the adsorbed overlayer induces nucleation (increase in the number of R-S interactions), which is accompanied by the self-sorting effect leading to a decrease in h. Such a transition is not observed when f is small (see the right-hand panel), and the nucleating aggregates preserve their mixed structure in the extended R-S domain formed at low T, which means that the director content is too low to induce the sorting process.

The above results indicate that, regardless of the surface coverage, the self-sorting process is the most effective when the majority of the molecules of the racemate are forced to adopt one planar orientation. On the contrary, the role of the director molecules is found to be much larger for the enantiopure assemblies, which turned out to be more sensitive to the guiding influence of the active dopant.

4. Conclusions

In summary, using the coarse-grained MC modelling we have demonstrated that the orientational confinement of adsorbed chiral molecules can lead to their on-surface resolution producing homochiral ordered patterns. The calculations performed for the hockey-stick-shaped molecules, resembling the

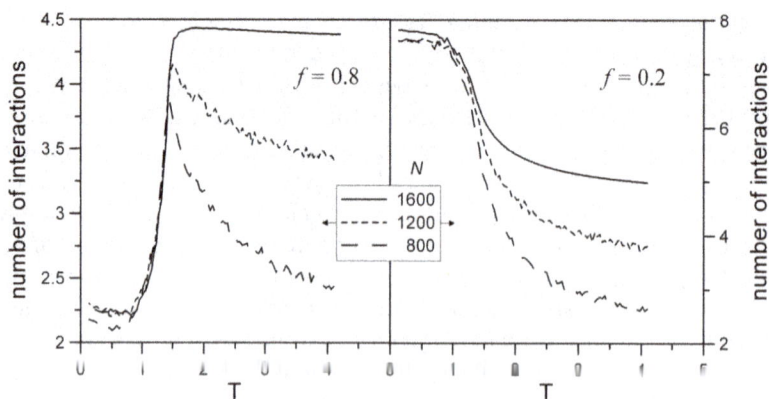

Figure 8. Effect of surface coverage (number of adsorbed molecules, N) on the average number of heterogeneous interactions per molecule, h, calculated for racemic mixtures differing in the amount fraction of the director molecules, f indicated in each panel. These results are averages over ten independent system replicas.

tetraphene, revealed the capability of this building block to form extended mirror-image domains upon 2D orientational confinement. It was found that the onset of phase transformation, in both enantiopure and racemic overlayers, shifts towards higher temperatures when the confinement is imposed on the molecules. However, this effect was more profound in the case of enantiopure systems highlighting the enhanced cooperativity of the like enantiomers in the structure formation. Separate simulations, in which only a part of the molecules was orientationally confined, showed that the chiral resolution (racemic) can be somewhat enhanced using the guiding dopant molecules with a fixed in-plane orientation. However, in this case, nearly all of the adsorbed molecules, R and S, should be capable of aligning uniaxially and preserving this orientation.

The results of our theoretical investigations can help understand the role of a directional bias in the chiral resolution of enantiomers in adsorbed overlayers. These findings can be also useful in developing the methods of fabrication of homochiral surfaces, such as in designing strategies to improve an adsorptive separation of chiral molecules.

References

1. Hu L., Zhang R., Chen Q., Nanoscale, 2014, **6**, 14064; doi:10.1039/c4nr05108d.
2. Bharti B., Velev O.D., Langmuir, 2015, **31**, 7897; doi:10.1021/la504793y.
3. Berg A.M., Patrick D.L., Angew. Chem. Int. Ed., 2005, **44**, 1821; doi:10.1002/anie.200461833.
4. Mougous J.D., Brackley A.J., Foland K., Baker R.T., Patrick D.L., Phys. Rev. Lett., 2000, **84**, 2742; doi:10.1103/PhysRevLett.84.2742.
5. Cun H., Wang Y., Yang B., Zhang L., Du S., Wang Y., Ernst K.-H., Gao H.J., Langmuir, 2010, **26**, 3402; doi:10.1021/la903193a.
6. Hanke F., Haq S., Raval R., Persson M., ACS Nano, 2011, **5**, 9093; doi:10.1021/nn203337v.
7. Klappenberger F., Zhang Y.-Q., Björk J., Klyatskaya S., Ruben M., Barth J.V., Acc. Chem. Res., 2015, **48**, 2140; doi:10.1021/acs.accounts.5b00174.
8. Fan Q., Gottfried J.M., Zhu J., Acc. Chem. Res., 2015, **48**, 2484; doi:10.1021/acs.accounts.5b00168.
9. Perusquía R., Peón J., Quintana J., Physica A, 2005, **345**, 130; doi:10.1016/j.physa.2004.05.089.
10. Martínez-González J.A., Pablo-Pedro R., Armas-Pérez J.C., Chapela G.A., Quintana-H J., RSC Adv., 2014, **4**, 20489; doi:10.1039/C4RA00617H.
11. Barnes B.C., Siderius D.W., Gelb L.D., Langmuir, 2009, **25**, 6702; doi:10.1021/la900196b.
12. Medved I., Trník A., Huckaby D.A., Phys. Rev. E., 2009, **80**, 011601; doi:10.1103/PhysRevE.80.011601.
13. Medved I., Trník A., Huckaby D.A., J. Stat. Mech., 2010, **12**, 12027; doi:10.1088/1742-5468/2010/12/P12027.
14. Martínez-González J.A., Chapela G.A., Quintana-H J., J. Chem. Phys., 2014, **140**, 194505; doi:10.1063/1.4876575.
15. Paci I., Szleifer I., Ratner M.A., J. Am. Chem. Soc., 2007, **129**, 3545; doi:10.1021/ja066422b.
16. Popa T., Paci I., Soft Matter, 2013, **9**, 7988; doi:10.1039/C3SM50312G.
17. Popa T., Paci I., J. Phys. Chem. C, 2015, **119**, 9829; doi:10.1021/acs.jpcc.5b00380.
18. Szabelski P., Woszczyk A., Langmuir, 2012, **28**, 11095; doi:10.1021/la301763k.
19. Woszczyk A., Szabelski P., Annales UMCS Chem., 2013, **68**, 133; doi:10.2478/umcschem-2013-0011.
20. Woszczyk A., Szabelski P., RSC Adv., 2015, **5**, 81933; doi:10.1039/C5RA15192A.
21. Romá F., Ramirez-Pastor A.J., Riccardo J.L., Langmuir, 2003, **19**, 6770; doi:10.1021/la0209785.
22. Ramirez-Pastor A.J., Nazarro M.S., Riccardo J.L., Zgrablich G., Surf. Sci., 1995, **341**, 249; doi:10.1016/0039-6028(95)00665-6.
23. Frenkel D., Smit B., Understanding Molecular Simulation from Algorithms to Applications, 2nd Edn., Academic Press, San Diego, 2002.

7

Profiles of electrostatic potential across the water-vapor, ice-vapor and ice-water interfaces

T. Bryk[1,2], A.D.J. Haymet[3]

[1] Institute for Condensed Matter Physics of the National Academy of Sciences of Ukraine,
1 Svientsitskii St., 79011 Lviv, Ukraine

[2] Institute of Applied Mathematics and Fundamental Sciences, Lviv Polytechnic National University,
12 Bandera St., 79013 Lviv, Ukraine

[3] Scripps Institution of Oceanography, UC San Diego, San Diego, California 92093, USA

Ice-water, water-vapor interfaces and ice surface are studied by molecular dynamics simulations with the SPC/E model of water molecules having the purpose to estimate the profiles of electrostatic potential across the interfaces. We have proposed a methodology for calculating the profiles of electrostatic potential based on a trial particle, which showed good agreement for the case of electrostatic potential profile of the water-vapor interface of TIP4P model calculated in another way. The measured profile of electrostatic potential for the pure ice-water interface decreases towards the liquid bulk region, which is in agreement with simulations of preferential direction of motion of Li^+ and F^- solute ions at the liquid side of the ice-water interface. These results are discussed in connection with the Workman-Reynolds effect.

Key words: *ice-water interface, ice surface, water-vapor interface, solute ions, profile of electrostatic potential, Workman-Reynods effect, molecular dynamics simulations*

1. Introduction

Fundamental presence of water, ice and their ice-water, ice-vapor(air) and water-vapor(air) interfaces in nature defines huge interest to exploration of their structural and dynamic properties [1, 2]. Microscopic structure and dynamics of the interfaces can be studied by atomistic molecular dynamics (MD) computer simulations using either models of water molecules with effective interactions or on *ab initio* level with explicit account for electron subsystem. Among the different ice and water interfaces, the ice-water interface has been studied much less, although by the date the issues of stability of the water-ice interface [3–5], its width [6–9] and interfacial free energy [10–13] are quite well elaborated.

Not only the pure interfaces are of great interest — even more interesting and fascinating problem is the behavior of different molecular and atomic solutes in the interfacial regions. One can mention a problem of the effect of solutes on the surface tension of aqueous solutions [14, 15], a problem of ionic transfer across aqueous and ice interfaces [16–19], the effects of salt on premelting the surface layers of ice [20–22], and the Workman-Reynolds effect [23–29] observed during the freezing of dilute aqueous solutions. The Workman-Reynolds effect of the emergence of freezing potential between the solid and liquid phases is connected with a charge separation occuring at a growing ice surface (moving ice-water interface). For simple aqueous solutions of NaCl, NaF, NaI, LiCl, etc., the Workman-Reynolds experiments gave evidence of positive potential of liquid with respect to ice, i.e., the charge separation with an excess of positive ions on the liquid side of the ice-aqueous solution interface. By date it is not clear what causes the observed charge separation at the ice-water interface.

Computer simulations support the observations of the charge separation occuring in Workman-Reynolds effect. Following a molecular dynamics study of the tendencies in ionic solute motion close to the water/dichloroethane interface [30], one may use the same methodology and place positive/negative ions on the liquid side of the ice-water interface. Upon proper equilibration, one can observe the preferable direction of motion of the ions towards/outwards the interface. Such a study with Na^+ and Cl^- solute ions at the liquid side of ice-water interface was reported in [10]. The positive sodium ions were moving towards the bulk water region of the two-phase system, while the negative Cl^- ions were trying to penetrate deeper into the interface heading towards the ice bulk region. These results for behaviour of positive/negative ions at the ice-water interface imply that the preferential direction of motion of solute ions can be caused by a difference in electrostatic potential between the solid and liquid sides of the pure interface. Therefore, the aim of this study was to check if the same effect is observed for the ions at the ice-water interface other than those reported in [10], and to estimate the profiles of electrostatic potential across three different interfaces: the water-vapor, ice-vapor and ice-water ones — and then compare them. The profile of electrostatic potential has already been calculated for water-vapor interface [31, 32], while for ice surface there are experimental estimates of the surface potential, as well as there were reported the profiles of the potential of the mean force for positive/negative ions [10, 22]. The rest of the paper is organized as follows: in the next section we supply the details of our molecular dynamics simulations as well as the methodology of calculations of the profile of electrostatic potential will be explained. In section 3, we report the results for the profiles of electrostatic potential for the water-vapor, ice-vapor and ice-water interfaces and discuss our findings. The last section contains a conclusion of this study.

2. Methodology of calculations

Molecular dynamic simulations for the three two-phase systems (containing water-vapor, ice-vapor and ice-water interfaces) were performed with the rigid SPC/E model [33] of water molecules. The simulated model ice-water system consisted of 2304 water molecules, ice-vapor and water-vapor systems — each of 1344 molecules. The average temperature in simulations of ice-water and ice-vapor systems was 225 K, i.e., nearly at the melting point for ice of the SPC/E model [4, 34], and for the water-vapor system the temperature was 298 K. For the case of ice-water and ice-vapor interfaces, we studied only the basal face of the interfaces, although the other orientations of ice I_h form ice-water interfaces with a bit smaller 10–90 widths than the basal face (see [9]), however, this should not change the general tendency of the behavior of ions at the inerface. Calculations of electrostatic profiles for the prism, $(20\bar{2}1)$ and $(2\bar{1}\bar{1}0)$ interfaces will be reported elsewhere.

The preparation of the interfaces for simulations at ambient pressure was described in detail in [10]. We used the same sequence of (NPT), (NP_zAT) and (NVT) ensembles in order to first prepare the bulk ice and the bulk water systems at the same ambient pressure, and then to bring water and ice into contact having the same area A and equilibrate the two-phase system by fluctuating only the z-box length at z-component of pressure tensor fixed to 1 bar. The size of equilibrated ice-water system of 2304 SPC/E water molecules was 26.8253 Å × 30.9903 Å × 85.3827 Å. The ultimate production runs of simulations with the purpose of calculating the electrostatic potentials were performed in NVT ensemble with Nose-Hoover thermostats. The electrostatics in simulations was treated by 3D Ewald method in all three cases of different interfaces. The cut-off distance for short-range part of potentials was 10 Å. All the simulations were performed by DL_POLY package [35].

We also performed $\sim 1.1 - 1.2$ nanosecond-long simulations of the behaviour of single Li^+ and F^- ions on the liquid side of the basal face of ice-water interface at $T = 225$ K. For ion interaction with the oxygens of SPC/E water molecules we used the following parameters for short-range Lennard-Jones potential: $\varepsilon_{LiO} = 0.160026$ kcal/mol, $\sigma_{LiO} = 2.337$ Å and $\varepsilon_{FO} = 0.167144$ kcal/mol, $\sigma_{FO} = 3.143$ Å. The ions at initial positions were slowly grown in at a fixed z-position first as a neutral particle with an increasing size every 500 timesteps (up to the mentioned Lennard-Jones parameter σ_{LiO}), and then gradually increasing its charge by ± 0.1 every 500 steps [10]. This procedure allowed us to grow in the solute ions without any structural damages for the ice-water interface.

Calculations of the electrostatic potential across the simulated two-phase system can be performed by means of the estimated charge-density profiles using their Fourier-transforms to solve the Poisson equation. For the case of ice-water and ice-vapor interfaces, the charge-density profiles contain positive and negative sharp peaks [9] in the solid phase which makes a direct application of the standard fast

Fourier-transform programs problematic. In fact, sharp peaks in the charge-density profiles come from the locations of the effective charges of oxygens and hydrogens in the atomic planes. However, the solute ions never reach the locations of these point charges on oxygens and hydrogens. Therefore, we used a simple methodology which is based on a trial neutral particle which moves in a plane with constrained z-coordinate and the single-particle electrostatic energy is calculated every few steps at the position of the trial particle (as if it had a charge +1). The trial particle interacts via soft-core potential $\sim (\sigma/r)^{12}$ with the oxygens and hydrogens and, therefore, it practically moves freely in the atomic plane with constrained z avoiding only the locations of the point charges on oxygens and hydrogens due to soft-core repulsion. In that case (having in mind that we can apply the Ewald methodology to calculate the single-particle electrostatic energy as if the trial particle had the charge +1) the electrostatic potential $V(z)$ is simply the average of single-particle electrostatic (Ewald) energies over configurations with the same z-coordinate of the trial particle minus the Ewald self-interaction. Note, that in order to avoid a shift of the interfaces in a system with a constrained particle, the motion of the simulation cell is constrained by removing the z-component of the velocity at every step during the molecular dynamics simulations, similarly as it was proposed in [36, 37] and was used earlier in our simulations reported in [10].

The z-coordinate of trial particle was changing with a step of 0.5 Å. In order to grow in the trial particle at a new z-position, we slowly increased the effective parameter σ every 500 time steps. After several thousand steps, the trial particle was grown in at a new z-position practically without any effect on the structure of the interface.

3. Results and discussion

Although it is known that the solute ions are expelled from the ice bulk phase, and free energy calculations [38, 39] and MD simulations [40] support this fact, the observations for the direction of preferential motion of positive/negative ions at the ice-water interface can shed light on the Workman-Reynolds effect [23]. We observed in MD simulations of the stable ice-water interface (the mass-density profile for the equilibrium relaxed basal face of ice I_h in contact with water simulated with the SPC/E model [33] at $T = 225$ K is shown in figure 1) the same tendencies in the preferential direction of motion for simple Li^+ and F^- ions (see figure 2) as were observed earlier for the larger in size Na^+ and Cl^- solute ions at the liquid side of ice-water interface [10]. It is seen from figure 2, that the negative F^- ion moves on the liquid side of the ice-water interface and is heading towars the bulk ice region, while the positive Li^+ ion shows a very similar behaviour with the Na^+ [10] having short-time trapping in cages on the liquid side of the interface, although with the long-time tendency of motion towards the water bulk region.

Before calculating the profile of electrostatic potential across the ice-water interface we tested our methodology on a simpler case of SPC/E water-vapor interface. In figure 3 we show the results for the density and electrostatic potential profile across the water-vapor interface. The interface is located be-

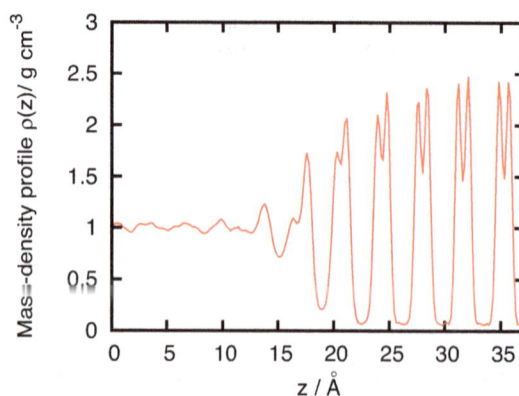

Figure 1. (Color online) Mass-density profile of the basal face ice-water interface obtained from simulations with the SPC/E model at $T = 225$ K.

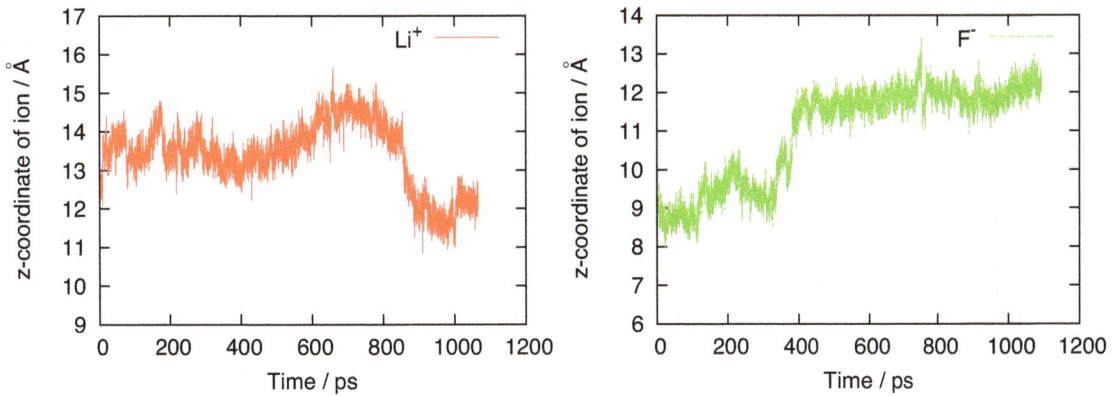

Figure 2. (Color online) Evolution of z-coordinate of Li^+ and F^- ions initially placed and equilibrated on the liquid side of the ice-water interface. The values of z-coordinate correspond to the mass-density profile shown in figure 1.

tween $z \approx 16$ Å and $z \approx 23$ Å, and namely in this region one observes a strong drop of the electrostatic potential, which shows a minimum nearly at the liquid side of the interface. In the literature there were reported calculations of the electrostatic potentials for the water-vapor interface of TIP4P water molecules [31, 32] and very similar (to our case of SPC/E model) profile of electrostatic potential was obtained. The minimum in profile of electrostatic potential was observed at the distance ~ 2.5 Å in the interface with respect to the Gibbs dividing plane, and the drop of potential between the vapor and minimum in the electrostatic potential in [32] was ~ -0.4 eV. In our case of the SPC/E water-vapor interface, we observed a minimum in the obtained profile of the electrostatatic potential also approximately at a distance 3 Å from the Gibbs dividing surface of the water-vapor density profile, and the drop in electrostatic potential was ~ 8.68 kcal/mol (figure 3), which corresponds to -376.4 meV, and is in good agreement for the estimate for TIP4P model [31], keeping in mind that the two water models (SPC/E and TIP4P) result in slightly different values of dielectric permittivity. Hence, our proposed methodology of estimation of the profile of electrostatic potential for the case of water-vapor interface gives a very reasonable agreement with the calculations by other methodology.

For the case of the pure ice surface, we used the same methodology of calculation of the profile of electrostatic potential. The obtained density profile and electrostatic potential are shown in figure 4. Qualitatively, the profile of eletrostatic potential for the ice surface is similar to the water-vapor interface, although the surface potential is smaller than for the water-vapor interface and is ~ 5.63 kcal/mol. There is a well-pronounced minimum in the electrostatic potential located in the top smeared-out atomistic

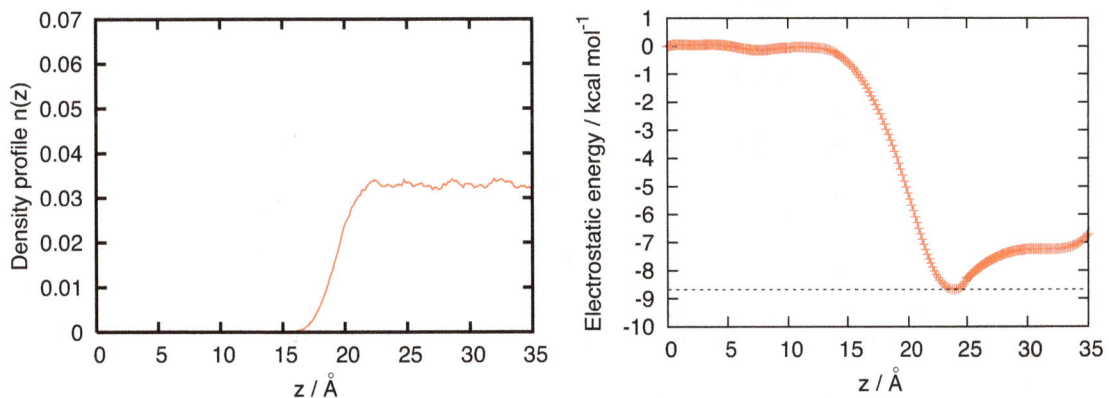

Figure 3. (Color online) Number density profile and profile of electrostatic potential for the water-vapor interface of SPC/E model at 298 K.

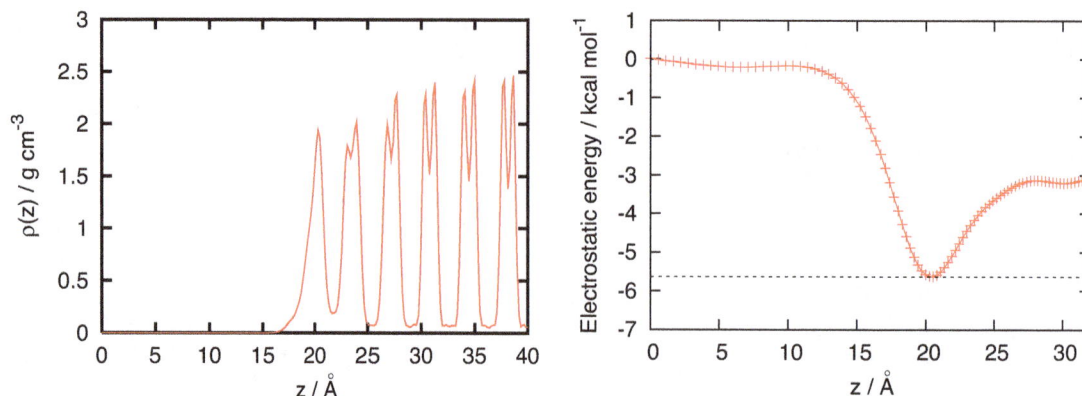

Figure 4. (Color online) Mass-density profile and profile of electrostatic potential for the ice (basal) surface at temperature 225 K.

layer which actually corresponds to the liquid-like layer on the ice surface. The MD simulations of the solute ions at the ice surface [22] gave evidence of localization of the positive solute ions in the liquid-like layer, while the negative ions were trying to penetrate deeper in the interface creating more defects and practically destroying the remaining order in the top surface layers. In the literature, there is not much information on the surface potential of pure ice surface. We were able to find some experimental estimates for the pure ice surface potential from measurements with gold electrodes at temperatures between $0°C$ and $-15°C$ [25]. The ice surface potential with respect to gold electrode was reported to be in the range $\sim 185 - 195$ meV, which, however, upon riming changed the sign and became ~ -215 meV. Our MD calculations for the pure ice surface potential (with an existing liquid-like layer) give the value of ~ -244 meV. It is obvious that the presence of the liquid-like layer on the surface of ice at melting temperature (at which the MD simulations were performed) should give a comparable value and the same sign of the surface potential as were obtained for the water-vapor interface.

Figure 5 reports the profile of electrostatic potential, that corresponds to the ice-water interface with the mass-density profile shown in figure 1. Our calculations show an increase of the electrostatic potential from the water to the ice phase with approximately 2.1 kcal/mol (~ 91 meV) net difference. This is in agreement with our observations of the tendencies of the preferential motion of the positive solute ion towards liquid phase and of the negative ion towards the bulk ice. Note that the ice-water interface has the smallest change in the profile of electrostatic potential in comparison with the water-vapor interface and ice surface. It is obvious that in the crystal bulk region the profile of electrostatic potential should

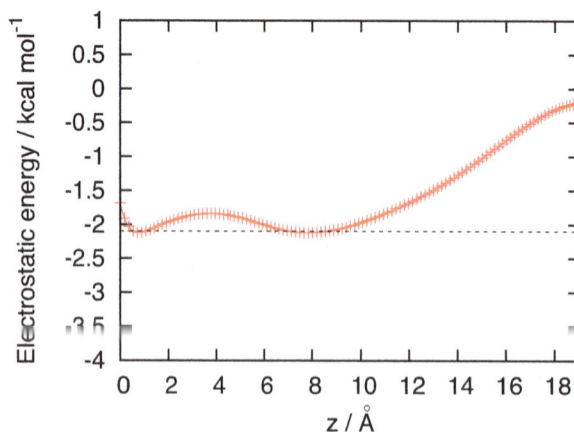

Figure 5. (Color online) Profile of electrostatic potential for the ice (basal face)-water surface of the SPC/E model, which corresponds to the mass-density profile shown in figure 1.

have the periodicity of the charge-density profile. We did not extend our calculations into the ice bulk region for the ice-water and ice-vapor interfaces. In order to recover the periodicity of the electrostatic potential in ice bulk region one has to use in our methodology much smaller step in constrained z-position for the trial particle.

4. Conclusions

In MD simulations of Li^+ and F^- ions at the ice-water interface we have observed different tendencies in the preferential direction of motion which were the same as in the earlier reported behavior of larger in size ions Na^+ and Cl^- in MD simulations [10]. In both cases, the positive ions had a preferential direction of motion towards the bulk water region, while the negative ions showed an opposite direction and were trying to penetrate deeper in the interface. These observations are in agreement with the Workman-Reynolds effect of emerging freezing potential at the growing ice in the aqueous solutions, which reflects the charge separation with the excess of positive ions in the liquid phase.

In order to get insight into electrostatic effects which occur at different interfaces, we calculated from MD simulations the profiles of electrostatic potential for the pure ice-water, water-vapor interfaces and pure ice surface. We have proposed a new methodology of calculating the profiles of electrostatic potentials by making use of a trial particle which moves in a plane with a fixed z-coordinate and due to soft-core repulsion avoids the positions of point charges. Calculations of a single-particle electrostatic energy by Ewald method at the instantaneous position of the trial particle is connected with the electrostatic potential at this point, which after the average over different configurations results in the profile of electrostatic potential. Our check performed for the water-vapor interface showed good agreement with the profile of electrostatic potential for the water-vapor interface simulated with TIP4P model [31].

Our results indicate that among the three interfaces, the pure water-vapor interface has the largest decrease of the profile of electrostatic potential in the liquid phase, while the ice-water interface has the smallest difference. The obtained profile of electrostatic potential for the pure ice-water interface decreases towards the bulk water region which can explain the observed difference in the preferential direction of motion of the positive/negative solute ions observed in MD simulations.

References

1. Howe J.M., Interfaces in Materials, Wiley, NewYork, 1997.
2. Laird B.B., Haymet A.D.J., Chem. Rev., 1992, **92**, 1819; doi:10.1021/cr00016a007.
3. Karim O.A., Haymet A.D.J., Chem. Phys. Lett., 1987, **138**, 531; doi:10.1016/0009-2614(87)80118-5.
4. Bryk T., Haymet A.D.J., Mol. Simul., 2004, **30**, 131; doi:10.1080/0892702031000152172.
5. Yoo S., Xantheas S.S., J. Chem. Phys., 2011, **134**, 121105; doi:10.1063/1.3573375.
6. Nada H., Furukawa Y., Jpn. J. Appl. Phys., 1995, **34**, 583; doi:10.1143/JJAP.34.583.
7. Bàez L., Clancy P., J. Chem. Phys., 1995, **103**, 9744; doi:10.1063/1.469938.
8. Hayward J.A., Haymet A.D.J., J. Chem. Phys., 2001, **114**, 3713; doi:10.1063/1.1333680.
9. Bryk T., Haymet A.D.J., J. Chem. Phys., 2002, **117**, 10258; doi:10.1063/1.1519538.
10. Haymet A.D.J., Bryk T., Smith E.J., In: Ionic Soft Matter: Modern Trends in Theory and Applications, Henderson D., Holovko M., Trokhymchuk A. (Eds.), NATO Science Series II, Vol. 206, Springer, Dordrecht, 2005, 333–359; doi:10.1007/1-4020-3659-0_13.
11. Handel R., Davidchack R.L., Anwar J., Brukhno A., Phys. Rev. Lett., 2008, **100**, 036104; doi:10.1103/PhysRevLett.100.036104.
12. Pirzadeh P., Beaudoin E.N., Kusalik P.G., Cryst. Growth Des., 2012, **12**, 124; doi:10.1021/cg200861e.
13. Davidchack R.L., Handel R., Anwar J., Brukhno A.V., J. Chem. Theory Comput., 2012, **8**, 2383; doi:10.1021/ct300193e.
14. Heydweiller A., Ann. Phys. (Leipzig), 1910, **33**, 145; doi:10.1002/andp.19103381108.
15. Levin Y., dos Santos A.P., Diehl A., Phys. Rev. Lett., 2009, **103**, 257802; doi:10.1103/PhysRevLett.103.257802.
16. Wick C.D., Dang L.X., Chem. Phys. Lett., 2008, **458**, 1; doi:10.1016/j.cplett.2008.03.097.
17. Lee A.J., Rick S.W., J. Phys. Chem. Lett., 2012, **3**, 3199; doi:10.1021/jz301411q.
18. Soniat M., Rick S.W., J. Chem. Phys., 2015, **143**, 044702; doi:10.1063/1.4874256.
19. Soniat M., Kumar R., Rick S.W., J. Chem. Phys., 2015, **143**, 044702; doi:10.1063/1.4926831.

20. Dash J.G., Rempel A.W., Wettlaufer J.S., Rev. Mod. Phys., 2006, **78**, 695; doi:10.1103/RevModPhys.78.695.
21. Kim J.S., Yethiraj A., J. Chem. Phys., 2008, **129**, 124504; doi:10.1063/1.2979247.
22. Bryk T., Haymet A.D.J., J. Mol. Liq., 2004, **112**, 47; doi:10.1016/j.molliq.2003.11.008.
23. Workman E.J., Reynolds S.E., Phys. Rev., 1950, **78**, 254; doi:10.1103/PhysRev.78.254.
24. Gross G.W., J. Geophys. Res., 1965, **70**, 2291; doi:10.1029/JZ070i010p02291.
25. Caranti J.M., Illingworth A.J., Nature, 1980, **284**, 44; doi:10.1038/284044a0.
26. Caranti J.M., Illingworth A.J., J. Geophys. Res., 1983, **88**, 8483; doi:10.1029/JC088iC13p08483.
27. Bronshteyn V.L., Chernov A.A., J. Cryst. Growth, 1991, **112**, 129; doi:10.1016/0022-0248(91)90918-U.
28. Wilson P.W., Haymet A.D.J., J. Phys. Chem. B, 2008, **112**, 11750; doi:10.1021/jp804047x;
29. Wilson P.W., Haymet A.D.J., J. Phys. Chem. B, 2008, **112**, 15260; doi:10.1021/jp807642s.
30. Benjamin I., Science, 1993, **261**, 1558; doi:10.1126/science.261.5128.1558.
31. Wilson M.A., Pohorille A., Pratt L.R., J. Chem. Phys., 1988, **88**, 3281; doi:10.1063/1.453923.
32. Wilson M.A., Pohorille A., Pratt L.R., J. Chem. Phys., 1989, **90**, 5211; doi:10.1063/1.456536.
33. Jorgensen W.L., Chandrasekhar J., Madura J.D., Impey R.W., Klein M.L., J. Chem. Phys., 1983, **79**, 926; doi:10.1063/1.445869.
34. Sanz E., Vega C., Abascal J.L.F., MacDowell L.G., Phys. Rev. Lett., 2004, **92**, 255701; doi:10.1103/PhysRevLett.92.255701.
35. http://www.ccp5.ac.uk/DL_POLY_CLASSIC/.
36. Dang L.X., Chang T.-M., J. Phys. Chem. B, 2002, **106**, 235; doi:10.1021/jp011853w.
37. Dang L.X., J. Phys. Chem. B, 2002, **106**, 10388; doi:10.1021/jp021871t.
38. Smith E.J., Bryk T., Haymet A.D.J., J. Chem. Phys., 2005, **123**, 034706; doi:10.1063/1.1953578.
39. Smith E.J., Bryk T., Haymet A.D.J., J. Chem. Phys., 2007, **126**, 237102; doi:10.1063/1.2738062.
40. Vrbka L., Jungwirth P., Phys. Rev. Lett., 2005, **95**, 148501; doi:10.1103/PhysRevLett.95.148501.

8

Maier-Saupe nematogenic fluid with isotropic Yukawa repulsion at a hard wall: Mean field approximation

M. Holovko[1], T. Patsahan[1]*, I. Kravtsiv[1], D. di Caprio[2]

[1] Institute for Condensed Matter Physics of the National Academy of Sciences of Ukraine,
1 Svientsitskii St., 79011 Lviv, Ukraine

[2] Institut de Recherche de Chimie Paris, CNRS — Chimie ParisTech,
11 rue Pierre et Marie Curie, 75005 Paris, France

The mean field approximation is formulated within the framework of the density field theory to study the properties of a Maier-Saupe nematogenic fluid near a hard wall. The density and the order parameter profiles are obtained using the analytical expressions derived in the linearized mean field approximation. The temperature dependencies of the contact values of the density and order parameter profiles are analyzed in detail. To estimate a validity of the applied approximations, the obtained theoretical results are compared with the original computer simulation data.

Key words: *Maier-Saupe nematogenic fluid, field theory, interface, hard wall, contact theorem, Yukawa potential*

1. Introduction

It is a great pleasure and a big honor for us to contribute this paper to the festschrift dedicated to Professor Stefan Sokołowski. Stefan is a well-known expert on the modelling of physico-chemical properties of complex fluids such as chemically reacting fluids in disordered porous media [1, 2], studies of the isotropic-nematic phase transition in confined nematic fluids [3], modelling of the properties of fluids in pores with walls decorated with tethered polymer brushes [4], treatment of phase behavior in confined functional colloids [5] and many other complex fluids in complex confinements.

In this paper, we study the influence of the surface on a nematic fluid near a hard wall. Due to orientational ordering, such systems have many unique properties which are very important in the display industry. The anchoring phenomena are among them, according to which the surface induces a specific orientation of the nematic director with respect to the surface [6]. In order to understand the connection between the anchoring phenomena and the interaction between nematic molecules and the surface, there the Henderson-Abraham-Barker (HAB) approach previously developed for isotropic fluids at the wall [7] has been employed [8, 9]. In this approach, the distribution of the fluid near the wall is described by the wall-particle Ornstein-Zernike (OZ) equation with the fluid distribution function in the bulk calculated in the mean spherical approximation (MSA) [10]. Using this approach, it is possible to evaluate the role of an orientation-dependent interaction of nematic molecules with the surface in the anchoring phenomena. However, in the MSA, the HAB approach does not correctly take into account the contribution from long-range intermolecular interactions. As a result, it does not satisfy an exact relation known as the contact theorem, which was formulated in [11, 12] for isotropic fluids near a wall, and recently in [13] it was reformulated for anisotropic fluids near a hard wall. According to this theorem, the contact value

---

*E-mail: tarpa@icmp.lviv.ua

of the density profile of a fluid near a hard wall is determined by the pressure of the fluid in the bulk. In [13], the contact theorem was also formulated for the order parameter profile.

An alternative way of describing fluids at a hard wall was developed within the framework of the density field theory. In this theory, the contributions from the mean field and from fluctuations are separated. The theory was successfully applied to ionic fluids at a hard wall [14–17] and to simple fluids with Yukawa-type interactions near a hard wall [18, 19]. It was shown that the mean field treatment of a Yukawa fluid at a hard wall reduces to the solution of a non-linear differential equation for the density profile, while the treatment of fluctuations reduces to the OZ equation with the Riemann boundary condition [20]. The density field theory was applied to the description of bulk properties of a nematic fluid in [21, 22]. The application of the density field theory to the description of a nematic fluid at a hard wall was initiated in [23].

In this paper, we use the mean field approximation to investigate the effect of a hard wall on the properties of a nematic fluid. We demonstrate the principal difference between the behavior of the order parameter profile obtained in the mean field approximation and its linearized version. In order to check the validity of both approaches we compare the obtained theoretical results with computer simulations data.

2. Theory

In this paper, we consider Maier-Saupe (MS) nematogenic fluid model [24, 25] as one of the simplest models that accounts for the isotropic-nematic phase transition. For simplification, we consider a fluid of point uniaxial nematogens interacting through the pair potential

$$v(r_{12},\Omega_1\Omega_2) = v_0(r_{12}) + v_2(r_{12})P_2(\cos\theta_{12}), \tag{1}$$

where the first term $v_0(r_{12}) = (A_0/r_{12})\exp(-\alpha_0 r_{12})$ describes isotropic repulsion and the second term with $v_2(r_{12}) = (A_2/r_{12})\exp(-\alpha_2 r_{12})$ describes anisotropic attraction between particles ($A_0 > 0$, $A_2 < 0$), r_{12} denotes the distance between particles 1 and 2, $\Omega = (\theta,\phi)$ are orientations of particles, $P_2(\cos\theta_{12}) = (3\cos^2\theta_{12} - 1)/2$ is the second order Legendre polynomial of the relative orientation θ_{12}.

It is necessary, in numerical calculations, to cut-off the potential $v(r_{12},\Omega_1\Omega_2)$ at some finite distance and due to this in expression (1), $v_0(r)$ and $v_2(r)$ are replaced by $\tilde{v}_i(r) = v_i(r)$ for $r \leqslant r_c$ and $\tilde{v}_i(r) = 0$ for $r > r_c$, where $i = 0, 2$ and r_c is the cut-off radius.

Within the field-theoretical formalism, the Hamiltonian is a functional of the density field and can be written as a sum of the entropic and the interaction terms

$$\beta H[\rho(\mathbf{r},\Omega)] = \int \rho(\mathbf{r},\Omega)\left\{\ln\left[\rho(\mathbf{r},\Omega)\Lambda_R\Lambda_T^3\right] - 1\right\}d\mathbf{r}d\Omega$$
$$+ \frac{\beta}{2}\int v(r_{12},\Omega_1\Omega_2)\rho(\mathbf{r}_1,\Omega_1)\rho(\mathbf{r}_2,\Omega_2)d\mathbf{r}_1 d\mathbf{r}_2 d\Omega_1 d\Omega_2, \tag{2}$$

where $\beta = 1/k_B T$ is the inverse temperature, $d\Omega = (1/4\pi)\sin\theta d\theta d\phi$ is the normalized angle element, $\rho(\mathbf{r},\Omega)$ is particle density per angle such that $\int \rho(\mathbf{r},\Omega)d\Omega = \rho(\mathbf{r})$, Λ_T is the thermal de Broglie wavelength of the molecules, the quantity Λ_R^{-1} is the rotational partition function for a single molecule [26].

2.1. Mean field approximation

In this paper, we restrict our consideration to the mean field (MF) approximation which is the lowest order approximation for the partition function. In the canonical formalism, it corresponds to fixing the Lagrange parameter λ such that the following relation is true for the singlet distribution function

$$\left.\frac{\delta\beta H[\rho(\mathbf{r},\Omega)]}{\delta\rho(\mathbf{r},\Omega)}\right|_{\rho^{MF}} = \lambda. \tag{3}$$

As a result, we have

$$\rho(\mathbf{r}_1,\Omega_1) = \rho^{bulk}(\Omega_1)\exp\left\{-\beta\int v(r_{12},\Omega_1\Omega_2)\left[\rho(\mathbf{r}_2,\Omega_2) - \rho^{bulk}(\Omega_2)\right]d\mathbf{r}_2 d\Omega_2\right\}, \tag{4}$$

where

$$\rho^{\text{bulk}}(\Omega) = \rho_{\text{b}} \frac{\exp\left[-(\kappa_2^2 S_{\text{b}}/\alpha_2^2) P_2(\cos\theta)\right]}{\int\limits_0^1 \text{d}\cos\theta \exp\left[-(\kappa_2^2 S_{\text{b}}/\alpha_2^2) P_2(\cos\theta)\right]} \tag{5}$$

is the singlet distribution function for the bulk nematic fluid in the MF approximation, defined within the framework of the Maier-Saupe theory [24, 25], $\kappa_2^2 = 4\pi\rho_{\text{b}}\beta A_2$, ρ_{b} is the bulk value of the fluid density, $S_{\text{b}} = (1/\rho_{\text{b}})\int_0^1 P_2(\cos\theta)\rho^{\text{bulk}}(\Omega)\text{d}\cos\theta$ is the bulk value of the orientational order parameter.

After integration with respect to orientation Ω_2, we obtain

$$\frac{\rho(\mathbf{r}_1,\Omega_{1n},\Omega_{wn})}{\rho^{\text{bulk}}(\Omega_1)} = \exp\left\{-\left[V_0(\mathbf{r}_1,\Omega_{wn}) - V_0^{\text{b}}\right] - \frac{1}{\sqrt{5}}\sum_m Y_{2m}(\Omega_{1n})\left[V_{2m}(\mathbf{r}_1,\Omega_{wn}) - V_{2m}^{\text{b}}\right]\right\}, \tag{6}$$

where Ω_{wn} denotes the angle between the nematic director and the surface, and the mean field potentials

$$V_0(\mathbf{r}_1,\Omega_{wn}) = \beta\int v_0(r_{12})\rho(\mathbf{r}_2,\Omega_{wn})\text{d}\mathbf{r}_2, \tag{7}$$

$$V_{2m}(\mathbf{r}_1,\Omega_{wn}) = \beta\int v_2(r_{12})S_{2m}(\mathbf{r}_2,\Omega_{wn})\text{d}\mathbf{r}_2. \tag{8}$$

The bulk values of these potentials are $V_0^{\text{b}} = \kappa_0^2/\alpha_0^2$, $V_{20}^{\text{b}} = \kappa_2^2 S_{\text{b}}/\alpha_2^2$, $V_{2m}^{\text{b}} = 0$ for $m \neq 0$, where $\kappa_0^2 = 4\pi\rho_{\text{b}}\beta A_0$,

$$\rho(\mathbf{r},\Omega_{wn}) = \int \rho(\mathbf{r},\Omega_{1n},\Omega_{wn})\text{d}\Omega_{1n} \tag{9}$$

is the density profile. The quantities

$$S_{2m}(\mathbf{r},\Omega_{wn}) = \frac{1}{\sqrt{5}}\int \rho(\mathbf{r},\Omega_{1n},\Omega_{wn})Y_{2m}(\Omega_{1n})\text{d}\Omega_{1n} = \rho(\mathbf{r},\Omega_{wn})S_{2m}^*(\mathbf{r},\Omega_{wn}), \tag{10}$$

where $S_{2m}^*(\mathbf{r},\Omega_{wn})$ are the order parameter profiles. Far from the wall we have $S_{20}^*(\mathbf{r},\Omega_{wn}) \to S_{\text{b}}$, $S_{2m}^*(\mathbf{r},\Omega_{wn}) \to 0$ for $m \neq 0$. Simple calculations show that, in order to take into account the cut-off radius r_{c}, one should substitute the quantities κ_i^2 by $\tilde{\kappa}_i^2$ such that

$$\tilde{\kappa}_i^2 = 4\pi\rho\beta\int\limits_0^{r_{\text{c}}} v_i(r)r^2\text{d}r = \kappa_i^2\left[1 - \exp(-\alpha_i r_{\text{c}}) - \alpha_i r_{\text{c}}\exp(-\alpha_i r_{\text{c}})\right]. \tag{11}$$

2.2. Linearized MF approximation

The gradient of equation (6) gives

$$\frac{1}{\rho(\mathbf{r},\Omega_{1n},\Omega_{wn})}\nabla\rho(\mathbf{r},\Omega_{1n},\Omega_{wn}) = \mathbf{E}_0(\mathbf{r},\Omega_{wn}) + \frac{1}{\sqrt{5}}\sum_m Y_{2m}(\Omega_{1n})\mathbf{E}_{2m}(\mathbf{r},\Omega_{wn}), \tag{12}$$

where we define an equivalent of the electric field as

$$\mathbf{E}_0(\mathbf{r},\Omega_{wn}) \equiv -\nabla V_0(\mathbf{r},\Omega_{wn}), \qquad \mathbf{E}_{2m}(\mathbf{r},\Omega_{wn}) \equiv -\nabla V_{2m}(\mathbf{r},\Omega_{wn}). \tag{13}$$

According to the properties of the Yukawa potential we can write

$$\left(\triangle - \alpha_0^2\right)V_0(\mathbf{r},\Omega_{wn}) = -4\pi\beta A_0\rho(\mathbf{r},\Omega_{wn}), \tag{14}$$

$$\left(\triangle - \alpha_2^2\right)V_{2m}(\mathbf{r},\Omega_{wn}) = -4\pi\beta A_2 S_{2m}(\mathbf{r},\Omega_{wn}). \tag{15}$$

Due to translational invariance parallel to the wall, the functions considered depend only on the distance z to the wall. Equations (10)–(15) make a set of six differential equations for the unknown functions

$\rho(\mathbf{r},\Omega_{1n},\Omega_{wn})$, $S_{2m}(\mathbf{r},\Omega_{wn})$, $E_0(\mathbf{r},\Omega_{wn})$, $E_{2m}(\mathbf{r},\Omega_{wn})$, $V_0(\mathbf{r},\Omega_{wn})$, $V_{2m}(\mathbf{r},\Omega_{wn})$. We note that in the case when the director is oriented perpendicularly to the wall, $\Omega_{wn} = 0$, the singlet distribution function is axially symmetric. Consequently, the equations considered will retain only the terms with $m = 0$. In this paper, we will restrict our further investigation to this special case.

As was shown in [23], the differential equations obtained can be solved analytically in the linear approximation for the expression (6)

$$\rho'(z,\Omega) = [E_0(z) + E_{20}(z)P_2(\cos\theta)]\,\rho^{\text{bulk}}(\Omega), \tag{16}$$

where the prime denotes derivative by z.

The resulting solutions of the linearized profile are as follows:

$$\frac{\rho(z)}{\rho_b} = 1 - \frac{\lambda_0^2 - \alpha_2^2 - \frac{1}{5}\kappa_2^2\left(\langle Y_{20}^2\rangle_\Omega - \langle Y_{20}\rangle_\Omega^2\right)}{\kappa_2^2 S_b}\,B_1\,e^{-\lambda_0 z}$$
$$-\frac{\lambda_2^2 - \alpha_2^2 - \frac{1}{5}\kappa_2^2\left(\langle Y_{20}^2\rangle_\Omega - \langle Y_{20}\rangle_\Omega^2\right)}{\kappa_2^2 S_b}\,B_2\,e^{-\lambda_2 z}, \tag{17}$$

$$\frac{S_{20}(z)}{\rho_b S_b} = 1 - \frac{\left(\lambda_0^2 - \alpha_2^2\right)}{\kappa_2^2 S_b}\,B_1\,e^{-\lambda_0 z} - \frac{\left(\lambda_2^2 - \alpha_2^2\right)}{\kappa_2^2 S_b}\,B_2\,e^{-\lambda_2 z}, \tag{18}$$

where

$$B_1 = \frac{\kappa_2^2 S_b}{2\left(\lambda_0^2 - \lambda_2^2\right)}\left[-\frac{\kappa_0^2}{\alpha_0^2} + \frac{\lambda_2^2 - \alpha_2^2 - (\kappa_2^2/5)\langle Y_{20}^2\rangle_\Omega}{\alpha_2^2}\right], \qquad B_2 = -\frac{\kappa_2^2 S_b}{2\alpha_2^2} - B_1, \tag{19}$$

$$\langle Y_{20}^k\rangle_\Omega = (1/\rho_b)\int_0^1 Y_{20}^k(\Omega)\rho^{\text{bulk}}(\Omega)\mathrm{d}\cos\theta. \tag{20}$$

Parameters λ_0 and λ_2

$$\lambda_{0,2}^2 = \frac{1}{2}\left\{\kappa_0^2 + \alpha_0^2 + \kappa_2^2\langle P_2^2(\cos\theta)\rangle + \alpha_2^2 \pm \sqrt{\left[\kappa_0^2 + \alpha_0^2 - \kappa_2^2\langle P_2^2(\cos\theta)\rangle - \alpha_2^2\right]^2 + 4\kappa_0^2\kappa_2^2 S_b^2}\right\} \tag{21}$$

are identical to the parameters found in the bulk phase when Gaussian fluctuations are taken into account [22] and characterize a decay of the isotropic repulsive and the anisotropic attractive interactions, respectively.

Hereafter, the approach presented in this subsection is referred to as the linearized mean field (LMF) approximation. The expressions for $\rho(z)$ and $S_{20}^*(z)$ obtained within this approximation correspond to the case of infinite cut-off radius $r_c \to \infty$.

2.3. Contact theorem

As it was shown in [13], the density and order parameter profiles satisfy some exact relations known as the contact theorems. According to these relations in the absence of wall-particle interactions, the contact values of the density profile $\rho(z = 0)$ and of the order parameter profile $S_{20}(z = 0)$ do not depend on the angle Ω_{wn} and are equal to

$$\rho(z = 0) = \beta\int \mathrm{d}\Omega_{1n}P(\Omega_{1n}) = \beta P, \tag{22}$$

$$S_{20}(z = 0) = \beta\int \mathrm{d}\Omega_{1n}P_2(\cos\theta)P(\Omega_{1n}), \tag{23}$$

where P is the bulk pressure and $P(\Omega_{1n})$ can be treated as the bulk partial pressure for molecules with a given orientation Ω_{1n}.

In the MF approximation for the model under consideration, relations (22) and (23) give

$$\frac{\rho(0^+)}{\rho_b} = 1 + \frac{\kappa_0^2}{2\alpha_0^2} + \frac{\kappa_2^2}{2\alpha_2^2} S_b^2, \tag{24}$$

$$\frac{S_{20}(0^+)}{S_b \rho_b} = 1 + \frac{\kappa_0^2}{2\alpha_0^2} + \frac{\kappa_2^2}{2\alpha_2^2} \langle P_2^2(\cos\theta) \rangle. \tag{25}$$

These relations are used as boundary conditions in the solution of differential equations of the LMF approximation. In order to take into account the cutoff distance r_c in relations (24)–(25), we should change κ_i^2 to $\tilde{\kappa}_i^2$ given by equation (11). However, we should note the principal difference between the exact results (22)–(23) and results (24)–(25) of the MF approximation. In paper [21], an invariant

$$\frac{\alpha_0^2}{2\kappa_0^2} V_0^2(z, \Omega_{wn}) - \frac{1}{2\kappa_0^2} E_0^2(z, \Omega_{wn}) + \sum_m \left[\frac{\alpha_2^2}{2\kappa_2^2} V_{2m}^2(z, \Omega_{wn}) - \frac{1}{2\kappa_2^2} E_{2m}^2(z, \Omega_{wn}) \right] \tag{26}$$

was found which was used to prove the contact theorem for the MF density profile of a nematogenic fluid in form (24). However, no similar proof of the contact theorem for the MF order parameter profile in form (25) exists.

3. Numerical calculation details

In order to verify the theoretical approaches presented in the previous section, a series of numerical calculations were carried out. To this end, for the model pair potential (1), the following parameters were chosen: $A_0/|A_2| = 3.0$ and $\alpha_0/\alpha_2 = 1.6$. It should be noted that all quantities marked by a star in our paper are considered as non-dimensional. For instance, all distances are reduced as $r^* = r\alpha_2$ or $z^* = z\alpha_2$, densities are reduced as $\rho^* = \rho/\alpha_2^3$ and temperature as $T^* = k_B T/(|A_2|\alpha_2)$. The potential (1) is characterized by a rather soft repulsive part (isotropic contribution) and a small attractive part dependent on the relative orientation between a pair of fluid particles (anisotropic contribution). It is worth noting that the considered fluid is mostly a repulsive one, and a small attractive contribution affects mainly an orientational properties of the fluid, at least at the conditions used in our study. In particular, we consider the fluid at the density $\rho_b^* = 1.0$, and the temperature interval $T^* = 0.5 - 3.5$ is chosen. We have found that, at these temperatures, the considered fluid is beyond its vapour-liquid phase transition region. Therefore, only a nematic-isotropic phase transition can be expected in our case.

To obtain a numerical solution of the integral equation (6) used in the MF approximation, the Picard iterative method was applied. The problem was considered in the cylindrical coordinates, which are set along z-axis normal to the wall surface. The integrations over z were performed using trapezoidal rule with a step $\Delta z = 0.02/\alpha_2$, while all integrations over r were done analytically. A step of integration over $\cos(\theta)$ was chosen as 0.0025. To take into account the confinement, we consider a fluid between two hard walls at a distance $L_z = 36/\alpha_2$ to each other. Two cut-off radii $r_c = 6.0/\alpha_2$ and $12.0/\alpha_2$ are used in our study. The chosen distance L_z appears to be sufficient to get a bulk-like region of a fluid in the middle between the two walls. The presence of the hard walls is introduced by the boundary conditions $\rho(z) = 0$ if $z < 0$ or $z > L_z$. Equation (6) is solved in combination with equation (10) leading to density $\rho(z)$ and order parameter $S_{20}^*(z)$ profiles. The precision of this solution expressed in terms of standard deviation is 10^{-5}.

The bulk order parameter S_b is used both in the MF and LMF calculations. To obtain this quantity, the integral equation (5) was applied. It was also solved numerically by the iterative method, but here with a precision 10^{-12} and with a step of integration over $\cos(\theta)$ taken equal to 0.00125. The cut-off radii $r_c = 6.0/\alpha_2$ and $12.0/\alpha_2$ were used to obtain S_b as well.

We compare the numerical results calculated from MF and LMF approaches with Monte-Carlo (MC) simulation results. For this purpose, a series of MC simulations in canonical ensemble [27] were performed to obtain S_b, $\rho(z)$ and $S_{20}^*(z)$. A system of N_p fluid particles interacting with the pair potential (1) were placed into a rectangular box of a size $L_x \times L_y \times L_z$, where $L_x = L_y = 24/\alpha_2$ and $L_z = 4r_c$ if $r_c = 6.0/\alpha_2$ and $L_z = 3r_c$ if $r_c = 12.0/\alpha_2$. For L_z taken in our simulations, a bulk-like region in the middle of the box is observable up to the temperature $T^* = 2.2$. Since the bulk fluid density is set to $\rho_b^* = 1.0$, a number of fluid particles N_p is set to 13824 for the case of $r_c = 6.0/\alpha_2$ and 20736 for the case of $r_c = 12.0/\alpha_2$. The simulations were carried out with the periodical boundary conditions applied in three dimensions in the case of a bulk fluid, and in X and Y dimensions in the case of a fluid between two hard walls. Each simulation procedure was performed at a constant temperature and volume and consisted of three stages: 1) equilibration of a fluid in a strong field applied along Z-axis to give the fluid particles a preferential orientation, which was normal to the wall surfaces; 2) equilibration of the system obtained at the previous stage with the field switched off; 3) production of the necessary characteristics for a system obtained at the second stage with the field switched off. A criterium defining an equilibrated system is a stabilization of the total order parameter in the system, i.e., when the order parameter fluctuates around one average value during an essential number of MC steps, usually it was over 20 000 steps at least. It should be noted that in our simulations, one MC step corresponds to N_p trial translational or rotational movements.

4. Results and discussion

4.1. Bulk fluid

The order parameter of the bulk MS fluid is calculated at temperatures in the range of $T^* = 0.5 - 3.5$ using the MF approximation with the different cut-off radii (figure 1). We have checked at which r_c^* the results tend to the case of $r_c^* \to \infty$ and have tested what the effect of the cut-off radius in general is. This information is valuable to make a comparison with the computer simulations in which the cut-off radius is used. As can be seen in the case of $r_c^* = 6.0$, the MF approximation leads to the values of the order parameter lower than in the cases of $r_c^* = 12.0$ and $r_c^* \to \infty$ for the whole stable nematic region up to the critical region of the nematic-isotropic phase transition, which appears at $S_b < 0.443$ [22] (figure 1, left-hand panel). Also, it is observed that the results for $r_c^* = 12.0$ totally coincide with the case of $r_c^* \to \infty$ (figure 1, right-hand panel). The values of the order parameter obtained from the simulations are systematically lower than in the MF approximations (figure 1, triangle symbols).

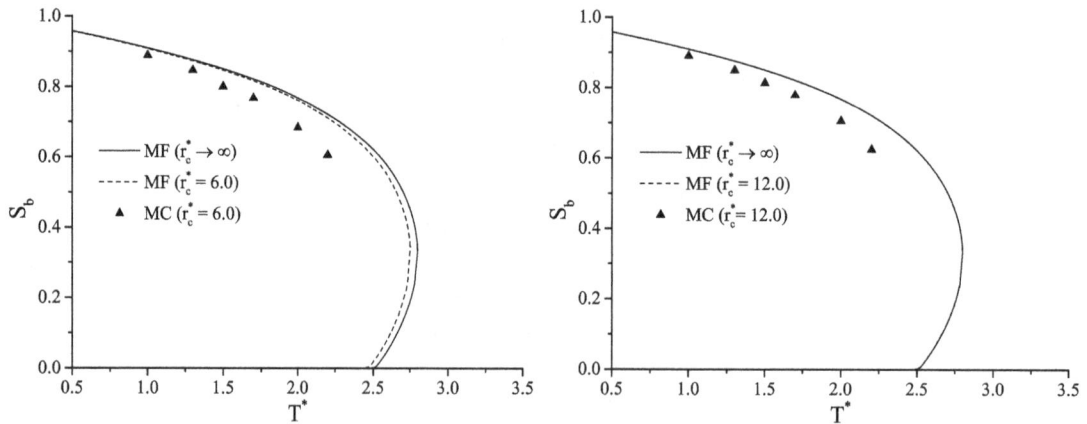

Figure 1. Temperature dependence of the order parameter S_b of a bulk MS fluid. The results obtained in the MF approximation as well as with the use of the Monte-Carlo (MC) simulation method.

4.2. Density and order parameter profiles

A series of density profiles are calculated for the MS fluid near a hard wall at different temperatures in the range $T^* = 0.5 - 3.5$. For this purpose, the LMF and MF approximations are applied and compared with the corresponding simulation results. In figure 2, we present two selected results at temperatures $T^* = 1.3$ and 2.0. The cut-off radius $r_c^* = 12.0$ is used in the MF approximation and in the computer

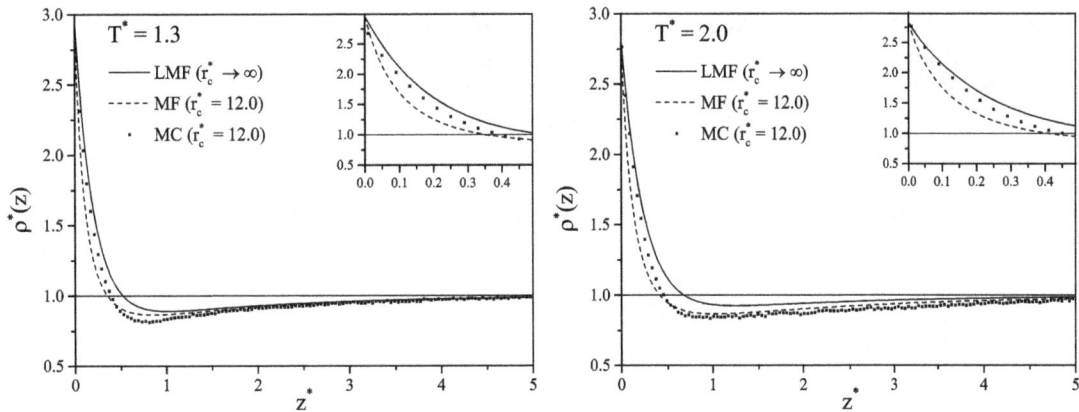

Figure 2. Density profiles of the MS fluid near a hard wall obtained in the LMF and MF approximations as well as with the use of the Monte-Carlo (MC) simulation method. The results were obtained at temperature $T^* = 1.3$ (left-hand panel) and $T^* = 2.0$ (right-hand panel).

simulations. As has been shown above, this cut-off should be long enough to give results comparable to the case $r_c^* \to \infty$. It is observed that all presented profiles have the same qualitative behavior, i.e., they have a distinct high maximum at the contact with the wall, then a minimum appears around $z^* = 0.5 - 1.0$ and at $z^* > 5$ one can see a convergence of $\rho^*(z)$ to ρ_b^* when the bulk-like region is reached. We note that the contact values obtained in the MF and LMF practically coincide — the cut off correction is negligible, while for $\rho^*(0)$ obtained from the simulations, we observe some small difference. The contact value of the obtained density profiles will be discussed more in detail later on. For this moment, we focus on $\rho^*(z)$ at distances $z^* > 0$ and on comparison of the theoretical approaches with the computer simulations. One can see that at small distances z^*, the LMF approximation agrees with the simulations better than the MF, while the MF approximation describes $\rho^*(z)$ better at distances z^* around the minimum of $\rho^*(z)$ and larger. Moreover, at all temperatures considered in our study, the z^*-position of the minimum of $\rho^*(z)$ is very close in the MF approximation to that in the computer simulations, while in the LMF approximation, the z^*-position of the minimum is notably shifted towards the higher values.

A different situation is observed for the order parameter profile $S_{20}^*(z)$ of the MS fluid near a hard wall, which is presented in figure 3 as a normalized quantity $\tilde{S}(z) = S_{20}^*(z)/S_b$. We take such a presentation because of an essential deviation of the theory from the simulations results for bulk order parame-

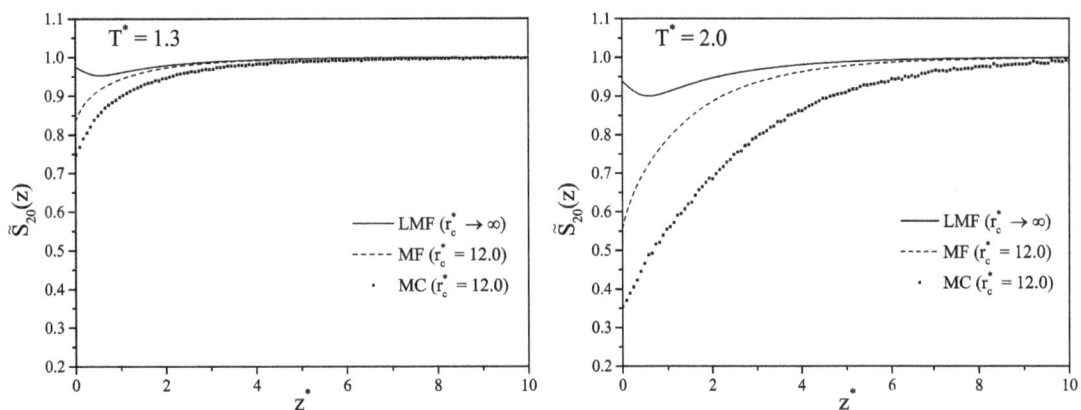

Figure 3. Profiles of the normalized order parameter $S_{20}^*(z)/S_b$ of the MS fluid near a hard wall obtained in the LMF and MF approximations and with the use of the MC simulation method. The results were obtained at temperature $T^* = 1.3$ (left-hand panel) and 2.0 (right-hand panel).

ters (figure 1). Using the normalization we can consider all profiles on the same plot. Nevertheless, even with this normalization, one can observe an important quantitative and qualitative difference between not only the theory and simulations, but also between the MF and LMF approximations (figure 3). In the LMF approximation, a minimum of $\tilde{S}(z)$ is found at $z^* = 0.54$ and $z^* = 0.58$ at $T^* = 1.3$ and $T^* = 2.0$, respectively, while in the MF approximation no minimum has been noticed at all. At the same time, the simulation results do not give any evidence of a $\tilde{S}(z)$ minimum either. In this context, the MF results are similar to those obtained from simulations. However, all the considered approaches give completely different contact values of $\tilde{S}(z)$. The contact value of the order parameter obtained from the simulation is the lowest one, then an essentially larger $\tilde{S}(z)$ is given by the MF approximation and the largest value of $\tilde{S}(0)$ close to 1.0 is calculated from the LMF approximation. In the bulk region (far enough from the wall), all the profiles converge to 1.0 as expected. The only difference is found in the rates of this convergence, which is lower in the MF approximation than in the LMF approximation, and in the simulations it is the lowest one. Also, it is worth noting that the rate of convergence of $S_{20}^*(z)$ to S_b [i.e., $\tilde{S}(z)$ to 1.0] decreases with the temperature.

Before discussing the contact values obtained in our study we will try to understand the behavior of the order parameter profile $S_{20}^*(z)$ and how it relates to the density profile $\rho^*(z)$. First of all, we have noticed that the order parameter of fluid particles next to the wall ($z^* \rightarrow 0$) is smaller than the bulk order parameter S_b. However, the density profiles (figure 2) for the same z^* is higher than the bulk density. From the knowledge of the bulk, for which the higher density leads to the higher order parameter, one can expect that the order parameter near the wall is also higher than S_b. Therefore, we encounter a contradiction, since both the MF approximations and simulation predict the values of $S_{20}^*(z)$ smaller than S_b everywhere except the bulk-like region where $S_{20}^*(z)$ is equal to S_b (figure 3). Apparently we are dealing with two competing effects: densification of fluid particles near the wall, which should increase $S_{20}^*(z)$ at small z^* and the absence of fluid particles beyond the wall, which should decrease $S_{20}^*(z)$. Following the obtained results, one can conclude that the latter effect is more essential for the system considered in our study.

4.3. Contact values

The density and order parameter profiles of a MS fluid near a hard wall are used to calculate the corresponding contact values $\rho^*(0)$ and $S_{20}^*(0)$ as functions of the temperature. These contact values can be calculated from the expressions of the contact theorem (CT) (22) and (23), which in the MF approximation reduce to the relations (24) and (25), respectively. In the LMF approximation, the contact values $\rho^*(0)$ and $S_{20}^*(0)$ satisfy the relations (24) and (25) automatically due to the boundary conditions applied in the solution of the corresponding differential equation. In the MF approximation, the situation is different for $\rho^*(0)$ and $S_{20}^*(0)$. As it was shown in [23], the contact value for the density profile $\rho^*(0)$ satisfies the relation (24), although it has not been proven that the contact value of the order parameter $S_{20}^*(0)$ obtained in the MF approximation should satisfy the relation (25).

First we consider the temperature dependencies of the contact value of density profiles obtained with the different cut-off radii $r_c^* = 6.0$ and 12.0 (figure 4). As can be seen, the CT leads to the same result as the MF approximation. A small difference between the LMF and the MF approximation (or the CT) appears when the cut-off radius is equal to $r_c^* = 6.0$ (figure 4, left-hand panel). If the cut-off radius is increased to $r_c^* = 12.0$, equivalent results are obtained in all of the approximations (figure 4, right-hand panel). At the same time, the agreement between the theoretical approaches and the simulations is mostly qualitative. We would like to draw attention to the non-monotonous behavior of the contact value $\rho^*(0)$ with a distinct minimum observed both in the simulations and in the theoretical predictions. To understand this interesting effect, we should analyze the temperature dependence of the density contact value more in detail.

As it is seen in figure 4, at low temperatures, $\rho^*(0)$ is large and lowers as the temperature increases. This is related to the reduction of the repulsive contribution of the pair potential (1). Since in our model a soft repulsive interaction is used, it becomes weaker if the temperature increases. It should be noted that the attractive part of the pair potential in our model is rather small and mainly affects orientational properties of fluid particles. At the same time, a relative orientation can indirectly strengthen the fluid-fluid repulsion contribution by reducing the attractive interaction term. Therefore, at some point in the

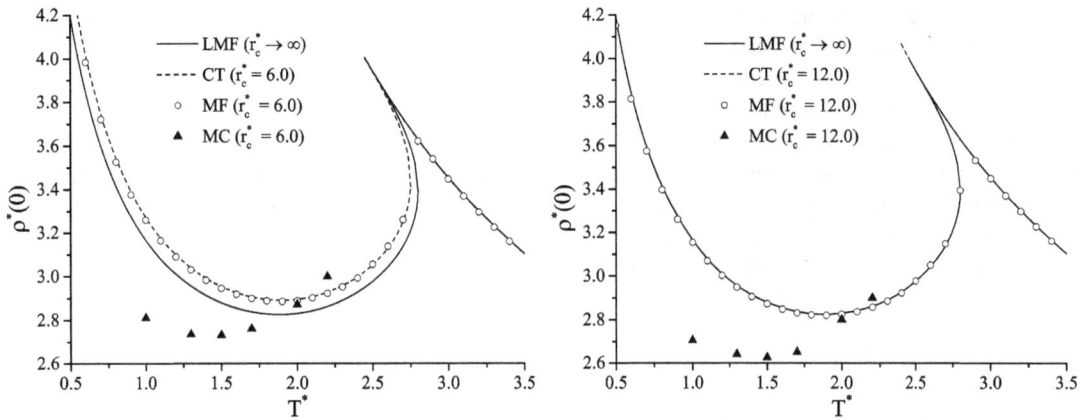

Figure 4. Contact value of the density profile as a function of the temperature obtained by different approaches. Two cut-off radii are used: $r_c^* = 6.0$ (left panel) and $r_c^* = 12.0$ (right-hand panel).

nematic phase, when the order parameter remains sufficiently small, the repulsion becomes stronger causing an increase of the contact value $\rho^*(0)$. This effect is observed in figure 4 at temperature $T^* = 1.885$, where $\rho^*(0)$ reaches its minimum and starts to increase rapidly until the fluid becomes totally isotropic [$S_{20}^*(z) = 0$]. In the isotropic phase, the fluid particles are totally orientationally disordered, and a further temperature increase leads to the weakening of the repulsion. Thus, a continuous decrease of $\rho^*(0)$ is obtained at high temperatures. It should be noted that the explanation presented here concerns solely, models with a pair potential consisting of a soft-core term combined with a Maier-Saupe attractive potential. In the case of a hard-core type of repulsive interaction, the temperature dependence can be opposite and the effect of orientational ordering can be completely different.

While the contact value of the density profile is rather understandable, the behavior of the contact value of the order parameter profile is not so clear. First of all, as it has been already shown, there is an essential inconsistency between the MF and LMF approximations (figure 3). There is also a problem if one compares the contact values $S_{20}^*(0)$ obtained in the MF and LMF approximations and those calculated from the CT theorem. As can be seen in figure 5, the MF results significantly differ from those of the LMF and the CT. A perfect agreement of the LMF and the CT appears due to the definition of the boundary conditions used in the LMF approximation, which are taken to fit the CT theorem (25). A deviation of the CT from the LMF is seen only for the case of $r_c^* = 6.0$ chosen in the CT (figure 5, left-hand panel). For

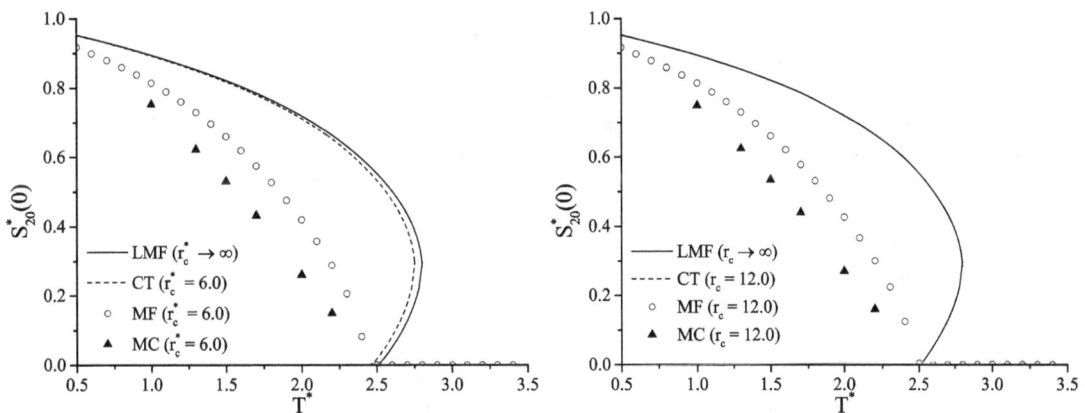

Figure 5. Contact value of the order parameter profile as a function of the temperature obtained by different approaches. Two cut-off radii are used: $r_c^* = 6.0$ (left-hand panel) and $r_c^* = 12.0$ (right-hand panel).

$r_c^* = 12.0$, the contact values of the order parameter of a MS fluid obtained from the CT and the LMF totally coincide. At the same time, the MF approximation gives much lower $S_{20}^*(0)$, although it is closer to the simulation results than the CT and LMF. We assume that the difference between the CT and the MF approaches is related to a poor description of the bulk partial pressure $P(\Omega_{1n})$. In order to improve the results obtained from the CT (or the LMF approximation), one should take into account higher order terms, for instance the Gaussian fluctuations [22]. Also, it should be noted that in contrast to the CT and LMF approximation, $S_{20}^*(0)$ obtained from the MF approximation and the simulations decays to zero before the critical region is reached. Thus, the convex part of $S_{20}^*(0)$ is absent for them. It means that in the critical region there is no orientational ordering of fluid particles at a contact with the wall.

5. Conclusions

Within the framework of the density field theory, we have formulated the mean field (MF) approximation as a starting point for the theoretical description of a nematic fluid at a hard wall. Using the developed approach we have investigated the density and order parameter profiles of a confined Maier-Saupe nematogenic fluid with an isotropic Yukawa-like repulsion. Theoretical predictions have been compared with analytical results obtained within the framework of the linearized mean field (LMF) approximation [23]. For the density profile, the results obtained in the MF and LMF approximations are in qualitative agreement with computer simulations data. For the order parameter profile, the results of the MF and LMF approximations have qualitatively different behaviors. In the LMF approximation, a minimum of the order parameter profile is present, while in the MF approximation no minimum has been observed. The results of computer simulations do not give any evidence of the existence of a minimum in the z-dependence of the order parameter profile either. In this context, the MF and computer simulations results are qualitatively similar. This contradiction in the description of the density and order parameter profiles within the framework of the LMF approximation is connected with the problem of the respective contact theorems in forms (24) and (25) which are used as boundary conditions in the solution of the corresponding system of differential equations. As it was shown in reference [23], the contact value of the density profile $\rho(0)$ satisfies the relation (24) but the validity of the relation (25) for the contact value of the order parameter $S_{20}(0)$ is not evident. Moreover, the comparison with the results of computer simulations in figure 5 shows that relation (25) is probably incorrect. As we can see from figure 5, the LMF approximation can give a better result if we change the boundary condition for the order parameter profile by correcting the contact value for $S(0)$ obtained within the framework of the MF approximation. We hope that the considered problem can be better understood by going beyond the MF approximation and including a contribution from fluctuations.

The temperature dependencies of the contact values of the density and order parameter profiles have been analyzed more in detail. We have found a non-monotonous behavior of the density contact value $\rho(0)$ as a function of the temperature with a distinct minimum observed in both theoretical predictions and computer simulations. This non-monotonous temperature dependence of $\rho(0)$ is explained by the competition of the soft isotropic repulsive and the soft anisotropic attractive contributions. For the contact value of the order parameter, we have observed a monotonous decrease with an increase of temperature. Both the simulations and the MF results indicate that there is no orientational ordering of fluid particles at the contact with the wall when the fluid is in the critical region.

The results presented in this paper have been obtained within the framework of the MF approximation. We note that the agreement between theoretical predictions and computer simulation data is mostly qualitative. For a better theoretical description one should take fluctuations into account. For the bulk properties of the model under consideration, the influence of the contribution from fluctuations was already discussed in reference [22]. It was shown that the singlet distribution function reduces to the form (5) with a change of κ_2^0 to $\kappa_2^0 l$, where $l - 1 = \rho A_0 u_2^2 / (\Lambda_0 + \Lambda_2)$. It was shown that the temperature trend of deviation between the order parameter S_b calculated with fluctuations included and the MF value is the same as that between computer simulations and the MF value presented in figure 1. In the next paper we plan to include the contribution from fluctuations in the description of a MS nematogenic fluid at a hard wall similar to the way it was done for isotropic Yukawa fluids [18, 19].

References

1. Trokhymchuk A.D., Pizio O., Holovko M.F., Sokołowski S., J. Phys. Chem., 1996, **100**, 17004; doi:10.1021/jp961443l.
2. Trokhymchuk A.D., Pizio O., Holovko M.F., Sokołowski S., J. Chem. Phys., 1997, **106**, 200; doi:10.1063/1.473042.
3. Ilnytskyi J., Sokołowski S., Pizio O., Phys. Rev. E, 1999, **59**, 4161; doi:10.1103/PhysRevE.59.4161.
4. Ilnytskyi J., Patsahan T., Sokołowski S., J. Chem. Phys., 2011, **134**, 204903; doi:10.1063/1.3592562.
5. Sokołowski S., Kalyuzhnyi Y.V., J. Phys. Chem. B, 2014, **118**, 9076; doi:10.1021/jp503826p.
6. Jerome B., Rep. Prog. Phys., 1991, **54**, 391; doi:10.1088/0034-4885/54/3/002.
7. Henderson D., Abraham F.F., Barker J.A., Mol. Phys., 1976, **31**, 1291; doi:10.1080/00268977600101021.
8. Sokolovska T.G., Sokolovskii R.O., Patey G.N., Phys. Rev. Lett., 2004, **92**, 185508; doi:10.1103/PhysRevLett.92.185508.
9. Sokolovska T.G., Sokolovskii R.O., Patey G.N., J. Chem. Phys., 2005, **122**, 034703; doi:10.1063/1.1825373.
10. Holovko M., Sokolovska T., J. Mol. Liq., 1999, **82**, 161; doi:10.1016/S0167-7322(99)00098-7.
11. Henderson D., Blum L., Lebowitz J.L., J. Electroanal. Chem., 1979, **102**, 315; doi:10.1016/S0022-0728(79)80459-3.
12. Holovko M., Badiali J.P., Di Caprio D., J. Chem. Phys., 2005, **123**, 234705; doi:10.1063/1.2137707.
13. Holovko M., Di Caprio D., J. Chem. Phys., 2015, **142**, 014705; doi:10.1063/1.4905239.
14. Di Caprio D., Stafiej J., Badiali J.P., Mol. Phys., 2003, **101**, 2545; doi:10.1080/0026897031000154293.
15. Di Caprio D., Stafiej J., Badiali J.P., J. Chem. Phys., 1998, **108**, 8572; doi:10.1063/1.476286.
16. Di Caprio D., Stafiej J., Borkowska Z., J. Electroanal. Chem., 2005, **41**, 582; doi:10.1016/j.jelechem.2005.02.008.
17. Di Caprio D., Valisko M., Holovko M., Boda D., J. Phys. Chem. C, 2007, **111**, 15700; doi:10.1021/jp0737395.
18. Di Caprio D., Stafiej J., Holovko M., Kravtsiv I., Mol. Phys., 2011, **109**, 695; doi:10.1080/00268976.2010.547524.
19. Kravtsiv I., Patsahan T., Holovko M., Di Caprio D., J. Chem. Phys., 2015, **142**, 194708; doi:10.1063/1.4921242.
20. Gahov F., Cherski Y., Convolution-Type Equations, Nauka, Moscow, 1978.
21. Holovko M., Di Caprio D., Kravtsiv I., Condens. Matter Phys., 2011, **14**, 33605; doi:10.5488/CMP.14.33605.
22. Kravtsiv I., Holovko M., Di Caprio D., Mol. Phys., 2013, **111**, 10023; doi:10.1080/00268976.2012.762615.
23. Holovko M., Kravtsiv I., Di Caprio D., Condens. Matter Phys., 2013, **16**, 14002; doi:10.5488/CMP.16.14002.
24. Maier W., Saupe A., Z. Naturforsch. A, 1959, **14**, 882; doi:10.1515/zna-1959-1005.
25. Maier W., Saupe A., Z. Naturforsch. A, 1960, **15**, 287; doi:10.1515/zna-1960-0401.
26. Gray C.G., Gubbins K.E., Theory of Molecular Fluids, Clarendon Press, Oxford, 1984.
27. Frenkel D., Smith B., Understanding Molecular Simulations, Academic, San Diego, 1995.

9

Pressure-driven flow of oligomeric fluid in nano-channel with complex structure. A dissipative particle dynamics study *

J.M. Ilnytskyi[1], P. Bryk[2], A. Patrykiejew[2]

[1] Institute for Condensed Matter Physics of the National Academy of Sciences of Ukraine,
1 Svientsitskii St., 79011 Lviv, Ukraine

[2] Department for the Modeling of Physico-Chemical Processes, Maria Curie-Skłodowska University,
20–031 Lublin, Poland

We develop a simulational methodology allowing for simulation of the pressure-driven flow in the pore with flat and polymer-modified walls. Our approach is based on dissipative particle dynamics and we combine earlier ideas of fluid-like walls and reverse flow. As a test case we consider the oligomer flow through the pore with flat walls and demonstrate good thermostatting qualities of the proposed method. We found the inhomogeneities in both oligomer shape and alignment across the pore leading to a non-parabolic velocity profiles. The method is subsequently applied to a nano-channel decorated with a polymer brush stripes arranged perpendicularly to the flow direction. At certain threshold value of a flow force we find a pillar-to-lamellar morphological transition, which leads to the brush enveloping the pore wall by a relatively smooth layer. At higher flow rates, the flow of oligomer has similar properties as in the case of flat walls, but for the narrower effective pore size. We observe stretching and aligning of the polymer molecules along the flow near the pore walls.

Key words: *Pouiseuille flow, polymer brush, oligomers, dissipative particle dynamics*

1. Introduction

Understanding the behavior of polymers attached to surfaces is of importance in many research areas including biophysics, polymer-induced effective interactions in colloidal suspensions, chromatographic separation, catalysis, and drug delivery [1]. Grafting polymer chains can significantly alter the properties of the surface and make it, for example, biocompatibile or responsive to external stimuli [2]. Due to the large field of potential applications, polymer brushes have been the subject of many theoretical studies. In the seminal works Alexander [3], de Gennes [4, 5] have calculated the brush profile and explored analytically the impact of grafting density and molecular weight. Since then, the properties of tethered chains have been investigated by means of self-consistent field theory [6–10], polymer density functional theory [11–16], and computer simulation [17–21]. Many theoretical predictions have been confirmed by experiment [22–25].

Polymer brushes can be used to tailor static properties of surfaces, such as wettability, as well as dynamical, such as hydrodynamic boundary conditions and friction. Fluid flow in polymer grafted nano-pores can be described via continuum hydrodynamic equations (e.g., the Brinkman equation [26]) with a priori assumed permeability related to the monomer density profile. The resulting velocity profile is sensitive to the assumed form of the monomer profile [27]. However, the continuum hydrodynamic description of a flow has not been firmly established on the nanoscale [28]. This is important in the context

---

* It is our pleasure to dedicate this paper to Professor Stefan Sokołowski, our Colleague and Mentor for many years.

of micro- and nanofluidic devices [29]. Downsizing a channel to the nanoscale, increases the surface-to-volume ratio and introduces new physical phenomena not observed in the macroscale [30]. Covering the surface by a polymer brush may introduce a pronounced reduction of friction, which lowers the pressure difference required to maintain the flow through a nanochannel [31]. Flow in polymer brushes has been the subject of numerous simulational studies in recent years [32–38].

Recently, the equilibrium properties of binary mixture confined in a slit-like pore decorated with polymer brush stripes were studied by means of dissipative particle dynamics (DPD) [39, 40]. It was found that, depending on the geometrical parameters characterizing the system (the size of the pore and the width of the stripes), several different structures (or morphologies) inside the pore can be formed. Such patterned brushes can be fabricated experimentally by means of electron beam litography [41]. In the present paper, we wish to study nonequilibrium properties of such system by considering the pressure driven oligomer flow inside a channel with either flat or brush-modified walls. In particular, we focus on three features such as: (i) the microstructure of a flow depending on the molecular mass of an oligomer and the magnitude of a bulk flow force; (ii) flow-induced morphology changes; and (iii) the effect of the patterned brush decoration of the walls on the properties of the flow. Our paper is arranged as follows: In section 2 we introduce a new simulation method which combines the ideas of fluid-like walls and reverse flow to minimise the near-wall artefacts and maintain constant temperature under flow condition. As a simple test, we apply the method to the case of oligomer flow through the pore with flat walls. In section 3, the analysis is extended to the case when the walls are modified by a polymer brush arranged in a form of stripes. Conclusions are provided in section 4.

2. Flow of oligomeric fluid through a channel with flat walls

Let us first consider the simulational approach employed in this study. We use the non-equilibrium extension of the DPD technique in a form discussed by Groot and Warren [42]. This is a mesoscopic method that operates at a level of coarse-grained beads, each representing either a fragment of a polymer chain or a collection of solvent particles. The force acting on ith bead due to its pairwise interaction with jth bead can be written as

$$\mathbf{F}_{ij} = \mathbf{F}_{ij}^{\mathrm{C}} + \mathbf{F}_{ij}^{\mathrm{D}} + \mathbf{F}_{ij}^{\mathrm{R}}, \tag{2.1}$$

where $\mathbf{F}_{ij}^{\mathrm{C}}$, $\mathbf{F}_{ij}^{\mathrm{D}}$ and $\mathbf{F}_{ij}^{\mathrm{R}}$ denote the conservative, dissipative and random contribution, respectively. These have the following form [42]

$$\mathbf{F}_{ij}^{\mathrm{C}} = \begin{cases} a(1 - r_{ij})\hat{\mathbf{r}}_{ij}, & r_{ij} < 1, \\ 0, & r_{ij} \geqslant 1, \end{cases} \tag{2.2}$$

$$\mathbf{F}_{ij}^{\mathrm{D}} = -\gamma w^{\mathrm{D}}(r_{ij})(\hat{\mathbf{r}}_{ij} \cdot \mathbf{v}_{ij})\hat{\mathbf{r}}_{ij}, \tag{2.3}$$

$$\mathbf{F}_{ij}^{\mathrm{R}} = \sigma w^{\mathrm{R}}(r_{ij})\theta_{ij}\Delta t^{-1/2}\hat{\mathbf{r}}_{ij}. \tag{2.4}$$

Here, $\mathbf{v}_{ij} = \mathbf{v}_i - \mathbf{v}_j$, $\mathbf{v}_i$ and $\mathbf{v}_j$ are the velocities of the beads, θ_{ij} is Gaussian random variable, $\langle \theta_{ij}(t) \rangle = 0$, $\langle \theta_{ij}(t)\theta_{kl}(t') \rangle = (\delta_{ik}\delta_{il} + \delta_{il}\delta_{jk})\delta(t - t')$ and Δt is the time-step of the integrator. As already discussed in references [28, 65], the effective range of friction between beads can be modified by adjusting the shape of the weight functions $w^{\mathrm{D}}(r_{ij})$ and $w^{\mathrm{R}}(r_{ij})$. We use the following general form for $w^{\mathrm{R}}(r_{ij})$:

$$w^{\mathrm{R}}(r_{ij}) = \begin{cases} (1 - r_{ij})^{\beta}, & r_{ij} < 1, \\ 0, & r_{ij} \geqslant 1, \end{cases} \tag{2.5}$$

where the exponent β is adjusted, and the weight function $w^{\mathrm{D}}(r_{ij})$ is set equal to $[w^{\mathrm{R}}(r_{ij})]^2$ according to Español and Warren [43] arguments. Likewise, it is required that $\sigma^2 = 2\gamma$.

The oligomers and tethered polymer chains (if any) are represented as necklaces of beads bonded together via harmonic springs, the force acting on ith bead from the interaction with its bonded neighbour, jth bead, is

$$\mathbf{F}_{ij}^{\mathrm{B}} = -k r_{ij}\hat{\mathbf{r}}_{ij}, \tag{2.6}$$

where $r_{ij} = |\mathbf{r}_{ij}|$, $\mathbf{r}_{ij} = \mathbf{r}_i - \mathbf{r}_j$ is the vector connecting the centers of i-th and j-th beads, $\hat{\mathbf{r}}_{ij} = \mathbf{r}_{ij}/r_{ij}$ and k is the spring constant. The same bonding force is used to tether the end polymer bead to the surface. The length, mass, time and energy (expressed via $T^* \equiv k_{\mathrm{B}}T$) units are all normally set equal to unity.

Let us now turn to the case where the fluid (or a mixture of fluids) is confined within a slit-like pore. In order to commence a simulation of the pressure-driven flow, it is required to provide a set of rules defining the behaviour of the fluid particles at walls, and a prescription for the construction of the walls. These rules should recover the well known cases of hydrodynamic flow such as the Poiseuille flow (i.e., a flow of a Newtonian fluid with no-slip boundary conditions and a parabolic velocity profile). On the other hand, for the flow of a polymeric fluid (i.e., a non-Newtonian fluid) the set of rules should lead to the slip boundary conditions. The simplest set of rules comprise elastic reflections off the wall [44, 45]. Unfortunately, they give rise to a hydrodynamic slip for Newtonian fluids, as well as suffer from near-wall density artifacts at higher density. This can be traced back to the fact that the atoms repelled each other strongly but did not interact with the wall until they attempted to cross [46].

A number of more sophisticated set of rules have been suggested. One option is to form the crystalline walls of a few layers of frozen (or having large mass) particles [47–53]. The interaction between the bulk fluid particles and those of the wall creates the near-wall drag which leads to the formation of the Poiseuille flow. A drawback of this approach is the propagation of the crystalline order into the near-wall regions of bulk fluid. This effect is perfectly physical for the atomic molecular dynamics simulation, where the solid wall mimics a real crystalline structure. However, for the mesoscopic DPD simulations, each soft bead is assumed to represent a meso-scale portion of the material, on which scale the atomic crystalline structure is smeared-out.

This drawback can be avoided by using the structureless fluid-like walls [46]. The walls in this case are made of the fluid confined in the slabs adjacent to the pore boundary, and the elastic reflections are applied on both sides of the boundary. Therefore, bulk and wall fluid particles are immiscible. Still, the interaction between near-wall beads on the opposite sides of the boundary creates a near-wall drag ensuring no-slip boundary condition for Newtonian fluids.

Another important issue in flow simulation is to avoid the system overheating due to the presence of the body force. This problem was addressed in several studies, cf. for example references [54, 55]. In the molecular dynamics simulation, the excessive energy is absorbed by an external thermostat, in either bulk or near-wall form [54]. In DPD simulations, the thermostat is "internal", provided by the balance between interparticle friction and random forces. For the case of a flow, some means for dissipation of additional energy related to the body force should be provided. One of the elegant ways to do this is the concept of a reverse flow [56–58]. In this approach, the simulation box contains two sub-flows driven oppositely. The total force applied to the system is equal to zero and a no-slip boundary is formed at the interface between two opposite flows of Newtonian fluids.

In our study, we combine both concepts by employing the fluid-like walls on both boundaries of a pore and initiating contraflows (reverse flows) within them. Separation between the main pore and contraflow-containing walls prevents intermixing between the beads from both regions. This is important both in the case when the flow of a mixture is considered, or in the case of polymer modified walls, where polymer chains are tethered to the boundary between the main pore and fluid-like wall. However, the existence of the reflective boundaries does not prevent a friction between the beads located on the opposite sides of the boundary, enabling the formation of the no-slip boundary condition for Newtonian fluids.

In this section, we consider the pressure-driven flow of oligomeric one-component fluid through the pore with flat walls. The oligomers of length $L_0 = 1, 4, 10$ and 20 beads are considered. The aim is twofold. Firstly, we would like to test to what typical values of bulk force the approach outlined above can be stretched without violation of temperature conservation. Secondly, we aim to study the flow microstructure depending on molecular length of the flowing oligomer and the magnitude of a flow force. The geometry of the system is illustrated in figure 1. Here, X-axis runs from left to right, Z-axis — from bottom to top, Y-axis coincides with the viewing direction. The simulation box is of dimensions $L_x = 80$, $L_y = 50$ and $L_z = 26.667$ with the periodic boundary conditions applied along X and Y axes, the pore size is $d = 13.333$, the size of the contraflow regions is $c = d/2 = 6.667$. The chains in contraflow regions are of the same length L_0 as in the main pore. Therefore, the total number of main and contraflow chains is the same. All beads are assumed to be of the same type, which is reflected in the fact that the parameter

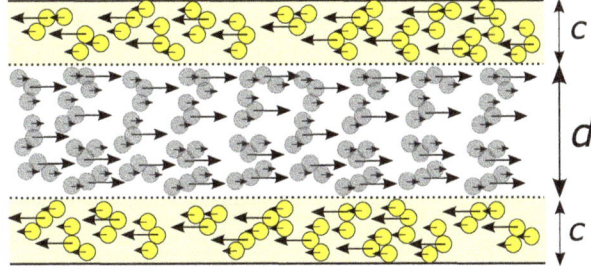

Figure 1. (Color online) Simulation box geometry showing the pressure-driven flow of the oligomer of $L_0 = 3$ beads in a slit-like pore of size d (gray beads). Contraflow containing an oligomer of the same length is contained in two fluid-like walls of size c (yellow beads). The direction of the applied flow force is shown via arrows of different length reflecting the force distribution biased towards the middle bead, see text for details.

a in equation (2.2) that controls the bead repulsion is the same for all pairwise interactions, $a = 25$. The pore and the fluid-like walls are separated via the internal boundaries (shown as dashed lines in figure 1), impenetrable for the beads on both sides by applying the reflection algorithm described in detail in reference [39]. The same reflection algorithm is used at the external walls (solid lines in the same figure) but, alternatively, the periodic boundary conditions can be used in Z direction, similarly to the original reverse flow setup [56–58].

Each ith bead within a pore is subjected to the flow force of certain amount f_i

$$\mathbf{F}_i^{\text{FL}} = f_i \hat{\mathbf{x}}, \qquad \hat{\mathbf{x}} = \{1, 0, 0\}, \tag{2.7}$$

applied along X-axis, where $f_i > 0$, this is indicated by the right-hand side directed arrows in figure 1. The beads in the contraflow regions are subjected to the force $\mathbf{F}_i^{\text{FL}} = -f_i \hat{\mathbf{x}}$, indicated as reversely directed arrows in the same figure. Several options are available for choosing the amount of f_i. The simplest one would be to choose $f_i \equiv f$, the same amount for each bead. However, such an algorithm could lead to less than optimal match of the micro-fluctuations of the applied pressure in real systems, since the polymer molecules tend to form coils with varying distribution of the density. Another, rather extreme option would be to apply the amount $f L_0$ to the middle bead only. The other beads feel this force indirectly and are delayed via the elastic spring forces. The latter approach might suffer from large fluctuations of bond lengths and slower relaxation of the intra-chain vibrations, due to the soft nature of the model. In our view, a reasonable compromise can be achieved by applying a fixed amount of the force $f L_0$ to each oligomer, but biasing it towards the middle bead of the chain. Namely, assuming that the beads of an oligomer are numbered sequentially as $l = 1, \dots, L_0$, then the amount of the force applied to the bead number l is found according to the Gaussian distribution:

$$f(l) = f w_{\text{G}}(l), \qquad w_{\text{G}}(l) = 2 \exp\left[-\frac{(l - \bar{l})^2}{\sigma^2}\right]. \tag{2.8}$$

Here, $\bar{l} = (L_0 + 1)/2$ is the mid-index of the chain, and the breadth of the distribution is given by $\sigma = L_0 / (2\sqrt{\pi})$. The distribution is normalized to L_0:

$$\sum_{l=1}^{N_0} w_{\text{G}}(l) = \int_{-\infty}^{+\infty} w_{\text{G}}(l) \mathrm{d}l = L_0. \tag{2.9}$$

The shape of the weight function $w_{\text{G}}(l)$ is shown in figure 2 for the cases of $L_0 = 4$, 10 and 20. As a result, the total force applied to the oligomer of L_0 beads is equal to $f L_0$, but it is biased towards the middle beads (illustrated by arrows of different length in figure 1).

The acceleration of the fluid beads due to applying the bulk force affects the accuracy of the integrator, as far as the expression for the coordinates at the time instance $t + \Delta t$ contains the term proportional to $v(t)\Delta t$, where $v(t)$ is the velocity of the particle at the time instance t. The only way to keep the same

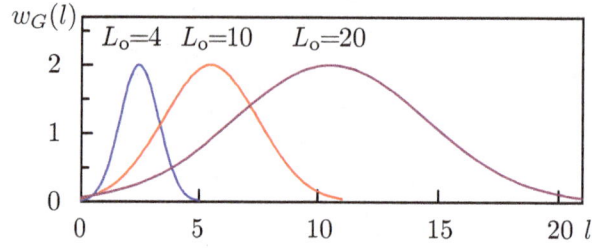

Figure 2. (Color online) Weight function $w_G(l)$ for the amount of bulk force applied to lth bead of the oligomer of length L_o, equation (2.8).

numerical accuracy of the integrator is to reduce the time-step Δt of the integrator. In our simulations, we use the value $\Delta t = 0.001$, about 40 times smaller compared to the values typically used in the case of equilibrium simulation. Temperature conservation is one of the most important indicators of the accuracy of the integrator. Following reference [55], we consider the transverse temperature, which is evaluated from the two components of the velocity perpendicular to the flow direction: $T_\perp^* = m\langle v_y^2 \rangle/2 + m\langle v_z^2 \rangle/2$. The profiles of $T_\perp^*$ for two extreme cases of oligomer length $L_o = 1$ (simple fluid) and 20 with respect to z coordinate are shown in figure 3 (a) and (b), respectively. We allow the maximum deviation of these profiles from the required value 1 not to exceed $3-4\%$. As follows from figure 3, this is achieved for all $L_o = 1-20$ if the flow force magnitude is restricted to $f \leqslant 0.2$. At larger values, $f > 0.2$, the system is prone to local heating near the internal walls, which signals a breakdown of this thermostatting method. We should also remark that for the setup with no contraflow regions, no thermostatting can be achieved at all: the temperature was found to rise monotonously even for the smallest considered values of f.

The profiles for the velocity components v_x of individual beads along the flow direction are built by binning the pore along the Z-axis. These are shown in figure 4 for the cases of $L_o = 1$ and $L_o = 20$ oligomer length obtained at various flow force amplitudes $f = 0.05, 0.1$ and 0.2. For the case of simple fluid (a), almost perfect parabolic shape is achieved inside the flow region indicating the properties of a Newtonian fluid. The velocity drops to zero exactly at the pore walls giving rise to the no-slip boundary condition. In this case, the Stokes formula can be used to estimate the viscosity of the fluid. With an increase of the oligomer length, L_o, the shape of the velocity profile gradually diverges from a parabolic one and turns into a bell-like shape at $L_o = 20$, as seen in (b). This indicates the non-Newtonian fluid behaviour. The models describing such non-parabolic profiles exist (see, e.g., reference [58]) and involve an analogue for the viscosity and a number of additional parameters. We found, however, a numerical fitting to these forms impractical. The set of rules defining the behaviour of the particles at walls, as imposed in our simulation, leads here to the slip boundary conditions. The discontinuity of the velocity profile at the wall boundary is clearly visible in figure 4 (c) and is a characteristic feature of the flows of polymeric fluids. In figure 4 (d), we compare two velocity profiles of the flows obtained with applying equation (2.8)–(2.9), i.e., the Gaussian distribution of the bulk force, and a uniform distribution of the bulk force. We note that even for such an extremely large value of the bulk force, the profiles are practically identical. We expect that for very long polymers, the Gaussian distribution of the bulk force would prove

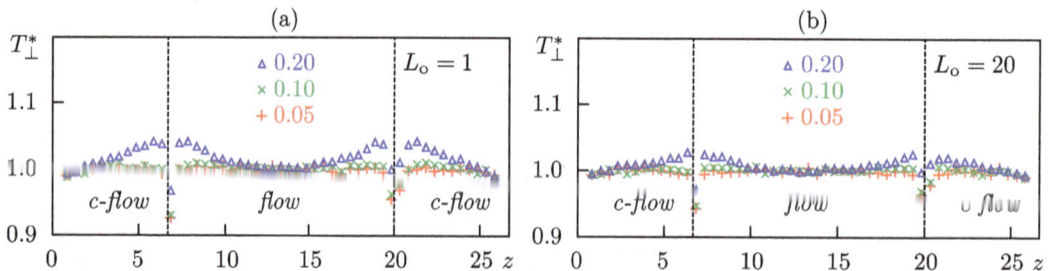

Figure 3. (Color online) $T_\perp^*$ profile at various amplitudes of bulk force f indicated in the figure. (a) simple liquid, $L_o = 1$; (b) longest oligomer, $L_o = 20$.

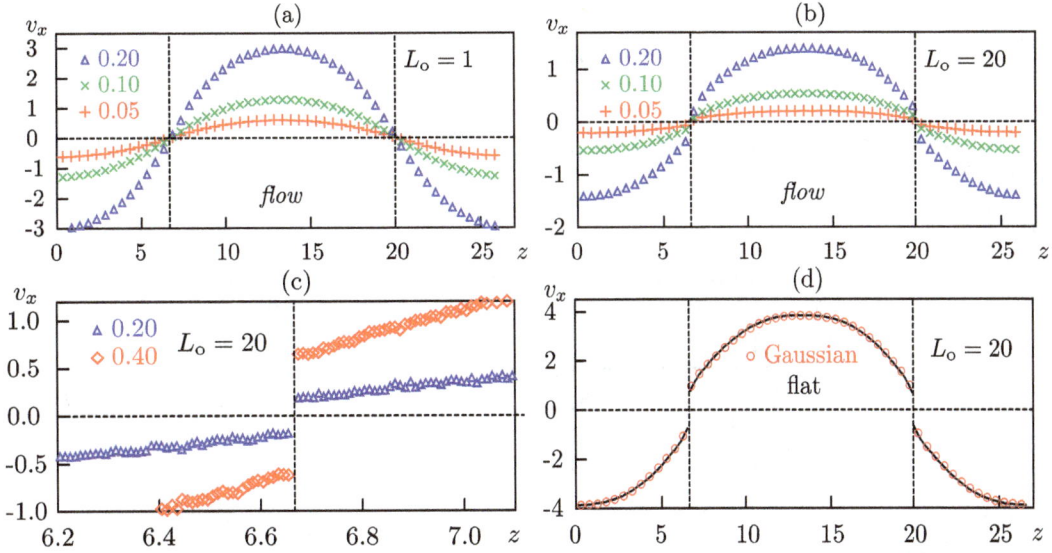

Figure 4. (Color online) Profiles for the velocity component along the flow direction, v_x, evaluated bead-wise at several values of f (indicated in the figure). (a) the case of simple fluid, $L_o = 1$; (b) the case of the longest oligomer considered, $L_o = 20$; (c) near-wall behaviour of the velocity profiles for $L_o = 20$; (d) the velocity profiles for $f = 0.4$ and for $L_o = 20$ evaluated with the Gaussian distribution of the bulk force, [cf. equation (2.8)–(2.9)], (symbols), and a uniform distribution of the bulk force (solid line).

beneficiary and could lead to better stability of the integration of the equations of motion.

An alternative route is to concentrate on the details of the microstructure of the oligomer flow, because these must be responsible for its non-Newtonian behaviour. In particular, comparing to the case of a simple fluid, oligomers have additional conformational degrees of freedom which will affect their flow properties. Therefore, we build the profiles for the average shape anisotropy and the molecular orientation for the oligomers in a flow. The components of the gyration tensor

$$G_{\alpha,\beta} = \frac{1}{L_o} \sum_{i=1}^{L_o} (r_{i,\alpha} - R_\alpha)(r_{i,\beta} - R_\beta) \tag{2.10}$$

are evaluated for each oligomer of length L_o at a given time instance t. Here, α, β denote the Cartesian axes, $r_{i,\alpha}$ are the coordinates of ith monomer, and R_α are the coordinates for the center of mass of the oligomer. In the equivalent ellipsoid representation, the eigenvalues $\lambda_1 > \lambda_2 > \lambda_3$ of this tensor provide the squared lengths of its semiaxes, whereas the respective eigenvectors $\mathbf{u}_1, \mathbf{u}_2$ and $\mathbf{u}_3$ — the orientation of these axes in space.

The shape anisotropy of an individual oligomer can be defined as

$$\kappa^2 = \frac{3}{2} \frac{\lambda_1^2 + \lambda_2^2 + \lambda_3^2}{[\lambda_1 + \lambda_2 + \lambda_3]^2} - \frac{1}{2}. \tag{2.11}$$

It is zero for a spherically symmetric body, where $\lambda_1 = \lambda_2 = \lambda_3 > 0$ and is equal to 1 for an infinitely long thin rod, where $\lambda_1 > 0$, $\lambda_2 = \lambda_3 = 0$. The average profile is built for the shape anisotropy in a steady state. It is obtained by first binning the system in Z-axis and averaging κ^2 for individual oligomers found in each bin. Then, time averaging within the steady state is performed.

The orientation of each oligomer in space is defined by that for the longest axis of its equivalent ellipsoid. The latter is characterised by the eigenvector $\mathbf{u}_1$ associated with the largest eigenvalue λ_1. The level of alignment of the oligomer along the flow axis X can be characterised by the order parameter:

$$S_x = P_2(\mathbf{u}_1 \cdot \hat{\mathbf{x}}), \tag{2.12}$$

where $\hat{\mathbf{x}}$ is defined in equation (2.7) and $P_2(x)$ is the second Legendre polynomial. The alignment profile is built then in a steady state by averaging S_x in each bin and then performing time averaging. It is obvious that both κ^2 and S_x can be defined for the case $L_o > 1$ only.

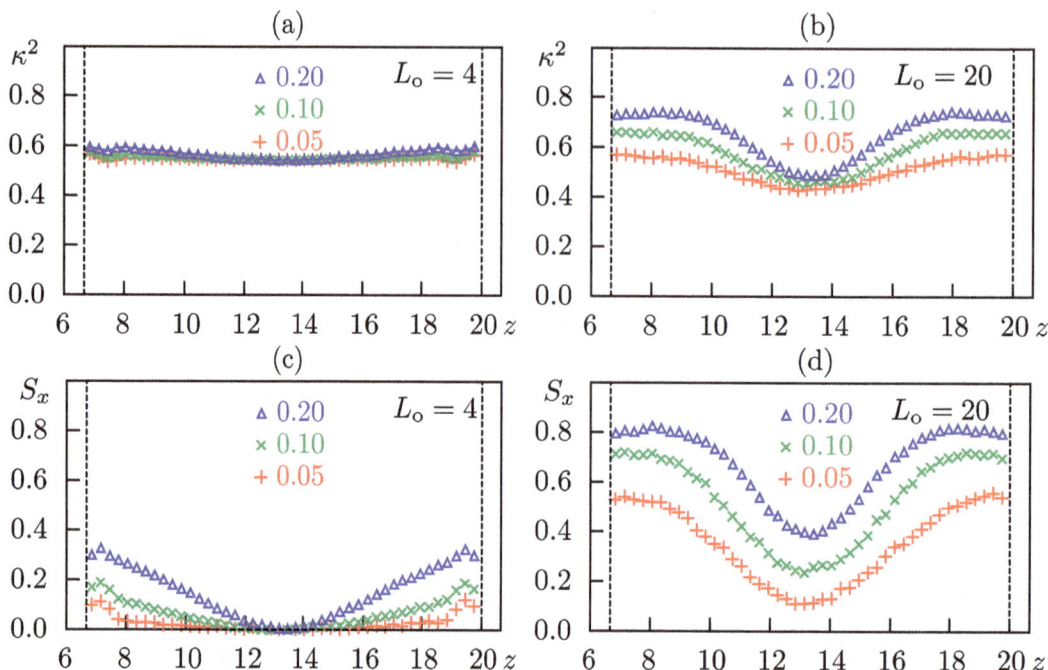

Figure 5. (Color online) Profiles for average shape anisotropy [(a) and (b)] and oligomer alignment [(c) and (d)]. Respective oligomer length L_o and the magnitude of the flow force are indicated in each plot.

Average shape anisotropy and oligomer alignment profiles are shown in figure 5 for the shortest $L_o = 4$ and longest $L_o = 20$ oligomer considered in this study. The case $L_o = 4$ is characterised by flat anisotropy profile with the value $\kappa^2 \approx 0.6$ independent of the magnitude of the flow force [see, frame (a)]. One can conclude that for the oligomer being this short, the flow does not change its shape (at least for the flow force magnitude range used here). For the longest oligomer, $L_o = 20$ [frame (b)], the average value of κ^2 over the profile is close to that for $L_o = 4$, but the profile exhibits distinct shoulders with higher κ^2 values near both channel edges and a well in its center. The channel edges, therefore, promote a stronger anisotropy for the adjacent oligomers, presumably due to entropic effects.

The alignment profile for the shortest oligomer [frame (c)] indicates no orientdensal order in the middle of the channel ($S_x \approx 0$), whereas a relatively weak alignment is observed near the channel edges, which rises to $S_x \sim 0.35$ with the increase of the flow force magnitude f. For the longest oligomer [frame (d)], the alignment profile has a cosine-like shape which moves up almost unchanged with an increase of f. It is non-zero in the middle of a channel for all f being considered. Therefore, at least for longer oligomers, $L_o > 4$, there is a variation of the oligomer shape and alignment across the channel: the molecules are found to be much more elongated and aligned near the edges as compared to the middle part.

The flow-induced deformation of the polymer molecules renders their shape to be more similar to liquid crystals. The effect is detected for longer oligomers $L_o > 4$ and stronger flows, where the effective length-to-breadth ratio of oligomer exceeds a certain threshold. Similar effect is well known for the systems of anisotropic hard bodies, where the orientationally ordered phases are also found above certain threshold length-to-breadth ratio [59–62]. Using this liquid crystal analogy, we recall the results obtained by Mazza et al. [63, 64] reporting the high self-diffusivity of the Gay-Berne-Kihara fluid along the director in the "supernematic" phase. Following these findings, one expects an essential reduction of the friction between the aligned oligomers near the channel edges, as compared to that in the central part. Larger friction between oligomers in the middle of a channel is seen as the reason for the suppression of the velocity profile here and, as a result, its non-parabolic, bell-like shape [cf. figure 4 (b)], and the appearance of the slip boundary condition [cf. figure 4 (c)].

3. Flow of oligomeric fluid through a channel with polymer modified walls

We turn now to the case when the pore walls are modified by polymer brushes arranged in the form of stripes (see, figure 6). Each chain of a brush is of length $L = 20$ beads of type A, the pore interior is filled with the oligomer fluid of the length L_o beads of type B, the contraflow regions contain an oligomer fluid of the length L_o beads of type A. The difference between the bead types is in the value of the repulsion amplitude a in equation (2.2) being set to $a_{AA} = a_{BB} = 25$ and $a_{AB} = 40$ for the interaction of similar and dissimilar beads, respectively. Therefore, the oligomer acts as a bad solvent for the brush. The good solvent case, $a_{AA} = a_{BB} = a_{AB} = 25$, is briefly discussed in the end of this section.

The equilibrium properties of the setup depicted in figure 6 for the case of $L_o = 1$ and no contraflow regions are studied in detail in reference [39, 40]. Equilibrium morphology was found to depend on the parameters d and w, and is formed as a result of an interplay between the enthalpy and the entropy of the system. In particular, at small $w \ll L$ and any d, the adjacent brush stripes belonging to the same wall merge and form a homogeneous "coat" on the wall resulting in the lamellar morphology. In this case, the chains are stretched and aligned along the X-axis. With an increase of w, the adjacent brush stripes are incapable of merging any more. Instead, they either stay separately (at relatively large $d \sim L$) or merge across the pore with their counterparts grafted to the opposite wall to form a pillar phase (at small enough $d < L$). In this case, the brush chains are stretched and aligned in Z direction. This demonstrates a strong correlation between the alignment direction of brush chains and the topology of the equilibrium morphology. Therefore, it looks plausible that the change of the alignment of the brush chain by means of an external stimulus could result in a morphology change in the system.

This is the case, indeed, when a flow force above certain threshold value is applied to the fluid in the pore. Let us consider first the visual representation of morphology changes in the form of a snapshot sequence. The case of $d = 13.333$, $w = 10$, $c = 4$, at various values of the force f is presented in figure 7. For this geometry of a pore, a stable pillar phase is observed when no or weak flow force is applied [cf. reference [39] and figure 7 (a)]. With an increase of f above the threshold value of $f \approx 0.04 - 0.06$, the pillars break and the morphology switches to the modulated lamellar morphology [see figure 7 (b)]. The layers, formed of brush chains bent along the flow, gradually flatten as f increases further, as shown in figure 7 (c). One should remark that a perfect stationary lamellar morphology is also aided by a microphase separation between the A beads of tethered chains and B beads of the flowing oligomer.

Let us check the quality of temperature conservation, similarly to the analysis performed in section 2 for the pore with flat boundaries. As follows from figure 8 (a), maximum deviation of the temperature profile from the required value 1 does not exceed 4% if the force amplitude is restricted to $f \leqslant 0.4$ for both cases of $L_o = 1$ and $L_o = 20$. It is worth mentioning that the maximum usable value for $f = 0.4$ here is twice as large as its counterpart for the case of flat boundaries, see figure 3. This relation can be attributed to the fact that the total amount of the force applied inside a pore with polymer modified

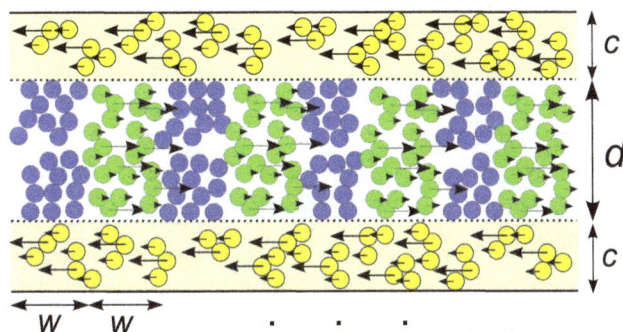

Figure 6. (Color online) Extension of the simulation box geometry of figure 1 to the case of a slit-like pore with its internal walls modified by stripes of polymer brushes (displayed in blue). The stripes are of width w and are periodic along the X-axis (direction of flow). Pore interior is filled with oligomeric fluid (displayed in green). The flow force is applied to the oligomers only.

Figure 7. (Color online) Sequence of snapshots showing the flow-induced transitions from pillar (a) through modulated lamellar (b) into flatten lamellar (c) morphology. System geometry: $d = 13.333$, $w = 10$, $c = 4$, $L_o = 1$, the flow force magnitude is $f = 0.02$, 0.1, and 0.4 for (a), (b) and (c), respectively. Colours follow these in figure 6, contraflow regions not shown.

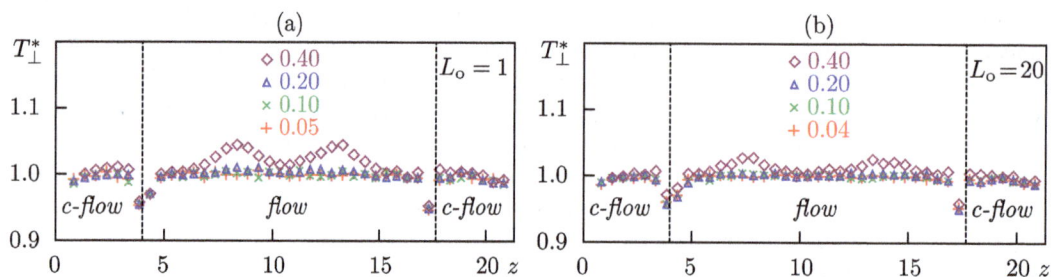

Figure 8. (Color online) $T_\perp^*$ profile at various amplitudes of the bulk force f indicated in the figure. Geometry $d = 13.333$, $w = 10$, $c = 4$ is considered with polymer-modified pore boundaries. (a) simple liquid, $L_o = 1$; (b) longest oligomer simulated, $L_o = 20$.

boundaries (figure 8) is twice less compared to the case of the pore with flat boundaries (figure 3). This is so due to the fact that no force is applied to the polymer brush beads (which are half of all the beads in the system).

The average profiles for the velocity component along the flow direction, v_x are shown in figure 9 at various force amplitudes $f = 0.05$, 0.1, 0.2 and 0.4 for the cases of $L_o = 1$ and $L_o = 20$. Comparing these profiles with their counterparts for the case of flat internal walls (figure 4), one can make the following observations. Firstly, at $f \geq 0.1$, the profiles exhibit two "shoulders" near each internal wall which are characterized by zero values for v_x. These are, obviously, the regions occupied by the polymer brush which "envelopes" the internal walls (see, figure 7). The flow is completely suppressed within these layers, rendering the walls thicker and reducing the pore size accessible to the flow. As a consequence, the maxima for v_x decrease compared to the case of flat walls. Secondly, the shape of the central part of

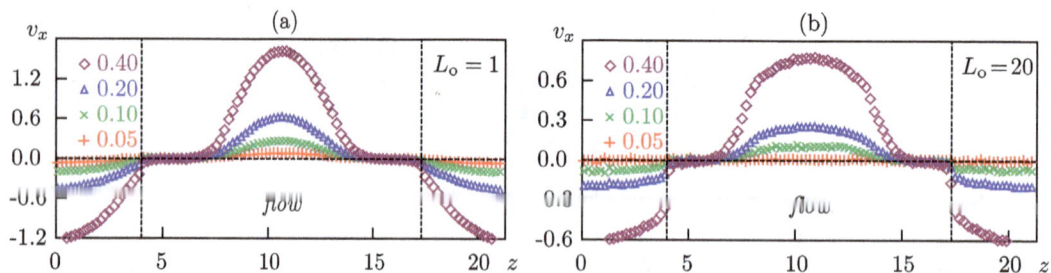

Figure 9. (Color online) Profile for velocity component along the flow direction, v_x, evaluated bead-wise at several values of f (indicated in the figure) for the setup depicted in figure 6, the other parameters are the same as in figure 8.

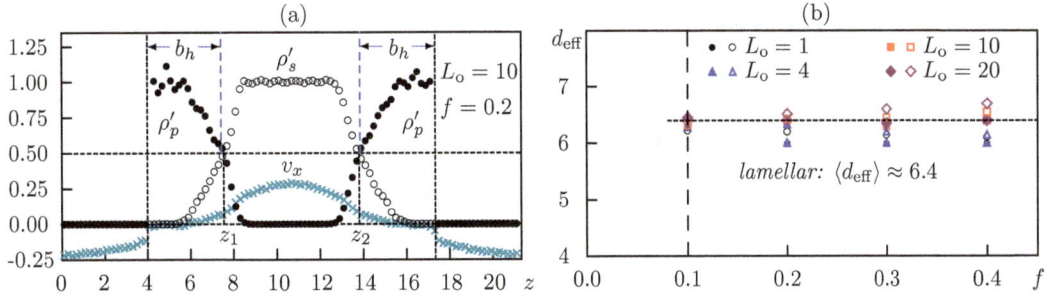

Figure 10. (Color online) (a) Schema explaining the estimates for the effective pore size d_{eff} in stationary lamellar morphology for $L_o = 10$, and $f = 0.2$, ρ'_p and ρ'_s are reduced density profiles for brush and oligomer beads, respectively. b_h is brush thickness (3.1), z_1 and z_2 are the intersection points of ρ'_p and ρ'_s. The velocity profile v_x is also shown (not in scale). (b) d_{eff} at various L_o and f. Solid legends: estimates from the intersection points z_1 and z_2, open legends: esimates from the brush height b_h. The dashed vertical line marks the formation of the lamellar morphology.

each velocity profile follows closely their counterparts in figure 4. It is of parabolic shape for $L_o = 1$ and bell-like for $L_o = 20$, suggesting qualitative similarities between both flows.

This interpretation brings up the possibility to treat a fluid flow within a stationary lamellar morphology similarly to the case of the pore with flat boundaries, discussed in section 2, except for the smaller effective pore size d_{eff} [65]. To evaluate the latter, one can use the expression for an average brush thickness

$$b_h = 2 \frac{\int \tilde{z} \rho_p(\tilde{z}) d\tilde{z}}{\int \rho_p(\tilde{z}) d\tilde{z}}, \tag{3.1}$$

where $\rho_p(\tilde{z})$ is the density profile of the beads that belong to the tethered chains and $\tilde{z}$ is the distance from the nearest pore boundary along the Z-axis. In this case, one obtains $d_{eff} = d - 2b_h$. Alternatively, the effective pore size can be estimated as the distance between the intersection points z_1 and z_2 for $\rho'_p = \rho_p/\rho$ and $\rho'_s = \rho_s/\rho$, the reduced density profiles for the polymer and the flowing oligomer beads, respectively. This is illustrated in figure 10 (a). It is evident that, for this particular case, both estimates for d_{eff} are extremely close. To check how this observation holds for other oligomer lengths L_o and flow forces f, we performed both types of estimates for d_{eff} in each case. The results are presented in figure 10 (b), where the estimates for d_{eff} made from the intersection points z_1 and z_2 are presented via solid legends, whereas the estimates performed via the evaluation of b_h are shown via open symbols. One can make several conclusions from this plot. First, the value of d_{eff}, evaluated by both approaches, are similar to each other for $f \geqslant 0.1$. This threshold correlates well with the value of f, at which the stationary lamellar morphology is formed (marked with the dashed line in figure 10). While d_{eff} can be also calculated at smaller values of f, these results would carry no physical significance due to the pillar morphology. Second, the difference between the values for d_{eff} estimated by means of two alternative methods at the same L_o and at the same f, does not exceed 4%. Therefore, either of the estimates for d_{eff} can be used. Third, there is a trend for an increase of d_{eff} with the growth of oligomer length L_o, although it is rather modest. For example, for the case of $f = 0.4$ the value of d_{eff} for oligomer length of $L_o = 20$ is only 10% higher than its counterparts for $L_o = 1$ and 4, and this increase is of the order of the error in the estimates of d_{eff} mentioned above. A slight increase of the brush height with an increasing flow can be attributed to the fact that there is some residual flow of oligomers inside the brush. As the flow increases, the flow-induced elongation of the oligomers leads to an increase of their effective size and this will lead to a slight increase of the brush height.

As was discussed in section 2, flat walls of the setup depicted in figure 1 act as effective "stretchers" and "aligners" for the adjacent oligomer molecules, which results in characteristic profiles for κ^2 and S_x shown in figure 5. It is, therefore, of interest to see whether or not the effective walls formed by a flattened polymer brush, as pictured in figure 7 (b) and (c), have a similar impact on the adjacent oligomer molecules. We examine the aligning capabilities of such flattened brushes more in detail, considering both cases of bad and good oligomer solvent. For the former case, the repulsion parameter a in equa-

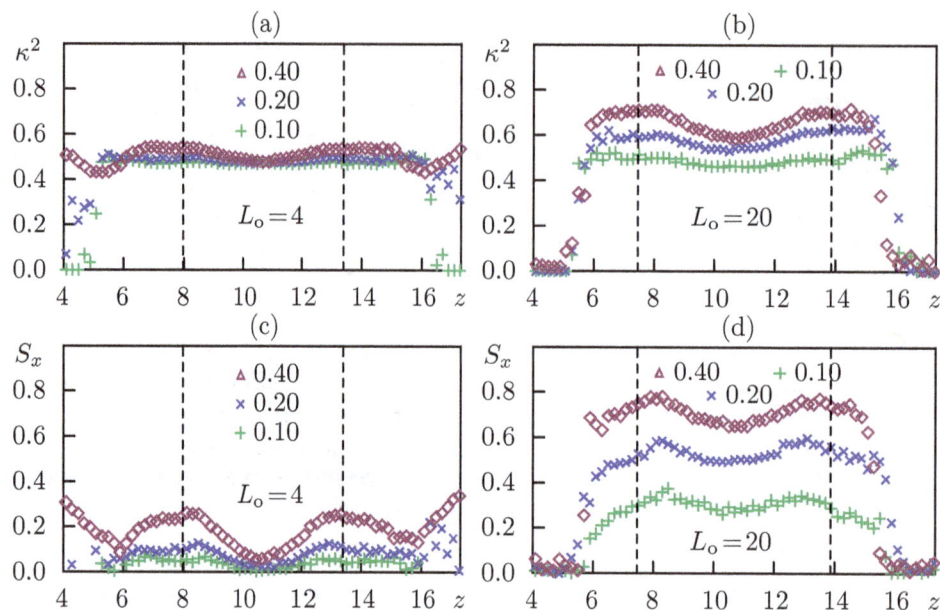

Figure 11. (Color online) Profiles of the average shape anisotropy [(a) and (b)] and oligomer alignment [(c) and (d)] for the geometry depicted in figure 6 and the case of bad oligomer solvent. Respective oligomer length L_o and the magnitude of the flow force are indicated in each plot.

tion (2.2) is set to $a_{AB} = 40$ for the interaction between oligomer and brush monomers, whereas for the latter case we set $a_{AB} = 25$. The repulsion parameter between similar beads is equal to $a_{AA} = a_{BB} = 25$ in both cases.

For the bad solvent case, the flattened brush and oligomer flow are strongly demixed with oligomers being expelled from the brush-rich regions. The profiles for κ^2 and S_x are shown in figure 11 for the oligomer lengths of $L_o = 4$ and $L_o = 20$. Here, we make use of our estimates for the effective pore size, d_{eff}, indicated on each plot by vertical dashed lines. If restricted to this region, then the profiles depicted

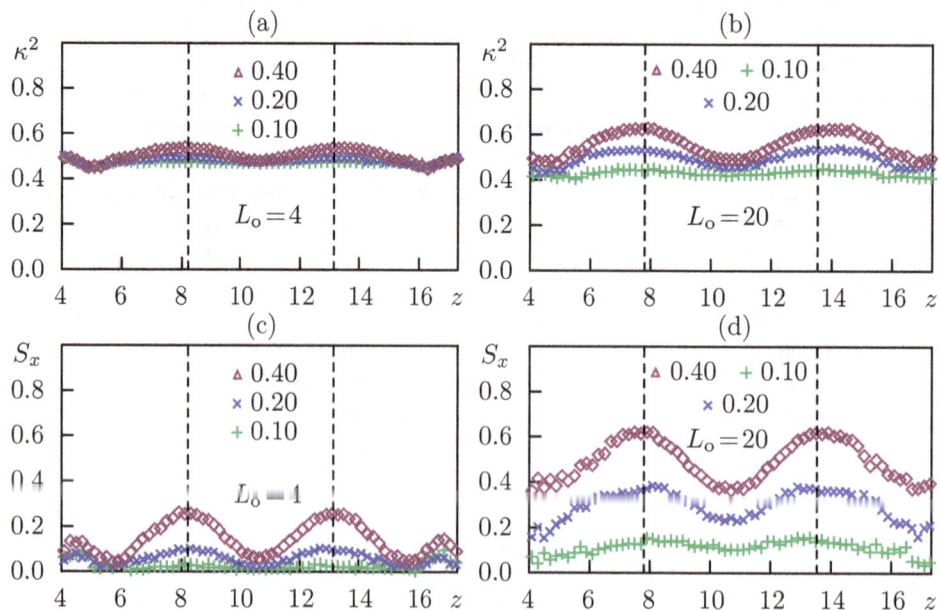

Figure 12. (Color online) The same as in figure 11 but for the case of good oligomer solvent.

in figure 11 are extremely close to their respective counterparts for the case of flat walls shown in figure 1 in both shape and absolute values, save for being "squeezed" into the smaller pore size d_{eff}. This indicates that the existing roughness of the flattened polymer wall does not reduce its impact on the adjacent oligomer molecules.

For the good solvent case, the brush and oligomer mix well if no flow force is applied, but a flow-driven lamellarization of the system takes place at about $f = 0.2$, similarly to the bad solvent case. The profiles for κ^2 and S_x are shown in figure 12 for the oligomer lengths of $L_o = 4$ and $L_o = 20$. The estimated effective pore size d_{eff} is of the same order but a fraction smaller than that for the bad solvent case. This is indicated in figure 12 by vertical dashed lines. One should remark that despite the strong alignment of the polymer brush and good mixing between the brush and the oligomers, the latter are found less elongated and less aligned along a flow compared to the bad solvent case shown in figure 11. The respective curves for κ^2 are lower by about 0.1 compared to their counterparts for the bad solvent case, whereas these for S_x are about 0.2 lower. One should attribute this to the fact that the flattening of the brush in the case of a good solvent requires a higher flow force compared to the case of a bad solvent. In the latter case, lamellarization is also aided by the microphase separation between the brush and the oligomer fluid.

Despite these quantitative differences, the qualitative picture emerging for both cases of the bad and good solvent is essentially the same. Namely, at a certain value of the flow force f, the stationary lamellar phase is formed with the flowing oligomer occupying the center part of the pore. The oligomer is found essentially elongated and aligned along the flow near the walls of this channel and much less in the center of the pore. This effect is detected for the oligomer lengths $L_o > 4$ and, due to its impact on the distribution of the local friction across the pore, affects the behaviour of the fluid turning it into a non-Newtonian one.

4. Conclusions

In this paper, we developed the simulation approach, which allows one to simulate the pressure-driven flow in the pore with flat and polymer-modified walls. It combines the earlier ideas of fluid-like walls and reverse flow. The former enables to avoid highly structured solid walls that usually lead to near-wall artefacts. The latter introduces friction between oppositely flowing streams which makes it possible to conserve the total momenta and keep the temperature constant. Our system geometry contains the central main pore "enveloped" by two fluid-like pores on each side. The flow force is introduced in the main pore and oppositely directed contra-flow of the same magnitude — in both fluid-like walls. Simulation of the oligomer flow through the pore with flat walls is used as a check for the credibility of the method and reproduction of the hydrodynamic boundary conditions. Good thermostatting of the system is achieved when the flow force magnitude does not exceed a certain threshold. For the case of the oligomer length $L_o > 4$, we found the molecules adjacent to the central pore boundaries essentially stretched and aligned along the flow, whereas their shape is more spherical and less aligned in the middle of a pore. This provides the basis for variation of the local friction across the pore and, as a consequence, the non-parabolicity of the velocity profile for the oligomer fluid and the slip boundary condition.

The case of polymer-modified walls is also considered when the polymer brush has the form of stripes arranged perpendicularly to the flow direction. In this case, at a certain threshold value of the flow force, one observes the pillar-to-lamellar transition induced by the flow which leads to the brush enveloping the pore wall with a relatively smooth layer. At higher flow rates, the flow of oligomer is similar to the case of flat walls, although for the narrower effective pore size. The latter is estimated both from the intersection of density profiles for the brush and the flowing oligomer and from the integral equation for the average brush thickness. The effect of local stretching and alignment of oligomers near the walls of the effective pore is detected the same as for the case of flat walls.

The method can be extended to more complex systems, namely: the flow of mixtures and their flow-induced separation; the flow of amphiphilic molecules; the flow of complex macromolecules or their solutions. Combined with the fine-tunable structure of the brush, this opens up a possibility to study various problems of transport of oligo- and macromolecules through a complex structured environment.

Acknowledgements

This work was supported by the EU under IRSES Project STCSCMBS 268498.

References

1. Zhao B., Brittain W.J., Prog. Polym. Sci., 2000, **25**, 677; doi:10.1016/S0079-6700(00)00012-5.
2. Minko S., J. Macromol. Sci., Polymer Rev., 2006, **46**, 397; doi:10.1080/15583720600945402.
3. Alexander S.J., J. Phys. (Paris), 1977, **38**, 983; doi:10.1051/jphys:01977003808098300.
4. De Gennes P.-J., J. Phys. (Paris), 1976, **37**, 1445; doi:10.1051/jphys:0197600370120144500.
5. De Gennes P.-J., Macromolecules, 1980, **13**, 1069; doi:10.1021/ma60077a009.
6. Milner S.T., Witten T.A., Cates M.E., Europhys. Lett., 1988, **5**, 413; doi:10.1209/0295-5075/5/5/006.
7. Scheutjens J.M.H.M., Fleer G.J., Cohen Stuart M.A., Colloids Surf., 1986, **21**, 285; doi:10.1016/0166-6622(86)80098-1.
8. Wijmans C.M., Scheutjens J.M.H.M., Zhulina E.B., Macromolecules, 1992, **25**, 2657; doi:10.1021/ma00036a016.
9. Netz R.N., Schick M., Macromolecules, 1998, **31**, 5105; doi:10.1021/ma9717505.
10. Carigano M.A., Szleifer I., J. Chem. Phys., 1995, **102**, 8662; doi:10.1063/1.468968.
11. McCoy J.D., Teixeira M.A., Curro J.G.J., J. Chem. Phys., 2002, **117**, 2975; doi:10.1063/1.1491242.
12. Cao D.P., Wu J., Langmuir, 2006, **22**, 2712; doi:10.1021/la0527588.
13. Xu X.F., Cao D.P., J. Chem. Phys., 2009, **130**, 164901; doi:10.1063/1.3119311.
14. Xu X.F., Cao D.P., Soft Matter, 2010, **6**, 4631; doi:10.1039/C0SM00034E.
15. Milchev A., Egorov S.A., Binder K., J. Chem. Phys., 2010, **132**, 184905; doi:10.1063/1.3414996.
16. Borówko M., Sokołowski S., Staszewski T., J. Phys. Chem. B, 2013, **117**, 10293; doi:10.1021/jp4027546.
17. Grest G.S., J. Chem. Phys., 1996, **105**, 5532; doi:10.1063/1.472395.
18. Weinhold J.D., Kumar S.K., J. Chem. Phys., 1994, **101**, 4312; doi:10.1063/1.467481.
19. Pastorino C., Binder K., Keer T., Müller M., J. Chem. Phys., 2006, **124**, 064902; doi:10.1063/1.2162883.
20. MacDowell L.G., Müller M., J. Chem. Phys., 2006, **124**, 084907; doi:10.1063/1.2172597.
21. Fouqueau A., Meuwly M., Bemish R.J., J. Phys. Chem. B, 2007, **111**, 10208; doi:10.1021/jp071721o.
22. Hadziioannou G., Patel S., Granick S., Tirrell M., J. Am. Chem. Soc., 1986, **108**, 2869; doi:10.1021/ja00271a014.
23. Taunton H.J., Toprahcioglu C., Fetters L.J., Klein J., Nature, 1988, **332**, 712; doi:10.1038/332712a0.
24. Taunton H.J., Toprahcioglu C., Fetters L.J., Klein J., Macromolecules, 1990, **23**, 571; doi:10.1021/ma00204a033.
25. Auroy P., Auvray L., Leger L., Phys. Rev. Lett., 1991, **66**, 719; doi:10.1103/PhysRevLett.66.719.
26. Brinkman H.C., Appl. Sci. Res., 1947, **1**, 27; doi:10.1007/BF02120313.
27. Milner S.T., Macromolecules, 1991, **24**, 3704; doi:10.1021/ma00012a036.
28. Servantie J., Müller M., Phys. Rev. Lett., 2008, **101**, 026101; doi:10.1103/PhysRevLett.101.026101.
29. Squires T.M., Quake A.R., Rev. Mod. Phys., 2005, 77, 977; doi:10.1103/RevModPhys.77.977.
30. Schoch R.B., Han J., Renaud P., Rev. Mod. Phys., 2008, **80**, 839; doi:10.1103/RevModPhys.80.839.
31. Klein J., Perahia D., Warburg S., Nature, 1991, **352**, 143; doi:10.1038/352143a0.
32. Lai P.Y., Binder K., J. Chem. Phys., 1993, **98**, 2366; doi:10.1063/1.464164.
33. Doyle P.S., Shaqfeh E.S.G., Gast A.P., Phys. Rev. Lett., 1997, **78**, 1182; doi:10.1103/PhysRevLett.78.1182.
34. Grest G.S., Phys. Rev. Lett., 1996, **76**, 4979; doi:10.1103/PhysRevLett.76.4979.
35. Adiga S.P., Brenner D.W., Nano Lett., 2005, **5**, 2509; doi:10.1021/nl051843x.
36. Huang J., Wang Y., Laradji M., Macromolecules, 2006, **39**, 5546; doi:10.1021/ma060628f.
37. Masoud H., Alexeev A., Chem. Commun., 2011, **47**, 472; doi:10.1039/c0cc02165b.
38. Dimitrov D.I., Milchev A., Binder K., Macromol. Theory Simul., 2008, **17**, 313; doi:10.1002/mats.200800038.
39. Ilnytskyi J.M., Patsahan T., Sokołowski S., J. Chem. Phys., 2011, **134**, 204903; doi:10.1063/1.3592562.
40. Ilnytskyi J.M., Sokołowski S., Patsahan T., Condens. Matter Phys., 2013, **16**, 13606; doi:10.5488/CMP.16.13606.
41. Paik M.Y., Xu Y., Rastogi A., Tanaka M., Yi Y., Ober C.K., Nano Lett., 2010, **10**, 3873; doi:10.1021/nl102910f.
42. Groot R.D., Warren P.B., J. Chem. Phys., 1997, **107**, 4423; doi:10.1063/1.474784.
43. Español P., Warren P., Europhys. Lett., 1995, **30**, 191; doi:10.1209/0295-5075/30/4/001.
44. Ashurst W.T., Hoover W.G., Phys. Rev. Lett., 1973, **31**, 206; doi:10.1103/PhysRevLett.31.206.
45. Levesque D., Ashurst W.T., Phys. Rev. Lett., 1974, **00**, 077; doi:10.1103/PhysRevLett.33.277
46. Ashurst W.T., Hoover W.G., Phys. Rev. A, 1975, **11**, 658; doi:10.1103/PhysRevA.11.658.
47. Koplik J., Banavar J.R., Willemsen J.F., Phys. Rev. Lett., 1988, **60**, 1282; doi:10.1103/PhysRevLett.60.1282.
48. Goujon F., Malfreyt P., Tildesley D.J., ChemPhysChem, 2004, **5**, 457; doi:10.1002/cphc.200300901.
49. Goujon F., Malfreyt P., Tildesley D.J., Mol. Phys., 2005, **103**, 2675; doi:10.1080/00268970500134706.
50. Visser D.C., Hoefsloot H.C.J., Iedema P.D., J. Comput. Phys., 2005, **205**, 626; doi:10.1016/j.jcp.2004.11.020.
51. Pivkin I.V., Karniadakis G.E., J. Comput. Phys., 2005, **207**, 114; doi:10.1016/j.jcp.2005.01.006.

52. Pivkin I.V., Karniadakis G.E., J. Chem. Phys., 2006, **124**, 184101; doi:10.1063/1.2191050.

53. Jian F., Yongmin H., Honglai L., Ying H., Front. Chem. Eng. China, 2007, **1**, 132; doi:10.1007/s11705-007-0025-5.

54. Ghosh A., Paredes R., Luding S., In: Proceedings of Congres on Particle Technology, PARTEC 2007, (Nürnberg, 2007), Peukert W., Schreglmann C. (Eds.), Nürnberg Messe, Nürnberg, 2007.

55. Pastorino C., Kreer T., Müller M., Binder K., Phys. Rev. E, 2007, **76**, 026706; doi:10.1103/PhysRevE.76.026706.

56. Müller-Plathe F., Phys. Rev. E, 1999, **59**, 4894; doi:10.1103/PhysRevE.59.4894.

57. Backer J.A., Lowe C.P., Hoefsloot H.C.J., Iedema P.D., J. Chem. Phys., 2005, **122**, 154503; doi:10.1063/1.1883163.

58. Fedosov D.A., Karniadakis G.E., Caswell B., J. Chem. Phys., 2010, **132**, 144103; doi:10.1063/1.3366658.

59. Vieillard-Baron J., Mol. Phys., 1974, **28**, 809; doi:10.1080/00268977400102161.

60. Frenkel D., Lekkerkerker H.N.W., Stroobants A., Nature, 1988, **332**, 822; doi:10.1038/332822a0.

61. McGrother S.C., Williamson D.C., Jackson G., J. Chem. Phys., 1966, **104**, 6755; doi:10.1063/1.471343.

62. Cuetos A., Dijkstra M., Phys. Rev. Lett., 2007, **98**, 095701; doi:10.1103/PhysRevLett.98.095701.

63. Mazza M.G., Greschek M., Valiullin R., Kärger J., Schoen M., Phys. Rev. Lett., 2010, **105**, 227802; doi:10.1103/PhysRevLett.105.227802.

64. Mazza M.G., Greschek M., Valiullin R., Schoen M., Phys. Rev. E, 2011, **83**, 051704; doi:10.1103/PhysRevE.83.051704.

65. Pastorino C., Binder K., Müller M., Macromolecules, 2009, **42**, 401; doi:10.1021/ma8015757.

10

Quasiparticle states driven by a scattering on the preformed electron pairs[*]

T. Domański

Institute of Physics, M. Curie-Skłodowska University, 20-031 Lublin, Poland

We analyze evolution of the single particle excitation spectrum of the underdoped cuprate superconductors near the anti-nodal region, considering temperatures below and and above the phase transition. We inspect the phenomenological self-energy that reproduces the angle-resolved-photoemission-spectroscopy (ARPES) data and we show that above the critical temperature, such procedure implies a transfer of the spectral weight from the Bogoliubov-type quasiparticles towards the in-gap damped states. We also discuss some possible microscopic arguments explaining this process.

Key words: *superconducting fluctutations, Bogoliubov quasiparticles, pseudogap*

1. Introduction

Superconductivity (i.e., dissipationless motion of the charge carriers) is observed at sufficiently low temperatures, when electrons from the vicinity of the Fermi surface are bound in the pairs and respond collectively (rather than individually) to any external perturbation such as electromagnetic field, pressure, temperature gradient, etc. Depending on specific materials, the pairing mechanism can be driven by phonons (in classical superconductors), magnons (in heavy fermion compounds) or by the antiferromagnetic exchange interactions originating from the Coulomb repulsion (in cuprate oxides). In most cases, the electron pairs are formed at the critical value T_c, marking the onset of superconductivity. There are, however, numerous exceptions to this rule. For instance, in the cuprate superconductors [1] or in the ultracold fermionic gasses [2], such pairs pre-exist well above T_c. To some extent, their presence causes the properties reminiscent of the superconducting state.

Early evidence for the preformed pairs existing above T_c has been indicated in the muon scattering experiments [3]. Later on, their existence was supported by the ultrafast (tera-Hertz) optical spectroscopy [4, 5] and the large Nernst effect [6, 7]. Spectroscopic signatures of the preformed pairs have been also detected directly in the ARPES measurements on yttrium [8] and lanthanum [9] cuprate oxides, revealing the Bogoliubov-type quasiparticle dispersion above T_c. Furthermore, the STM imaging provided clear fingerprints of such dispersive Bogoliubov quasiparticles (by the unique octet patterns) being unchanged from temperatures deep in the superconducting region up to $1.5T_c$ [10]. Superconducting fluctuations above T_c have been also reported by the Josephson-like tunneling [11] and the proximity effect induced in the nanosize metallic slabs deposited on $La_{2-x}Sr_xCuO_4$ [12]. More recently, the residual Meissner effect has been experimentally observed above the transition temperature T_c by the torque magnetometry [13] and other measurements [14, 15]. Additional evidence for the superconducting-like behaviour above T_c has been seen in the high resolution ARPES measurements [16], the superfluid fraction observed in the c-axis optical measurements $Re\{\sigma_c(\omega)\}$ [17], the Josephson spectroscopy for YBaCuO-LaSrCuO-YBaCuO junction using LaSrCuO in the pseudogap state well above T_c [18], optical conductance in the pseudogap state of YBaCuO superconductor [19] and the photo-enhanced antinodal conductivity in pseudogap state of the high T_c superconductors [20].

---

[*]This work is dedicated to professor Stefan Sokołowski on the occasion of his 65-th birthday.

Preformed pairs are correlated above T_c only on some short temporal τ_ϕ and spatial l_ϕ scales [21–23]. For this reason, the superconducting fluctuations are manifested in very peculiar way [24]. Their influence on the single particle spectrum is manifested by: a) two Bogoliubov-type branches and b) additional in-gap states that are over-damped (have a short life-time). Temperature has a strong effect on the transfer of the spectral weight between these entities. In the underdoped cuprate superconductors, such a transfer is responsible for filling-in the energy gap [16, 25], instead of closing it (as in the classical superconductors). Some early results concerning superconducting fluctuations have been known for a long time [26, 27], but they attracted much more interest in the context of cuprate superconductors [26–36] and ultracold fermion superfluids [37, 38].

In this work, we study qualitative changeover of the single particle electronic spectrum of the underdoped cuprate oxides for temperatures varying from below T_c (in the superconducting state) to above T_c , where the preformed pairs are not long-range coherent. In the superconducting state, the usual Bogoliubov-type quasiparticles are driven by the Bose-Einstein condensate of the (zero-momentum) Cooper pairs. We find that above T_c, the Bogoliubov quasiparticles are still preserved, but the scattering processes driven by the finite momentum pairs contribute the in-gap states whose life-time substantially increases with increasing temperature. We discuss this process on the phenomenological as well as microscopic arguments. Roughly speaking, a feedback of the electron pairs on the unpaired electrons resembles the long-range translational and orientational order that develops between the amphiphilic particles in presence of the ions at solid state surfaces studied by S. Sokołowski with coworkers [39].

2. Microscopic formulation of the problem

To account for the coherent/incoherent pairing we consider the Hamiltonian

$$\hat{H} = \sum_{\mathbf{k},\sigma} \left(\varepsilon_\mathbf{k} - \mu\right) \hat{c}_{\mathbf{k}\sigma}^\dagger \hat{c}_{\mathbf{k}\sigma} + \frac{1}{N} \sum_{\mathbf{k},\mathbf{k}',\mathbf{q}} V_{\mathbf{k},\mathbf{k}'}(\mathbf{q}) \hat{c}_{\mathbf{k}'\uparrow}^\dagger \hat{c}_{\mathbf{q}-\mathbf{k}'\downarrow}^\dagger \hat{c}_{\mathbf{q}-\mathbf{k}\downarrow} \hat{c}_{\mathbf{k}\uparrow} \tag{2.1}$$

describing the mobile electrons of kinetic energy $\varepsilon_\mathbf{k}$ (where μ is the chemical potential) interacting via the two-body potential $V_{\mathbf{k},\mathbf{k}'}(\mathbf{q})$. We assume a separable form $V_{\mathbf{k},\mathbf{k}'} = -g\,\eta_\mathbf{k}\,\eta_{\mathbf{k}'}$ of this pairing potential (with $g > 0$). In the nearly two-dimensional cuprate superconductors with the prefactor $\eta_\mathbf{k} = \frac{1}{2}\left[\cos(ak_x) + \cos(ak_y)\right]$ (where a is the unit length in CuO_2 planar structure), such pairing potential induces the d-wave symmetry order parameter [40, 41].

The Hamiltonian (2.1) can be recast in a more compact form, by introducing the pair operators

$$\hat{b}_\mathbf{q} = \frac{1}{\sqrt{N}} \sum_\mathbf{k} \eta_\mathbf{k} \hat{c}_{\mathbf{q}-\mathbf{k}\downarrow} \hat{c}_{\mathbf{k}\uparrow} \tag{2.2}$$

and $\hat{b}_\mathbf{q}^\dagger = (\hat{b}_\mathbf{q})^\dagger$, when the two-body interactions simplify to

$$\frac{1}{N} \sum_{\mathbf{k},\mathbf{k}',\mathbf{q}} V_{\mathbf{k},\mathbf{k}'}(\mathbf{q}) \hat{c}_{\mathbf{k}\uparrow}^\dagger \hat{c}_{\mathbf{q}-\mathbf{k}\downarrow}^\dagger \hat{c}_{\mathbf{q}-\mathbf{k}'\downarrow} \hat{c}_{\mathbf{k}'\uparrow} = -\sum_\mathbf{q} g\, \hat{b}_\mathbf{q}^\dagger \hat{b}_\mathbf{q}. \tag{2.3}$$

Using the Heisenberg equation of motion ($\hbar = 1$)

$$i\frac{d}{dt}\hat{c}_{\mathbf{k}\uparrow} = \left(\varepsilon_\mathbf{k} - \mu\right) \hat{c}_{\mathbf{k}\uparrow} - g\,\eta_\mathbf{k} \frac{1}{\sqrt{N}} \sum_\mathbf{q} \hat{c}_{\mathbf{q}-\mathbf{k}\downarrow}^\dagger \hat{b}_\mathbf{q} \tag{2.4}$$

we immediately notice that the single-particle properties of this model (2.1):

a) are characterized by the mixed particle and hole degrees of freedom (because the annihilation operators $\hat{c}_{\mathbf{k}\uparrow}$ couple to the creation operators $\hat{c}_{\mathbf{q}-\mathbf{k}\downarrow}^\dagger$),

b) depend on the pairing field $\hat{b}_\mathbf{q}$ (appearing in the equation of motion $\frac{d}{dt}\hat{c}_{\mathbf{k}\uparrow}$).

Both these features manifest themselves in the superconducting state, when there exists the Bose-Einstein (BE) condensate $\langle \hat{b}_{\mathbf{q}=0} \rangle \neq 0$ of the Cooper pairs. They also survive in the normal state, as long as the preformed (finite-momentum) pairs are present below the same characteristic temperature T^* marking an onset of the electron pairing. In the next section we explore their role in the superconducting state ($T \leq T_c$) and in the pseudogap region ($T_c < T < T^*$).

3. Pairs as the scattering centers

The Heisenberg equation of motion (2.4) indicates that the electronic states are affected by the pairing field $\hat{b}_{\mathbf{q}}$. Let us consider the generic consequences of such Andreev-type scattering, separately considering: the BE condensed $\mathbf{q} = \mathbf{0}$ and the finite-momentum $\mathbf{q} \neq \mathbf{0}$ pairs.

3.1. The effect of the Bose-Einstein condensed pairs

We start by considering the usual BCS approach, when only the zero momentum pairs are taken into account. This situation has a particularly clear interpretation within the path integral formalism, treating the pairing field via the Hubbard-Stratonovich transformation and determining it from the minimization of action (the saddle point solution). The same result can be obtained using the equation of motion (2.4), focusing on the effect of $\mathbf{q} = \mathbf{0}$ pairs

$$i\frac{d}{dt}\hat{c}_{\mathbf{k}\uparrow} \simeq \left(\varepsilon_{\mathbf{k}} - \mu\right)\hat{c}_{\mathbf{k}\uparrow} - g\,\eta_{\mathbf{k}}\,\hat{c}^{\dagger}_{-\mathbf{k}\downarrow}\frac{\hat{b}_0}{\sqrt{N}}, \tag{3.1}$$

$$i\frac{d}{dt}\hat{c}^{\dagger}_{\mathbf{k}\downarrow} \simeq -\left(\varepsilon_{\mathbf{k}} - \mu\right)\hat{c}^{\dagger}_{\mathbf{k}\downarrow} - g\,\eta_{\mathbf{k}}\frac{\hat{b}^{\dagger}_0}{\sqrt{N}}\,\hat{c}_{\mathbf{k}\uparrow}. \tag{3.2}$$

Macroscopic occupancy of the $\mathbf{q} = \mathbf{0}$ state implies that the bosonic operators $\hat{b}_0^{(\dagger)}$ can be treated as complex numbers $b_0^{(*)}$. By introducing the order parameter

$$\Delta_{\mathbf{k}} = g\eta_{\mathbf{k}}\frac{\langle\hat{b}_0\rangle}{\sqrt{N}} = g\eta_{\mathbf{k}}\frac{1}{N}\sum_{\mathbf{k}'}\eta_{\mathbf{k}'}\langle\hat{c}_{-\mathbf{k}'\downarrow}\hat{c}_{\mathbf{k}'\uparrow}\rangle \tag{3.3}$$

the equations (3.1), (3.2) can be solved exactly using the standard Bogoliubov-Valatin transformation. In such BCS approach, the classical superconductivity has close analogy with the superfluidity of weakly interacting bosons, whose collective sound-like mode originates from the interaction between the finite-momentum bosons and the BE condensate.

In the present context, the zero-momentum Copper pairs substantially affect the single particle excitation spectrum (and the two-body correlations as well). The single-particle Green's function $G(\mathbf{k}, \tau) = -\hat{T}_{\tau}\langle\hat{c}_{\mathbf{k}\uparrow}(\tau)\hat{c}^{\dagger}_{\mathbf{k}\uparrow}\rangle$, where $\hat{T}_{\tau}$ is the time ordering operator, obeys the Dyson equation

$$[G(\mathbf{k}, \omega)]^{-1} = \omega - \varepsilon_{\mathbf{k}} + \mu - \Sigma(\mathbf{k}, \omega), \tag{3.4}$$

with the BCS self-energy

$$\Sigma(\mathbf{k}, \omega) = \frac{|\Delta_{\mathbf{k}}|^2}{\omega + (\varepsilon_{\mathbf{k}} - \mu)}. \tag{3.5}$$

The self-energy (3.5), accounting for the Andreev-type scattering of the $\mathbf{k}$-momentum electrons on the Cooper pairs, can be alternatively obtained from the bubble diagram. The related spectral function $A(\mathbf{k}, \omega) = -\pi^{-1}\mathrm{Im}\,G(\mathbf{k}, \omega + i0^+)$ is thus characterized by the two-pole structure

$$A(\mathbf{k}, \omega) = u_{\mathbf{k}}^2\,\delta(\omega - E_{\mathbf{k}}) + v_{\mathbf{k}}^2\,\delta(\omega + E_{\mathbf{k}}) \tag{3.6}$$

with the Bogoliubov-type quasiparticle energies $E_{\mathbf{k}} = \pm\sqrt{\left(\varepsilon_{\mathbf{k}} - \mu\right)^2 + \Delta_{\mathbf{k}}^0}$ and the spectral weights $u_{\mathbf{k}}^2 = \frac{1}{2}\left[1 + (\varepsilon_{\mathbf{k}} - \mu)/E_{\mathbf{k}}\right]$ and $v_{\mathbf{k}}^2 = 1 - u_{\mathbf{k}}^2$. Let us remark that these quasiparticle branches are separated by the (true) energy gap $|\Delta_{\mathbf{k}}|$. In classical superconductors, $\Delta_{\mathbf{k}}$ implies the off-diagonal-long-range-order (ODLRO) that quantitatively depends on concentration of the BE condensed Cooper pairs. ODLRO is responsible for a dissipationless motion of the charge carriers and simultaneously causes the Meissner effect via the spontaneous gauge symmetry breaking.

3.2. The effect of the non-condensed pairs

In this section we shall study effect of the finite-momentum pairs existing above T_c, which no longer develop any ODLRO because there is no BE condensate. Nevertheless, according to (2.4), the single and paired fermions are still mutually dependent. This fact suggests that the previous BCS form (3.5) should be replaced by some corrections originating from the finite momentum pairs. Let us denote the pair propagator by $L(\mathbf{q}, \tau) = -\hat{T}_\tau \langle \hat{b}(\mathbf{r}, \tau)\hat{b}^\dagger(\mathbf{0}, 0) \rangle$ and assume its Fourier transform in the following form

$$L(\mathbf{q}, \omega) = \frac{1}{\omega - E_\mathbf{q} - i\Gamma(\mathbf{q}, \omega)}, \tag{3.7}$$

where $E_\mathbf{q}$ stands for the effective dispersion of pairs and $\Gamma(\mathbf{q}, \omega)$ describes the inverse life-time. Taking into account the equation (2.4), we express the self-energy via the bubble diagram

$$\Sigma(\mathbf{k}, \omega) = -T \sum_{i\nu_n, \mathbf{q}} \frac{1}{\omega - \xi_{\mathbf{q}-\mathbf{k}} - i\nu_n} L(\mathbf{q}, i\nu_n), \tag{3.8}$$

where $\xi_{\mathbf{q}-\mathbf{k}} = \varepsilon_{\mathbf{q}-\mathbf{k}} - \mu$ and $i\nu_n$ is the bosonic Matsubara frequency. Since above T_c the preformed pairs are only short-range correlated [21–23], we impose

$$\langle \hat{L}^\dagger(\mathbf{r}, t)\hat{L}(\mathbf{0}, 0) \rangle \propto \exp\left(-\frac{|t|}{\tau_\phi} - \frac{|\mathbf{r}|}{\xi_\phi}\right). \tag{3.9}$$

Following T. Senthil and P.A. Lee [22, 23], one can estimate the single particle Green's function $\mathscr{G}(\mathbf{k}, \omega)$ using the following interpolation

$$\Sigma(\mathbf{k}, \omega) = \Delta^2 \frac{\omega - \xi_\mathbf{k}}{\omega^2 - \left(\xi_\mathbf{k}^2 + \pi\Gamma^2\right)}, \tag{3.10}$$

where Δ is the energy gap due to pairing and the other parameter Γ is related to damping of the subgap states. In the low energy limit (i.e., for $|\omega| \ll \Delta$) the dominant contribution comes from the in-gap quasiparticle whose residue is $Z \equiv \left[1 + \Delta^2/(\pi\Gamma^2)\right]^{-1}$, whereas at higher energies the BCS-type quasiparticles are recovered. This selfenergy (3.10) can be derived from the microscopic considerations [42] within the two-component model, describing itinerant fermions coupled to the hard-core bosons [43–50].

The other (closely relative) phenomenological ansatz [31, 32]

$$\Sigma(\mathbf{k}, \omega) = \frac{\Delta^2}{\omega + \xi_\mathbf{k} + i\Gamma_0} - i\Gamma_1 \tag{3.11}$$

has been inferred considering the "small fluctuations" regime [26]. Experimental lineshapes of the angle resolved photoemission spectroscopy obtained for the cuprate superconductors at various doping levels and temperatures (including the pseudogap regime) amazingly well coincide with this simple formula (3.11). The gap and the phenomenological parameters Γ_0, Γ_1 are in general momentum-dependent, but for a given direction in the Brillouin zone one can restrict only to their temperature and doping variations. From now onwards we shall focus on such antinodal region.

In the overdoped samples, Γ_0 can be practically discarded from (3.11) and the remaining parameter Γ_1 simply accounts for T-dependent broadening of the Bogoliubov peaks until they disappear just above T_c. Physical origin of Γ_1 is hence related to the particle-particle scattering. On the contrary, in the underdoped regime, there exists a pseudogap up to temperatures T^* which by far exceed T_c. To reproduce the experimental lineshapes, one must then incorporate the other parameter Γ_0 (nonvanishing only above T_c) which is scaled by $T - T_c$ as shown in figure 1 reproduced from references [31, 32]. Since Γ_0 enters the self-energy (3.11) through the BCS-type structure, its origin is related to the particle-hole scatterings.

We now inspect some consequences of the parametrization (3.11) applicable for the pseudogap regime $T > T_c$ in the underdoped cuprates. Since neither the magnitude of Γ_1 nor Δ seem to vary over a large temperature region above T_c, it is obvious that the qualitative changes are there dominated by scatterings in the particle-hole channel, i.e., due to Γ_0. Roughly speaking, these processes are responsible for filling-in the low energy states upon increasing T as has been evidenced by ARPES [16] and STM [25]

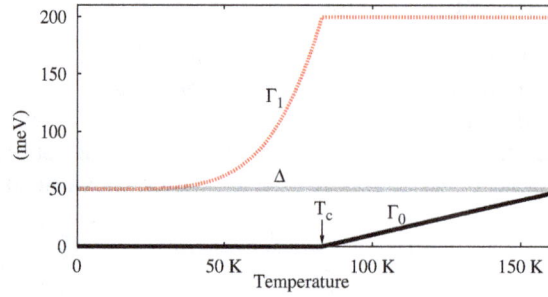

Figure 1. (Color online) Temperature dependence of the phenomenological parameters Δ, Γ_0 and Γ_1 which, through the self-energy (3.11), reproduce the experimental profiles of the underdoped Bi2212 ($T_c = 83$ K) sample. This fitting is adopted from references [31, 32].

measurements. On a microscopic level, such changes can be assigned to scattering on the preformed pairs.

For analytical considerations, let us rewrite the complex self-energy (3.11) as

$$\Sigma(\mathbf{k},\omega) = (\omega + \xi_{\mathbf{k}}) \frac{\Delta^2}{(\omega + \xi_{\mathbf{k}})^2 + \Gamma_0^2} - i\Gamma_{\mathbf{k}}(\omega), \qquad (3.12)$$

where the imaginary part is

$$\Gamma_{\mathbf{k}}(\omega) = \Gamma_1 + \Gamma_0 \frac{\Delta^2}{(\omega + \xi_{\mathbf{k}})^2 + \Gamma_0^2}. \qquad (3.13)$$

In what follows we indicate that above T_c the excitation spectrum can consist of altogether three different states, two of them corresponding to the Bogoliubov modes (signifying particle-hole mixing characteristic for the superconducting state) and another one corresponding to the single particle fermion states which form inside the pseudogap. These states start to appear at $T = T_c^+$ and initially represent heavily overdamped modes containing infinitesimal spectral weight (see reference [22, 23] for a more detailed discussion). Upon increasing temperature, their life-time gradually increases and simultaneously the in-gap states absorb more and more spectral weight at the expense of the Bogoliubov quasiparticles. Finally (in the particular case considered here, this happens roughly near $2T_c$) the single particle fermion states become dominant.

Anticipating the relevance of (3.11) to the strongly correlated cuprate materials, one can determine the single particle Green's function $G(\mathbf{k},\omega)$ and the corresponding spectral function $A(\mathbf{k},\omega)$. Quasiparticle energies are determined by poles of the Green's function, i.e.,

$$\omega - \xi_{\mathbf{k}} - \text{Re}\{\Sigma(\mathbf{k},\omega)\} = 0 \qquad (3.14)$$

provided that the imaginary part $\Gamma_{\mathbf{k}}(\omega)$ disappears. We clearly see that the latter requirement cannot be satisfied for $\Gamma_1 \neq 0$ regardless of Γ_0. Formally this means that the life-time of herein discussed quasiparticles is not infinite. Let us check these eventual (finite life-time) quasiparticle states determined through (3.14). Using the self-energy (3.11), the condition (3.14) is equivalent to

$$(\omega - \xi_{\mathbf{k}}) - (\omega + \xi_{\mathbf{k}}) \frac{\Delta^2}{(\omega + \xi_{\mathbf{k}})^2 + \Gamma_0^2} = 0. \qquad (3.15)$$

In general, there appear three solutions (figure 2) depending on temperature via the parameter Γ_0.

Superconducting region. The fitting procedure [31, 32] has estimated that the parameter Γ_0 vanishes in the superconducting state $T \leqslant T_c$. Under such conditions, (3.15) yields the standard BCS poles at $E_{\mathbf{k}} = \pm \sqrt{\xi_{\mathbf{k}}^2 + \Delta^2}$. Due to $\Gamma_1 \neq 0$, they show up in the spectral function $A(\mathbf{k},\omega)$ as Lorentzians whose broadening corresponds to the inverse life-time of the Bogoliubov modes. Owing to T-dependence of Γ_1 (see

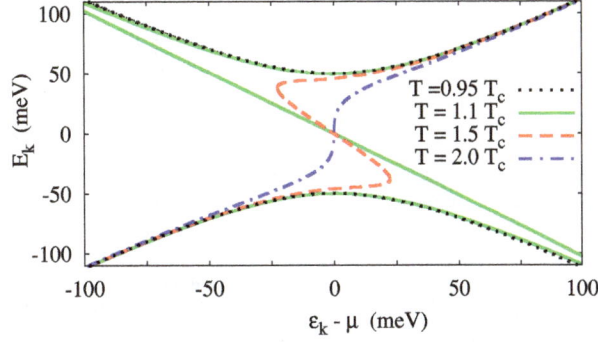

Figure 2. (Color online) Dispersion of $\omega = E_{\mathbf{k}}$ representing the solutions of equation (3.14) for $T/T_c = 0.95$ (dotted line), 1.1, 1.5 and 2 (solid curves as described) obtained for the parameters used in references [31, 32]. We notice three different crossings (3.15), two of them related to the Bogoliubov modes and additional one appearing in between.

figure 1), the broadening of these peaks increases upon approaching T_c from below, albeit $A(\mathbf{k}, \omega = 0) = 0$. Experimentally this process can be observed as the smearing of the coherence peaks [1].

Pseudogap regime. With the appearance of $\Gamma_0 \neq 0$ above T_c, the real part of the self-energy becomes a continuous function of ω (see figure 3). Consequently, besides the Bogoliubov modes, we now obtain an additional crossing located in-between. Figure 2 shows the representative dispersion curves obtained for $1.1 T_c$, $1.5 T_c$, $2 T_c$ and compared with the superconducting state (dotted line). We observe either the three branches or just the single one at sufficiently high temperatures when the spectral function $A(\mathbf{k}, \omega)$ evolves to a single peak structure.

As some useful example, let us study the Fermi momentum $\mathbf{k}_F$, when (3.15) simplifies to

$$\omega \left(1 - \frac{\Delta^2}{\omega^2 + \Gamma_0^2} \right) = 0. \tag{3.16}$$

In this case, we obtain: a) two symmetric quasiparticle energies at $\omega_{\pm} = \pm \tilde{\Delta}$, where $\tilde{\Delta} \equiv \sqrt{\Delta^2 - \Gamma_0^2}$, and b) the in-gap state at $\omega_0 = 0$. The corresponding imaginary parts $\Gamma_{\mathbf{k}}(\omega)$ are

$$\Gamma_{\mathbf{k}}(\omega_{\pm}) = \Gamma_1 + \Gamma_0, \tag{3.17}$$

$$\Gamma_{\mathbf{k}}(\omega_0) = \Gamma_1 + \frac{\Delta^2}{\Gamma_0}. \tag{3.18}$$

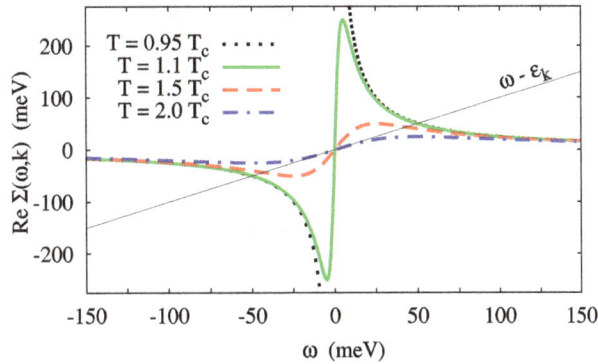

Figure 3. (Color online) The real part of the self-energy $\Sigma(\mathbf{k}, \omega)$ for $\varepsilon_{\mathbf{k}} = \mu$ at several representative temperatures $T/T_c = 0.95$ (dotted line) and 1.1, 1.5, 2.0 (as denoted). Below T_c there exist two poles at $\omega = \pm\Delta$ whereas for $T > T_c$, we obtain altogether three crossings which at higher temperature merge into a single one.

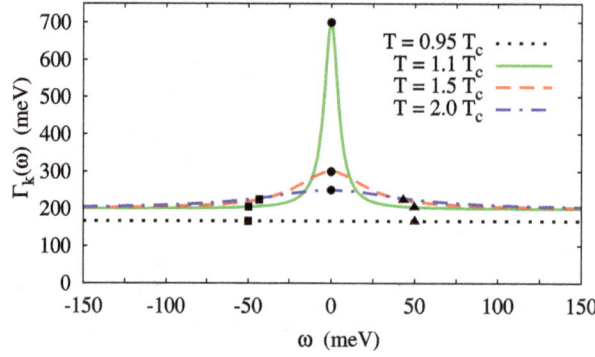

Figure 4. (Color online) The imaginary part $\Gamma_{\mathbf{k}}(\omega)$ for the same set of parameters as used in figure 3. The filled symbols indicate the value of $\Gamma_{\mathbf{k}}(\omega)$ and position of the crossings $\omega = E_{\mathbf{k}}$ of the lower Bogoliubov mode (squares), in-gap state (circles) and the upper Bogoliubov branch (triangles).

Since Γ_1 does not vary above T_c, the temperature dependence of $\Gamma_{\mathbf{k}}^{-1}(\omega_i)$ is controlled by Γ_0. Using the experimental estimations [31, 32], we thus find the qualitatively opposite temperature variations of $\Gamma_{\mathbf{k}}(\omega_\pm)$ and $\Gamma_{\mathbf{k}}(\omega_0)$ shown in figure 5. These quantities correspond to the life-times of quasiparticles and, therefore, we conclude that:

a) in-gap quasiparticles are forbidden for the superconducting state due to vanishing $\Gamma_{\mathbf{k}}^{-1}(\omega_0) = 0$ (in other words, spectrum consists of just the Bogoliubov modes typical of the BCS theories),

b) in the pseudogap state above T_c, where $\Delta \neq 0$ and $\Gamma_0 \neq 0$, besides the Bogoliubov branches there emerge in-gap states which initially at T_c^+ represent the heavily overdamped modes.

At a first glance, our conclusions seem to be in conflict with the ARPES data, which have not reported any pronounced in-gap features. Nevertheless, various studies of the pseudogap clearly revealed a rather negligible temperature dependence of $\Delta(T)$ upon passing T_c (at least for the anti-nodal areas). Instead of closing this gap, the low energy states are gradually filled-in [25]. Such a behavior can be thought as an indirect signature of the in-gap states, which for increasing temperatures absorb more and more spectral weight. To support this conjecture, we illustrate in figure 6 an ongoing transfer of the spectral weights between the Bogoliubov quasiparticles and in-gap states. Using (3.11), we show the spectral function $A(\mathbf{k}, \omega)$ subtracting its value at T_c in analogy to the detailed experimental discussion by T. Kondo et al. [16]. In-gap states emerge around ω_0 and gradually gain the spectral weight (figure 7) simultaneously increasing their life-time.

Intrinsic broadening of the in-gap states [53] unfortunately obscures their observation by the spectroscopic tools at temperatures close to T_c. These states might be, however, probed indirectly. T. Senthil and P.A. Lee [22, 23] suggested that such states could be responsible for the magnetooscillations observed ex-

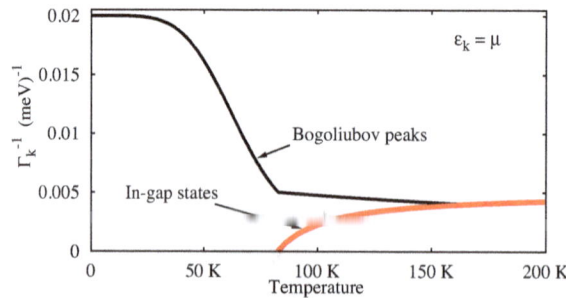

Figure 5. (Color online) Temperature dependence of the inverse broadening $\Gamma_{\mathbf{k}}^{-1}$ which corresponds to the effective life-time of the Bogoliubov modes (thin line) and the in-gap states (thick curve) obtained for $\varepsilon_{\mathbf{k}} = \mu$.

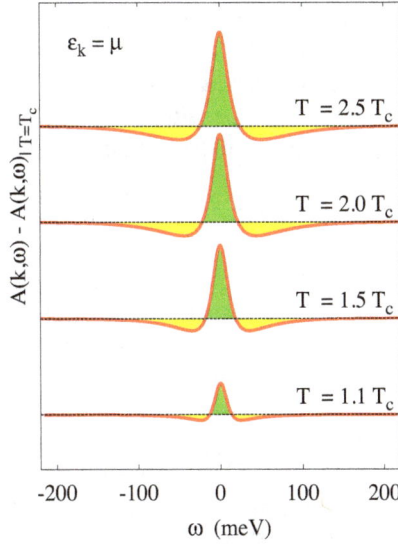

Figure 6. (Color online) Transfer of the spectral weight from the Bogoliubov quasiparticle peaks towards the in-gap states obtained using (3.11) for $\varepsilon_{\mathbf{k}} = \mu$. Temperature dependence of the total transferred spectral weight is shown in figure 7.

perimentally by N. Doiron-Leyraud et al. [51, 52]. They indicated that pair-coherence extending only over short spatial- and temporal length naturally implies the pair decay (scattering) into the in-gap fermion states. This line of reasoning has been also followed by some other groups [53, 54].

4. Microscopic toy model

Pairing of the cuprate superconductors occurs on a local scale, practically between the nearest neighbor lattice sites. To account for an interplay between the paired and unpaired charge carriers taking place in the pseudogap regime we consider here the following simplified picture

$$\hat{H}_{\text{loc}} = \varepsilon_0 \sum_{\sigma} \hat{c}_{\sigma}^{\dagger} \hat{c}_{\sigma} + E_0 \hat{b}^{\dagger} \hat{b} + \left(\Delta \hat{b}^{\dagger} \hat{c}_{\downarrow} \hat{c}_{\uparrow} + \text{h.c.} \right), \tag{4.1}$$

where $\hat{c}_{\sigma}^{(\dagger)}$ correspond to the unpaired fermions and $\hat{b}^{(\dagger)}$ to the pairs (hard-core bosons). We assume that in the pseudogap state, neither the fermions nor the hard-core boson pairs are long-living because of their mutual scattering by the Andreev charge exchange term. The same type of scattering, although in

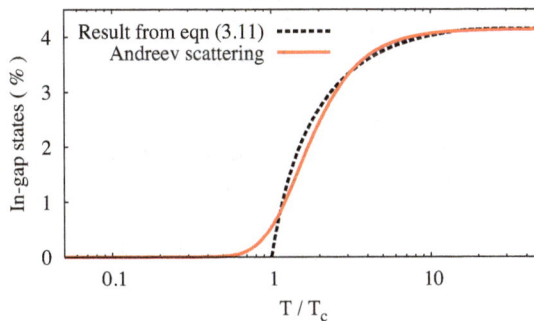

Figure 7. (Color online) Spectral weight corresponding to the in-gap states obtained from the phenomenological ansatz (3.11) for $\varepsilon_{\mathbf{k}} = \mu$ (dotted curve) and solution of the toy model (4.1) for $\varepsilon_0 = 0 = E_0$ (solid line).

the momentum space, has been considered in reference [22, 23] within the lowest order diagrammatic treatment. On a microscopic footing, the Hamiltonian (4.1) can be regarded as the effective low energy description of the plaquettized Hubbard model [47, 48].

Neglecting the itinerancy of the charge carriers, we can obtain a rigorous solution for a given local cluster (not to be confused with the individual copper sites in CuO_2 planes [50]). Exact diagonalization of the Hilbert space yields the following single particle Green's function [42]

$$G(\omega) = \frac{\mathcal{Z}_{QP}}{\omega - \varepsilon_0} + \frac{1 - \mathcal{Z}_{QP}}{\omega - \varepsilon_0 - \frac{|\Delta|^2}{\omega + \varepsilon_0 - E_0}}.$$ (4.2)

Let us notice that the second term on rhs of (4.2) acquires the same structure as imposed by (3.11). In the present case, no imaginary terms appear but the structure of the Green's function (4.2) mimics the role of Γ_0. Formally, it describes the bonding and antibonding states originating from the Andreev scattering and besides that we also have a remnant of the non-interacting propagator $[\omega - \varepsilon_0]^{-1}$ whose spectral weight is given by $\mathcal{Z}_{QP}$.

The quasiparticle weight $\mathcal{Z}_{QP}$ depends on occupancies of the fermion and boson levels. As an example, we explore here the symmetric (i.e., half-filled) case with $\varepsilon_0 = 0$ and $E_0 = 0$ when $\mathcal{Z}_{QP} = 2/[3 + \cosh(|\Delta|/k_B T)]$. Assuming the typical ratio $|\Delta|/k_B T_c = 4$, we plot in figure 7 the temperature dependence of the unpaired states contribution $\mathcal{Z}_{QP}$ to the spectrum. We find a very good agreement between our simple treatment and the estimations using the self-energy (3.11). It means that the parameter Γ_0 introduced in references [31, 32] and the local Andreev-type scattering considered here account for the very same particle-hole processes inducing the in-gap states. Transfer of the spectral weight from the paired to unpaired states (figure 7) confirms the qualitative agreement over a broad temperature region and the indication for the same critical point.

For more realistic comparison of the present study with the experimental data [16], one obviously has to consider the itinerant charge carriers. As a natural improvement of the local solution (4.2) we would expect the following type of Green's function

$$G(\mathbf{k}, \omega) = \frac{\mathcal{Z}_{QP}(\mathbf{k})}{\omega - \varepsilon_\mathbf{k}} + \sum_\mathbf{q} \frac{[1 - \mathcal{Z}_{QP}(\mathbf{k})] f(\mathbf{q}, \mathbf{k})}{\omega - \varepsilon_\mathbf{k} - \frac{|\Delta_\mathbf{k}|^2}{\omega + \varepsilon_{\mathbf{q}-\mathbf{k}} - E_\mathbf{q}}},$$ (4.3)

where T-dependent coefficients $f(\mathbf{q}, \mathbf{k})$ should be determined via the many-body techniques. Approaching T_c from above the predominant influence comes from $\mathbf{q} \to \mathbf{0}$ bosons and then we notice that (4.3) reduces to the ansatz (3.11). Such results have been recently reported from the dynamical mean field calculations for the Hubbard model [55, 56].

5. Conclusions and outlook

We have shown that the pairing ansatz (3.11), widely used for fitting the experimental ARPES profiles, above T_c corresponds to the pair scattering inducing the single particle fermion states inside the pseudogap. Temperature dependent phenomenological parameter Γ_0 is found to control the transfer of the spectral weight from the Bogoliubov quasiparticles to the unpaired in-gap states. To model such a process on a microscopic level, we have considered the scenario in which the local pairs are scattered into single fermions via the Andreev conversion [22, 23, 47, 48, 50]. We have found a unique relation between the transferred spectral weight (from the paired to unpaired quasiparticles) with the non-bonding state $\mathcal{Z}_{QP}$. It would be instructive to extend the present analysis onto the case of $\mathbf{k}$-dependent energy gap. Such a problem would be closely related to the issue of Fermi arcs, i.e., partially reconstructed pieces of the Fermi surface, and to nontrivial angular dependence of the pseudogap [1].

Acknowledgements

Author is indebted for the fruitful discussions with Adam Kamiński, Roman Micnas, Julius Ranninger, and Karol I. Wysokiński. This work is supported by the National Science Centre in Poland through the projects DEC-2014/13/B/ST3/04451.

References

1. Kaminski A., Kondo T., Takeuchi T., Gu G., Philos. Mag. B, 2014, **95**, 453; doi:10.1080/14786435.2014.906758.
2. Gaebler J.P., Stewart J.T., Drake T.E., Jin D.S., Perali A., Pieri P., Strinati G.C., Nat. Phys., 2010, **6**, 569; doi:10.1038/nphys1709.
3. Uemura Y.J., Le L.P., Luke G.M., Sternlieb B.J., Wu W.D., Brewer J.H., Riseman T.M., Seaman C.L., Maple M.B., Ishikawa M., Hinks D.G., Jorgensen J.D., Saito G., Yamochi H., Phys. Rev. Lett., 1991, **66**, 2665; doi:10.1103/PhysRevLett.66.2665.
4. Orenstein J., Corson J., Oh S., Eckstein J.N., Ann. Phys. (Leipzig), 2006, **15**, 596; doi:10.1002/andp.200510202.
5. Corson J., Mallozzi R., Orenstein J., Eckstein J.N., Bozovic I., Nature, 1999, **398**, 221; doi:10.1038/18402.
6. Wang Y., Li L., Ong N.P., Phys. Rev. B, 2006, **73**, 024510; doi:10.1103/PhysRevB.73.024510.
7. Xu Z.A., Ong N.P., Wang Y., Takeshita T., Uchida S., Nature, 2000, **406**, 486; doi:10.1038/35020016.
8. Kanigel A., Chatterjee U., Randeria M., Norman M.R., Koren G., Kadowaki K., Campuzano J., Phys. Rev. Lett., 2008, **101**, 137002; doi:10.1103/PhysRevLett.101.137002.
9. Shi M., Bendounan A., Razzoli E., Rosenkranz S., Norman M.R., Campuzano J.C., Chang J., Mansson M., Sassa Y., Claesson T., Tjernberg O., Patthey L., Momono N., Oda M., Ido M., Guerrero S., Mudry C., Mesot J., Eur. Phys. Lett., 2009, **88**, 27008; doi:10.1209/0295-5075/88/27008.
10. Lee J., Fujita K., Schmit A.R., Kim C.K., Eisaki H., Uchida S., Davis J.C., Science, 2009, **325**, 1099; doi:10.1126/science.1176369.
11. Bergeal N., Lesueur J., Aprili M., Faini G., Contour J.P., Leridon B., Nat. Phys., 2008, **4**, 608; doi:10.1038/nphys1017.
12. Yuli O., Asulin I., Kalchaim Y., Koren G., Millo O., Phys. Rev. Lett., 2009, **103**, 197003; doi:10.1103/PhysRevLett.103.197003.
13. Li L., Wang Y., Komiya S., Ono S., Ando Y., Gu G.D., Ong N.P., Phys. Rev. B, 2010, **81**, 054510; doi:10.1103/PhysRevB.81.054510.
14. Iye T., Nagatochi T., Ikeda R., Matsuda A., J. Phys. Soc. Jpn., 2010, **79**, 114711; doi:10.1143/JPSJ.79.114711.
15. Bernardi E., Lascilfari A., Ragimonti A., Romano L., Scavini M., Oliva C., Phys. Rev. B, 2010, **81**, 064502; doi:10.1103/PhysRevB.81.064502.
16. Kondo T., Khasanov R., Takeuchi T., Schmalian J., Kaminski A., Nature, 2009, **457**, 296; doi:10.1038/nature07644.
17. Dubroka A., Rössle M., Kim K.W., Malik V.K., Munzar D., Basov D.N., Schafgans A.A., Moon S.J., Lin C.T., Haug D., Hinkov V., Keimer B., Wolf Th., Storey J.G., Tallon J.L., Bernhard C., Phys. Rev. Lett., 2011, **106**, 047006; doi:10.1103/PhysRevLett.106.047006.
18. Kirzhner T., Koren G., Sci. Rep., 2014, **4**, 6244; doi:10.1038/srep06244.
19. Moon S.J., Lee Y.S., Schafgans A.A., Chubukov A.V., Kasahara S., Shibauchi T., Terashima T., Matsuda Y., Tanatar M.A., Prozorov R., Thaler A., Canfield P.C., Bud'ko S.L., Sefat A.S., Mandrus D., Segawa K., Ando Y., Basov D.N., Phys. Rev. B, 2014, **90**, 014503; doi:10.1103/PhysRevB.90.014503.
20. Cilento F., Dal Conte S., Coslovich G., Peli S., Nembrini N., Mor S., Banfi F., Ferrini G., Eisaki H., Chan M.K., Dorow C.J., Veit M.J., Greven M., van der Marel D., Comin R., Damascelli A., Rettig L., Bovensiepen U., Capone M., Giannetti C., Parmigiani F., Nat. Commun., 2014, **5**, 4353; doi:10.1038/ncomms5353.
21. Franz M., Nat. Phys., 2007, **3**, 686; doi:10.1038/nphys739.
22. Senthil T., Lee P.A., Phys. Rev. Lett., 2009, **103**, 076402; doi:10.1103/PhysRevLett.103.076402.
23. Senthil T., Lee P.A., Phys. Rev. B, 2009, **79**, 245116; doi:10.1103/PhysRevB.79.245116.
24. Mamedov T.A., de Llano M., Philos. Mag. B, 2014, **94**, 4102; doi:10.1080/14786435.2014.979903.
25. Fischer O., Kugler M., Maggio-Aprile I., Berthod Ch., Renner Ch., Rev. Mod. Phys., 2007, **79**, 353; doi:10.1103/RevModPhys.79.353.
26. Abrahams E., Red M., Woo J.W.F., Phys. Rev. B, 1970, **1**, 208; doi:10.1103/PhysRevB.1.208.
27. Schmid A., Z. Phys., 1970, **231**, 324; doi:10.1007/BF01397514.
28. Ranninger J., Robin J.M., Physica C, 1995, **253**, 279; doi:10.1016/0921-4534(95)00515-3.
29. Emery V.J., Kivelson S.A., Nature, 1995, **374**, 434; doi:10.1038/374434a0.
30. Tchernyshyov O., Phys. Rev. B, 1997, **56**, 3372; doi:10.1103/PhysRevB.56.3372.
31. Norman M.R., Randeria M., Ding H., Campuzano J.C., Phys. Rev. B, 1998, **57**, 11093; doi:10.1103/PhysRevB.57.R11093.
32. Franz M., Millis A.J., Phys. Rev. B, 1998, **58**, 14572; doi:10.1103/PhysRevB.58.14572.
33. Fujimoto S., J. Phys. Soc. Jpn., 2002, **71**, 1230; doi:10.1143/JPSJ.71.1230.
34. Domański T., Ranninger J., Phys. Rev. Lett., 2003, **91**, 255301; doi:10.1103/PhysRevLett.91.255301.
35. Domański T., Ranninger J., Phys. Rev. B, 2001, **63**, 134505; doi:10.1103/PhysRevB.63.134505.
36. Chubukov A.V., Norman M.R., Millis A.J., Abrahams E., Phys. Rev. B, 2007, **76**, 180501; doi:10.1103/PhysRevB.76.180501.
37. Chen Q.J., Stajic J., Tan S.N., Levin K., Phys. Rep., 2005, **412**, 1; doi:10.1016/j.physrep.2005.02.005.

38. Cichy A., Micnas R., Ann. Phys. (New York), 2014, **347**, 207; doi:10.1016/j.aop.2014.04.014.

39. Pizio O., Sokołowski S., Sokołowska Z., J. Chem. Phys., 2014, **140**, 174706; doi:10.1063/1.4873438.

40. Matsui H., Sato T., Takahashi T., Wang S.-C., Yang H.-B., Ding H., Fujii T., Watanabe T., Matsuda A., Phys. Rev. Lett., 2003, **90**, 217002; doi:10.1103/PhysRevLett.90.217002.

41. Chatterjee U., Shi M., Ai D., Zhao J., Kanigel A., Rosenkranz S., Raffy H., Li Z.Z., Kadowaki K., Hinks D.G., Xu Z.J., Wen J.S., Gu G., Lin C.T., Claus H., Norman M.R., Randeria M., Campuzano J., Nat. Phys., 2009, **5**, 1456; doi:10.1038/nphys1200.

42. Domański T., Phys. Rev. A, 2011, **84**, 023634; doi:10.1103/PhysRevA.84.023634.

43. Friedberg R., Lee T.D., Phys. Rev. B, 1989, **40**, 6740; doi:10.1103/PhysRevB.40.6745.

44. Micnas R., Ranninger J., Robaszkiewicz S., Rev. Mod. Phys., 1990, **62**, 113; doi:10.1103/RevModPhys.62.113.

45. Micnas R., Phys. Rev. B, 2007, **76**, 184507; doi:10.1103/PhysRevB.76.184507.

46. Geshkenbein V.B., Ioffe L.B., Larkin A.I., Phys. Rev. B, 1997, **55**, 3173; doi:10.1103/PhysRevB.55.3173.

47. Altman E., Auerbach A., Phys. Rev. B, 2002, **65**, 104508; doi:10.1103/PhysRevB.65.104508.

48. Mihlin A., Auerbach A., Phys. Rev. B, 2009, **80**, 134521; doi:10.1103/PhysRevB.80.134521.

49. Le Hur K., Rice T.M., Ann. Phys. (New York), 2009, **324**, 1452; doi:10.1016/j.aop.2009.02.004.

50. Ranninger J., Domański T., Phys. Rev. B, 2010, **81**, 014514; doi:10.1103/PhysRevB.81.014514.

51. Doiron-Leyraud N., Proust C., LeBoeuf D., Levallais J., Bonnemaison J.-B., Liang R., Bonn D.A., Hardy W.N., Taillefer L., Nature, 2007, **447**, 565; doi:10.1038/nature05872.

52. Taillefer L., J. Phys.: Condens. Matter, 2009, **21**, 164212; doi:10.1088/0953-8984/21/16/164212.

53. Ranninger J., Romano A., Phys. Rev. B, 2010, **82**, 054508; doi:10.1103/PhysRevB.82.054508.

54. Micklitz T., Norman M.R., Phys. Rev. B, 2009, **80**, 220513; doi:10.1103/PhysRevB.80.220513.

55. Imada M., Yamaji Y., Sakai S., Motome Y., Ann. Phys. (Berlin), 2001, **523**, 629; doi:10.1002/andp.201100028.

56. Sakai S., Civelli M., Nomura Y., Imada M., Phys. Rev. B, 2015, **92**, 180503(R); doi:10.1103/PhysRevB.92.180503.

11

Modelling and measurements of fibrinogen adsorption on positively charged microspheres

P. Żeliszewska[1], A. Bratek-Skicki[1], Z. Adamczyk[1], M. Cieśla[2]

[1] J. Haber Institute of Catalysis and Surface Chemistry Polish Academy of Sciences, Niezapominajek 8, 30-239 Cracow, Poland

[2] M. Smoluchowski Institute of Physics, Jagiellonian University, Łojasiewicza 11, 30-348 Cracow, Poland

Adsorption of fibrinogen on positively charged microspheres was theoretically and experimentally studied. The structure of monolayers and the maximum coverage were determined by applying the experimental measurements at pH = 3.5 and 9.7 for NaCl concentration in the range of $10^{-3} - 0.15$ M. The maximum coverage of fibrinogen on latex particles was precisely determined by the AFM method. Unexpectedly, at pH = 3.5, where both fibrinogen molecule and the latex particles were positively charged, the maximum coverage varied between 0.9 mg m^{-2} and 1.1 mg m^{-2} for 10^{-2} and 0.15 M NaCl, respectively. On the other hand, at pH = 9.7, the maximum coverage of fibrinogen was larger, varying between 1.8 mg m^{-2} and 3.4 mg m^{-2} for 10^{-2} and 0.15 M NaCl, respectively. The experimental results were quantitatively interpreted by the numerical simulations.

Key words: *fibrinogen adsorption, positively charged microspheres*

1. Introduction

Protein adsorption on various surfaces has received a considerable attention due to its importance in biomedical fields. New biotechnological methods of protein production, purification and separation depend on their interfacial properties. Furthermore, the protein adsorption at solid/liquid interfaces enabled the development of diverse biomedical applications, such as biosensors, immunological tests, and drug delivery systems. On the other hand, in biomaterial field, protein adsorption is undesirable because it can cause adverse responses such as blood coagulation and complement activation [1].

Fibrinogen (Fb) is a major serum blood protein that plays an essential role in the clotting cascade initiated by thrombin. Fibrinogen molecules exhibit a strong tendency to adsorb on various surfaces under broad range of conditions [2, 3] mediating cellular interactions that are a key event in determining biocompatibility of these materials [4]. On the other hand, fibrinogen monolayers adsorbed on various synthetic materials may promote platelet adhesion that often leads to fouling of artificial organs [6, 7].

Due to its fundamental significance, the chemical structure of fibrinogen and its bulk physicochemical properties have been extensively studied [8–15]. It was established [9] that the fibrinogen molecule is a symmetric dimer composed of three identical pairs of polypeptide chains, referred to as Aα, Bβ and γ chains [10]. They are coupled in the middle of the molecule through a few disulfide bridges forming a central E nodule. The longest Aα chain is composed of 610 amino acids, the Bβ chain comprises 460 amino-acids and the γ chain 411 amino acids. Accordingly, the molar mass of the fibrinogen molecule is equal to 338 kDa [11]. It is interesting to mention that a considerable part of the Aα chains extends from the core of the molecule forming two polar appendages (arms) called the αC domains [12].

Information on fibrinogen's molecule shape and dimensions was derived from electron [13–16] and atomic force microscopy (AFM) [17–22]. It was established that the molecule has a co-linear, trinodular shape with a total length of 47.5 nm [13]. The two equal end domains of rather irregular shape are approximated by spheres having a diameter of 6.5 nm; and the middle domain has a diameter of 5 nm.

Additionally, in references [23, 24] it was predicted that the length of the side arms is equal to 18 nm and that they are positively charged in the αC domains for the broad range of pHs from 3.5 to 9.7.

Due to its significance, the adsorption of fibrinogen at solid/electrolyte interfaces has been investigated by various experimental methods such as ellipsometry [25], Total Internal Reflection (TIRF) [26], AFM [27], radiolabeling [28], Quartz Crystal Microbalance (QCM) [29], streaming potential [23], etc. However, only in the work of Brash et al. [4] positively charged surfaces were investigated, prepared by controlled adsorption of polycations on a glass substrate. The radiolabeled fibrinogen was used to monitor the kinetics of adsorption. It is postulated that three populations of fibrinogen appear in the adsorbed monolayers: non- exchanging, slowly, and rapidly exchanging. It was shown that for positively charged surfaces, the amount of fast and slow exchanging molecules is oppositely different from the neutral and negatively charged surfaces.

In contrast to the vast literature dealing with other proteins, only a few studies focused on fibrinogen adsorption on polymeric microspheres have been reported [30, 31]. In reference [32], the effect of ionic strength in the adsorption of fibrinogen on polystyrene microspheres (latex) at pH = 7.4 was studied. The electrophoretic mobility measurements were performed to control the progress of protein adsorption under in situ conditions. The coverage of the protein was determined by a concentration depletion method involving AFM imaging of residual fibrinogen. It was revealed that the maximum coverage of fibrinogen on latex varied between 1.9 and 3.2 mg m^{-2} for 10^{-3} and 0.15 M NaCl concentration, respectively. In a recent publication [31], adsorption of fibrinogen at positively charged amidine microspheres was studied at pH of 7.4. Quite unexpectedly, the maximum coverage of the protein was much smaller than for negatively charged latex, equal to 0.6 and 1.3 mg m^{-2} for 10^{-3} and 0.15 M NaCl, respectively. This anomalous result was interpreted in terms of the random sequential adsorption model, by postulating a side-on adsorption of fibrinogen molecules at the latex surface.

The main goal of this work is to systematically study fibrinogen adsorption at positively charged microspheres at pH = 3.5 and 9.7 in order to quantitatively determine the maximum coverage as a function of this parameter. These results, supplemented by electrophoretic mobility measurements furnishing the charge density data, allow one to elucidate fibrinogen adsorption mechanisms for a broad range of pH that has a significance for basic science. This is also of a practical interest for developing a robust procedure of preparing stable fibrinogen monolayers at latex particles of a well-controlled coverage and a known orientation of molecules. Such latex/fibrinogen complexes can be exploited to study interactions with other proteins (antibodies) and low molar mass ligands.

2. Materials and methods

2.1. Experimental

Fibrinogen from human blood plasma, fraction I, type IV used for our study was purchased from Sigma (F4753) in the form of a powder. The sample contains 65% protein, 15% sodium citrate, and 25% sodium chloride. The dynamic surface tension method was used to check the purity of fibrinogen solution. The bulk concentration of the fibrinogen in electrolyte solutions was determined by the BCA method (Bicinchoninic acid Protein Assays) [33].

High-Performance Liquid Chromatography (HPLC) experiments were performed using an Knauer system. The column was packed with a composite of cross-linked agarose and dextran. The flow rate was 0.5 ml/min. The absorptiometric detection was monitored at 280 nm.

The water which was used in all experiments was purified by the Milipore Elix 5 instruments. All other reagents were purchased from Sigma–Aldrich and used without any purification.

Amidine microspheres (latex) used in our measurements were purchased from Invitrogen. This latex was positively charged, surfactant free with concentration equal to 3.7% and nominal size of 800 nm. The pH of Fb solutions and latex suspension was regulated within the range by the addition of HCl or NaOH. Buffers were not used for experiments, due to their specific adsorption on monolayers.

The dynamic light scattering (DLS) was used to determine diffusion coefficients of fibrinogen and latex particles. On the other hand, the electrophoretic mobility of fibrinogen and fibrinogen-covered microspheres was measured by the Laser Doppler Velocimetry (LDV) technique. The diffusion coefficient and the elctrophoretic mobility were measured using the Zetasizer Nano ZS Malvern instruments. The concentration depletion method described in reference [34] was used to determine the excess of fibrinogen after adsorption on the latex. This method consisted of several stages: the experiment was started by transferring the fibrinogen latex mixture after the adsorption step to the diffusion cell. Then, a few mica sheets were immersed in the suspension for 30 minutes (adsorption time). Subsequently, the fibrinogen covered mica sheets were rinsed using a pure electrolyte with the same pH and ionic strength as for the adsorption of fibrinogen on latex particles. The AFM imaging was used to determine the average number of fibrinogen molecules adsorbed over equal sized surface areas randomly selected over the mica sheets.

2.2. Theoretical modelling

The modelling of fibrinogen monolayer formation on microparticle surfaces was carried out by applying the random sequential adsorption (RSA) approach developed in references [35, 36] for quantifying irreversible adsorption proteins (ferritin) on flat interfaces. In these calculations, the specific interactions among protein molecules were neglected and their shape was approximated by a circular disk. Later on, the RSA model was extensively used for calculating the kinetics, the maximum (jamming) coverage and the monolayer structure of non-spherical particles of various shapes [37–40]. However, all these results were obtained for hard (non-interacting) particles and flat interfaces of infinite extension, by neglecting the curvature effect that can influence both the structure and the maximum coverage of monolayers.

The general rules of the Monte-Carlo simulation scheme based on the RSA approach are as follows [36, 38]:

(i) a virtual particle (molecule) is created, whose position within the simulation domain and orientation are selected at random with a probability depending on the interaction energy,

(ii) if the particle fulfills the pre-defined adsorption criteria it becomes irreversibly deposited and its position remains unchanged during the entire simulation process,

(iii) if the deposition criteria are violated, a new adsorption attempt is made, uncorrelated with the previous attempts.

Usually, two major deposition criteria are defined: (i) there should be no overlapping of any part of the virtual particle with the previously adsorbed particles and (ii) there should be a physical contact of the particle with the interface. Despite the simplicity of the governing rules, the RSA modelling is a powerful tool for efficiently producing populations of a large number of molecules N_p (often exceeding 10^5). It is also flexible because adsorption processes of anisotropic particles at interface of various geometry, for example on spherical microparticles, can be efficiently treated. In this work, adsorption of fibrinogen on microparticle surfaces is simulated by using the bead Model B (see figure 1), previously used for describing adsorption at flat and negatively charged interfaces [23]. In this model, the shape of the fibrinogen molecule is approximated by a string of 23 co-linear touching spheres of various diameters. The two external spheres have diameters of 6.7 nm and the central sphere has a diameter of 5.3 nm. The side arms are approximated as straight sequences of 12 beads of equal size, having also the diameter of 1.5 nm. These side chains form the angle φ with the core part of the fibrinogen molecule. Electric charges denoted by q_1, q_2, q_3 (see table 1) are attributed to various parts of the fibrinogen molecule as

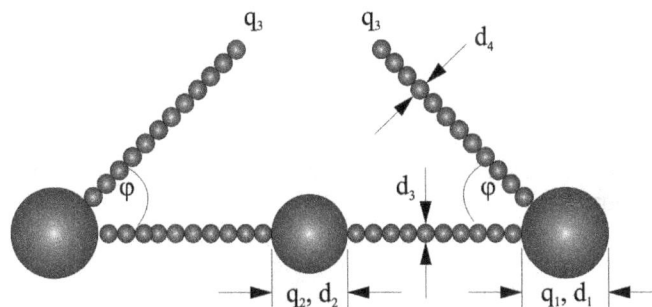

Figure 1. The bead model of the fibrinogen molecule [24], $d_1 = 6.7$ nm, $d_2 = 5.3$ nm, $d_3 = d_4 = 1.5$ nm. $\varphi = 56°$ (pH = 7.4).

Table 1. Model parameters used in numerical simulations. Diameters of the spheres: $d_1 = 6.7$ nm, $d_2 = 5.3$ nm, $d_3 = d_4 = 1.5$ nm were the same for all pH's and ionic strengths.

pH	Ionic strenght [M]	q_1 [e]	q_2 [e]	q_3 [e]	ϕ deg
3.5	10^{-2}	−6	0	0	95
	0.15	−8	0	0	
9.7	10^{-3}	−10	0	0	
	10^{-2}	−20	0	0	30
	0.15	−31	0	0	

shown in figure 1. The magnitude of these charges can be estimated from the electrophoretic mobility measurements as discussed later on.

The lateral electrostatic interactions between the adsorbed fibrinogen molecules were accounted for by using the Yukawa (screened Coulomb) pair potential ϕ given by

$$\phi(r_{ij}) = \frac{q_i q_j}{4\pi\epsilon r_{ij}} \exp\left[-\kappa\left(r_{ij} - \frac{d_i}{2} - \frac{d_j}{2}\right)\right],$$ (2.1)

where q_i, q_j are the effective (electrokinetic) charges of the two beads i and j, r_{ij} is the distance between the bead centers having the diameters d_i and d_j, and $\kappa^{-1} = \left(\epsilon k_B T / 2e^2 I\right)^{1/2}$ is the electrical double-layer thickness, ϵ is the permittivity of the medium, k_B is the Boltzmann constant, T is the absolute temperature, e is the elementary charge and I is the ionic strength.

In order to calculate the net interaction between the molecules, the pair potentials given by equation (2.1) were evaluated over the interaction area, significantly exceeding the double-layer thickness (by excluding interaction of the beads belonging to the same fibrinogen molecule).

The model molecules were adsorbed according to the above RSA scheme on a spherically shaped interface whose radius of curvature exactly matched the dimension of the latex particles used in adsorption experiments. The electrostatic interactions of the protein with the latex surface were assumed to be of the square well (perfect sink) type.

Typically, in one simulation run, 1500 molecules were generated. Therefore, in order to improve the statistics, the averages from ca. 50 independent runs were taken, with the total number of particles exceeding 70 000. This ensures a relative precision of the simulation better that 0.5%. The primary parameter derived from these simulations was the average number of molecules adsorbed on latex particles calculated as a function of time. By extrapolating this dependence to infinite time using the procedure previously described [40] one obtains the maximum number of molecules N_p adsorbed on the latex particles. Consequently, the surface concentration of the protein is $N_p/\Delta S$ (where $\Delta S = \pi d_1^2$ is the surface area of the latex particle of the diameter d_1) and the dimensionless coverage is calculated from the formula

$$\Theta = \frac{S_g N_p}{\Delta S},$$ (2.2)

where S_g is the characteristic cross-section area of the protein molecule.

Knowing N_p, the dimensional coverage commonly used for the interpretation of experimental data is calculated from the following dependence:

$$\Gamma_f = \frac{M_w}{A_v} N_p,$$ (2.3)

where M_w is the molar mass of fibrinogen and A_v is the Avogadros's constant.

3. Results and discussion

3.1. Fibrinogen and latex characteristics in the bulk

As mentioned before, the diffusion coefficient of the microspheres was measured by the DLS for various ionic strengths and pHs. Knowing the diffusion coefficient, the hydrodynamic diameter d_l [nm] was calculated using the Stokes-Einstein relationship. In this way, the hydrodynamic diameter of microspheres was determined to be 860 ± 15 nm, for 10^{-3} M, 830 ± 10 nm, for 10^{-2} M, and 815 ± 10 nm for 0.15 M.

The electrophoretic mobility μ_e of microspheres was determined by using the laser Doppler velocimetry (LDV) technique. Zeta potential values of latex particles for ionic strength 10^{-3} M, 10^{-2} M, 0.15 M NaCl and pH = 3.5 are equal to 78 mV, 87 mV and 40 mV, respectively. Knowing the zeta potential of latex particles, one can calculate its uncompensated charge using the Gouy-Chapman relationship for a symmetric 1:1 electrolyte [34]

$$\sigma_0 = \frac{(8\epsilon k_B T n_b)^{\frac{1}{2}}}{0.160} \sinh\left(\frac{e\zeta}{2k_B T}\right), \tag{3.1}$$

where σ_0 is the electrokinetic charge density of latex particles expressed in e nm^{-2} and n_b is the number concentration of the salt (NaCl) expressed in m^{-3}.

By using the experimental zeta potential values one obtains for pH = 3.5 from equation (3.1) $\sigma_0 = 0.057, 0.19$ and 0.25 e nm^{-2} for the NaCl concentration of $10^{-3}, 10^{-2}$ and 0.15 M, respectively. Analogously, for pH = 9.7, $\sigma_0 = 0.042$ and 0.14 for the NaCl concentration of 10^{-3} and 0.15 M, respectively (see table 2).

Table 2. Electrophoretic mobility, zeta potential (calculated from the Henry's model) and charge density of latex and human serum fibrinogen.

pH	Ionic strength [M]	Latex A800			Human serum fibrinogen			
		μ_e [μm cm / Vs]	ζ_l [mV]	σ_0 [nm^{-2}]	μ_e [μm cm / Vs]	ζ_f [mV]	N_c^*	N_c^{**}
3.5	1.3×10^{-3}	5.8	78	0.057	1.4	25	15	33
	10^{-2}	6.7	87	0.19	0.94	16	10	42
	0.15	3.1	40	0.25	0.52	9.0	5.7	63
7.4[†]	10^{-3}	5.3	71	0.044	−0.94	−18	−10	−21
	10^{-2}	6.2	81	0.17	−0.56	−9.8	−6.2	−25
	0.15	2.6	34	0.20	−0.30	−4.4	−3.3	−37
9.7	10^{-3}	5.2	70	0.042	−0.91	−17	−10	−20
	10^{-2}	5.5	72	0.14	−0.91	−16	−10	−40
	0.15	1.8	24	0.14	−0.51	−7.6	−5.6	−62

Footnotes:
$^*N_c = Q_c^0/1.602 \times 10^{-19}$, $^{**}N_c = Q_c/1.602 x \times 10^{-19}$,   [†] Previous results obtained in reference [31].

The electrophoretic mobility of fibrinogen molecules was determined previously [31]. These data are also given in table 2. As can be seen, the electrophoretic mobility of fibrinogen was positive for pH = 3.5 and negative for pH = 7.4 and 9.7. By using the electrophoretic data, one can calculate the electrokinetic charge of the fibrinogen molecule Q_c^0 (expressed in Coulombs) from the Lorenz-Stokes relationship [23, 41]

$$Q_c^0 = 3\pi d_H \eta \mu, \tag{3.2}$$

where d_H is the hydrodynamic diameter of fibrinogen, η is the dynamic viscosity of the solvent (water).

It should be mentioned that equation (3.2) is valid for molecules of arbitrary shape but its accuracy is limited for higher ionic strengths. Therefore, in this case, the following equation was used, valid for an arbitrary ionic strength and spherical particles [34]

$$Q_c = \frac{2}{3} Q_c^0 \frac{1 + \kappa d_H}{f_H(\kappa d_H)}, \tag{3.3}$$

where $f_H(\kappa d_H)$ is the Henry's function.

The results calculated from equations (3.1)–(3.3), converted to the number of charges N_c are collected in table 2. As can be noticed, at pH of 3.5, fibrinogen molecule acquires a net positive charge, whereas at higher pHs of 7.4 and 9.7, the charge is highly negative. It should also be mentioned that from the electrophoretic mobility measurements alone one cannot predict in a unique way the charge distribution among various parts of fibrinogen molecule. However, from the diffusion coefficient and dynamic viscosity measurements reported in references [24] it is predicted that the positive charge is concentrated in the end parts of the side arms, which means that q_3 remains positive for pH range 3.5–9.7. Accordingly, q_1 and q_2 should be negative. However, a precise charge distribution at this pH is not known. This can only be empirically determined form the thorough adsorption experiments reported below.

3.2. Fibrinogen adsorption on microspheres

Fibrinogen adsorption on polymeric microspheres was monitored by measuring the changes in electrophoretic mobility (zeta potential) induced by this process. The steps of the experiment were as follows:

(i) measurement of the zeta potential of bare latex particles whose concentration c_l was equal to 60 mg L^{-1},

(ii) adsorption of fibrinogen on latex particles by filling the cell with the fibrinogen solution of an opportune concentration c_f (0.1–5 mg L^{-1}) for 600 seconds,

(iii) purification of the latex suspension by using a membrane filtration and measurement of the electrophoretic mobility of latex using the electrophoretic method.

This procedure is reproducible and allows one to determine the changes of zeta potential as a function of fibrinogen concentration added to the latex suspension.

The fibrinogen coverage on latex particles was calculated by using the following formula

$$\Gamma_f = 10^{-3} \frac{v_s c_f}{S_l}, \tag{3.4}$$

where Γ_f is the fibrinogen coverage on microspheres (expressed in mg m^{-2}), v_s is the volume of suspension mixture, c_f is the initial concentration of fibrinogen in the suspension (after mixing with the latex), and S_l is the surface area of latex expressed in m^2, given by $S_l = 6c_l v_s / d_l \rho_l$, where ρ_l is the latex specific density, equal to 1.05×10^3 kg m^{-3}.

Experimental results expressed as the dependence of the microsphere zeta potential on the fibrinogen coverage calculated from equation (3.4) are shown in figure 2 (parts a, b, c) for pH = 3.5, and figure 3 (parts a, b, c) for pH = 9.7 and various ionic strengths $10^{-3} - 0.15$ M NaCl. As can be seen, in figure 2 (a), at pH = 3.5 and $I = 10^{-3}$ M, the zeta potential of latex remains constant for the entire range of the fibrinogen concentration during the adsorption experiments. This indicates that there is no fibrinogen adsorption under these conditions. However, for higher NaCl concentrations, ζ_f significantly decreases with Γ_f attaining limiting values that are higher than the zeta potential of fibrinogen in the bulk. Note, that ζ_f becomes constant where the maximal coverage is approached. Therefore, one can roughly estimate it from the dependence of ζ_f on fibrinogen coverage on latex particles Γ_f. Analysing figure 2, the limiting fibrinogen coverages (where the zeta potential does not change) are approximately 0.9 and 1.1 mg m^{-2} for NaCl concentration of 10^{-2}, 0.15 M, respectively. In an analogous way, the limiting coverages for pH = 9.7 and the NaCl concentration 10^{-3}, 10^{-2}, 0.15 M (figure 3) were estimated to be 2.0, 2.3, 3.4 mg m^{-2}, respectively. These results indicate that significant adsorption of fibrinogen on latex particles occurred for the applied ionic strength.

It should be mentioned, however, that the maximum coverages of fibrinogen derived from the zeta potential measurements are of a limited precision. Therefore, in order to more exactly determine the maximum coverage, a quantitative depletion method described in reference [30] was applied. According to this procedure, the residual fibrinogen remaining in the suspension after the adsorption step is adsorbed on mica and then imaged by AFM. Fibrinogen adsorption is carried out under diffusion transport

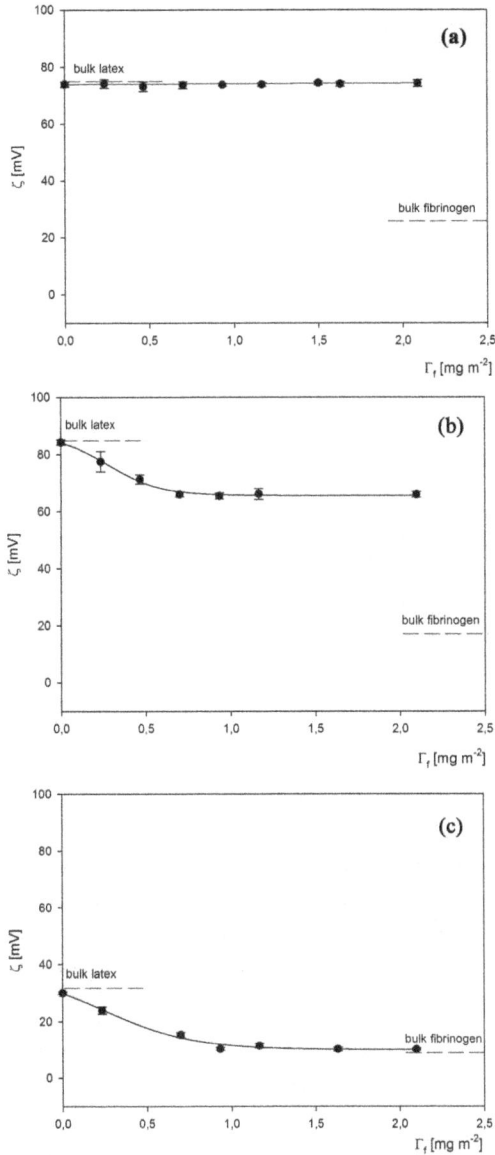

Figure 2. The dependence of the zeta potential of latex on the nominal fibrinogen coverage. The points denote experimental results obtained for pH = 3.5 and various NaCl concentration: 10^{-3} M (a), 10^{-2} M (b) and 0.15 M (c). The lines are guides for the eye for the experimental data.

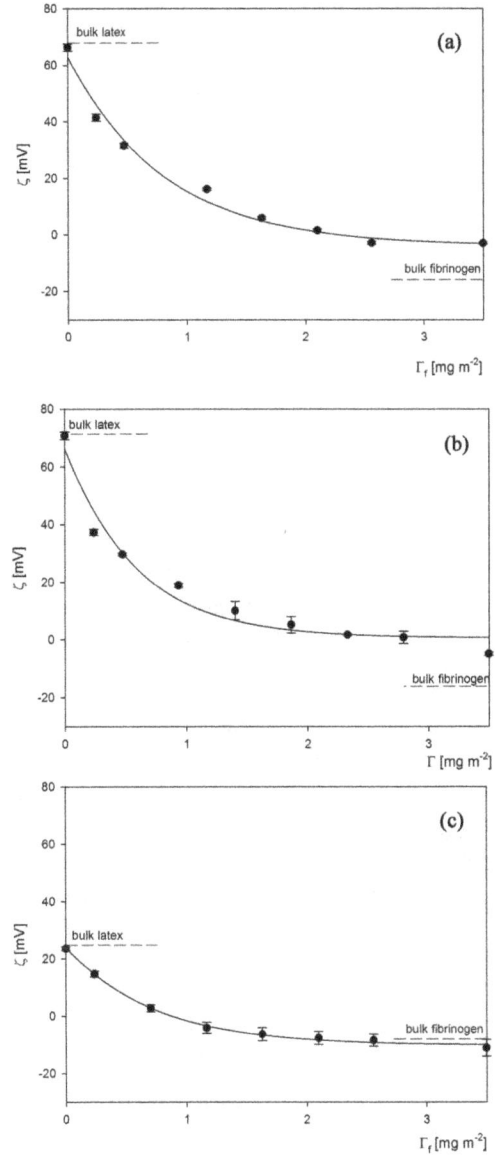

Figure 3. Same as in figure 2, but at pH = 9.7.

conditions using the latex suspension without applying any filtration or centrifugation. In these measurements, the average number of fibrinogen molecules per unit area of mica N_{fm} is determined as a function of time. The principle of this method is that under diffusion-controlled transport, N_{fm} increases proportionally to the concentration of fibrinogen, i.e., [34]

$$N_{fm} = \left(\frac{Dt}{\pi}\right)^{\frac{1}{2}} c_{fr} = C_f c_{fr}, \tag{3.5}$$

where $C_f = (Dt/\pi)^{\frac{1}{2}}$ is a constant, which can be calculated if the adsorption time t and the diffusion coefficient of fibrinogen D are known. This constant was also determined in calibrating measurements of the protein adsorption on bare mica sheets from solutions of defined concentrations. It is also important

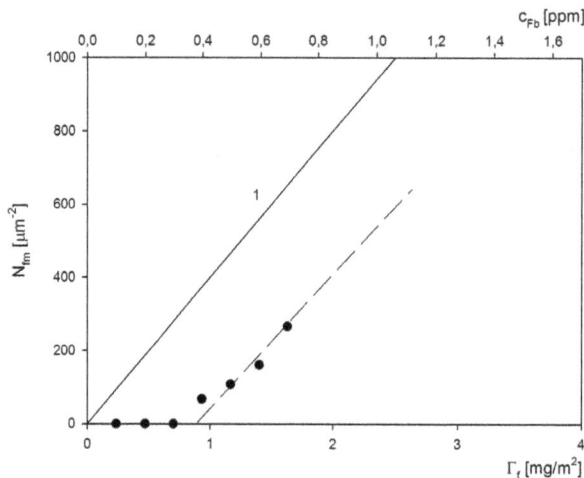

Figure 4. The dependence of the surface concentration of fibrinogen on mica (N_{fm}) determined by AFM imaging on the fibrinogen coverage Γ_f (lower axis) and on the fibrinogen concentration in the suspension c_{Fb}, bulk latex concentration after mixing $c_l = 60$ mg L^{-1}, pH = 3.5, and NaCl concentrations equal to 10^{-2} M. The solid line 1 represents the reference results predicted for diffusion-controlled transport of fibrinogen and the dashed line is a guide for eye for the experimental data.

to mention that by measuring N_{fm} and exploiting equation (3.5), one can uniquely calculate the unknown fibrinogen concentration in the suspension.

It should be mentioned that the latex particle deposition on mica during the fibrinogen adsorption runs is negligible due to their low concentrations and diffusion coefficients. The results obtained in these experiments are shown in figure 4 for pH = 3.5 and figure 5 for pH = 9.7 (NaCl concentrations equal to 10^{-2} M) as the dependence of N_{fm} on the initial concentration of fibrinogen in the suspension (the final latex particle concentration was 60 mg L^{-1}). Additionally, in figures 4 and 5, the reference data for fibrinogen solutions of the same concentration without contacting the latex are plotted as straight lines number 1. As can be seen, for low fibrinogen concentration in the initial solution, there was no adsorption on mica and N_{fm} fluctuated within error bounds near zero. Afterward, if c_{Fb}, exceeds threshold values, a liner increase in N_{fm} is observed with slopes similar to the reference data. This confirms that fibrinogen adsorption on latex was irreversible for the pH and ionic strength conditions presented.

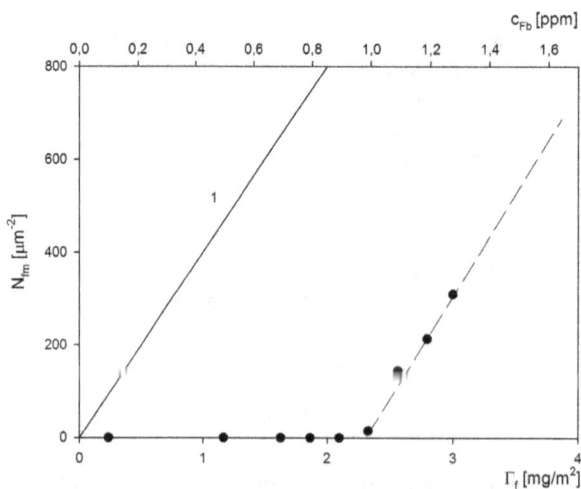

Figure 5. Same as in figure 4, but at pH = 9.7.

By extrapolating the linear experimental dependencies shown in figures 4 and 5 to zero, one can obtain the threshold fibrinogen concentrations denoted by c_{fmax}. Knowing the concentrations, the maximum coverage of fibrinogen on latex can be calculated directly from equation (3.4) by substituting c_f by c_{fmax} In this way, one obtains precise values of Γ_{max} that are listed in table 3. As can be noticed, the maximum coverages agree with the previous results derived from the zeta potential measurements (see figures 2–3). Thus, for pH = 3.5, the maximum coverage is negligible for NaCl concentrations equal to 10^{-3} M. However, for higher NaCl concentrations of 10^{-2} and 0.15 M, the maximum fibrinogen coverage assumes quite appreciable values of 0.9 and 1.1 mg m^{-2}, respectively. The latter results prove that the classical mean-field approach, where it is assumed that the effective charge on protein molecules is uniformly distributed and the molecule is characterized by an average value of zeta potential, is inadequate. This is so, because both the latex particles and fibrinogen molecule exhibit a large (positive) zeta potential at pH = 3.5 for NaCl concentration of 10^{-2} and 0.15 M (see table 2). This discrepancy can be explained in terms of a heterogeneous charge distribution over the fibrinogen molecule with a negative charge located in its core part whereas the rest of the molecule remains positively charged. This assumption is supported by the fact that the core part of the fibrinogen molecule comprises aspartic and glutamic acids residues characterized by pH value below 3 [36].

Table 3. Maximum coverage of fibrinogen on positively and negatively charged latex particles expressed in mg m^{-2} for various pHs and NaCl concentrations.

pH	NaCl concentration [M]	Maximum coverage of fibrinogen on latex [mg m^{-2}]			
		Γ_{max}		theoretical (RSA modelling)	
		negative latex*	positive latex		remarks
3.5	10^{-3}	1.8 ± 0.2	< 0.1	negligible	side-on adsorption
	10^{-2}	2.5 ± 0.2	0.9 ± 0.1	1.13	side-on adsorption
	0.15	3.6 ± 0.2	1.1 ± 0.1	1.22	side-on adsorption
7.4	10^{-3}	1.9 ± 0.2	$0.6 \pm 0.1^*$	0.65	side-on adsorption
	10^{-2}	2.7 ± 0.2	$1.2 \pm 0.1^*$	1.12	side-on adsorption
	0.15	3.2 ± 0.2	$1.3 \pm 0.1^*$	1.29	side-on adsorption
9.7	10^{-3}	–	2.0 ± 0.2	1.8	unoriented adsorption
	10^{-2}	–	2.3 ± 0.2	2.8	unoriented adsorption
	0.15	–	3.4 ± 0.2	3.7	unoriented adsorption

*Previous results obtained in references [30–32].

Therefore, our experimental studies show that even at pH = 3.5, the fibrinogen molecule contains negatively charged fragments which is in conflict with the previous reports where it was assumed that the entire molecule is positively charged at this pH [24, 34, 42].

The presence of a negative charge at the core part of the molecule suggests that fibrinogen adsorption at pH = 3.5 occurs exclusively in the side-on orientation with the αC domains of the Aα chains pointing toward the bulk solution. This assumption is supported by the fact that the theoretically predicted coverage derived by applying the random sequential adsorption modelling [30] is equal to 1.13 and 1.22 mg m^{-2} for NaCl concentration of 10^{-2} and 0.15 M, which is slightly lower than with experimental values. Snapshots of fibrinogen monolayers on latex particles derived from these simulations are shown in figure 6 for the ionic strength 10^{-2} and 0.15 M NaCl.

As can be seen, it is theoretically predicted that fibrinogen molecules adsorb in the side-on orientation exposing the positively charged side arms into the electrolyte solution. Additionally, due to the lateral electrostatic repulsion among the beads forming the core part of molecules, the monolayer obtained for 10^{-2} M, NaCl is characterized by a significantly smaller density compared to the jamming coverage for hard (non-interacting) spheres predicted to be 1.29 mg $^{-2}$ [24].

For pH = 7.4, the experimental maximum coverages are 1.2 and 1.3 mg m^{-2} that are only slightly larger than at pH = 3.5 but considerably lower than for fibrinogen adsorption at negatively charged latex determined in reference [32] and equal to 2.7 and 3.2 mg m^{-2}. This seems to be paradoxical since at pH =

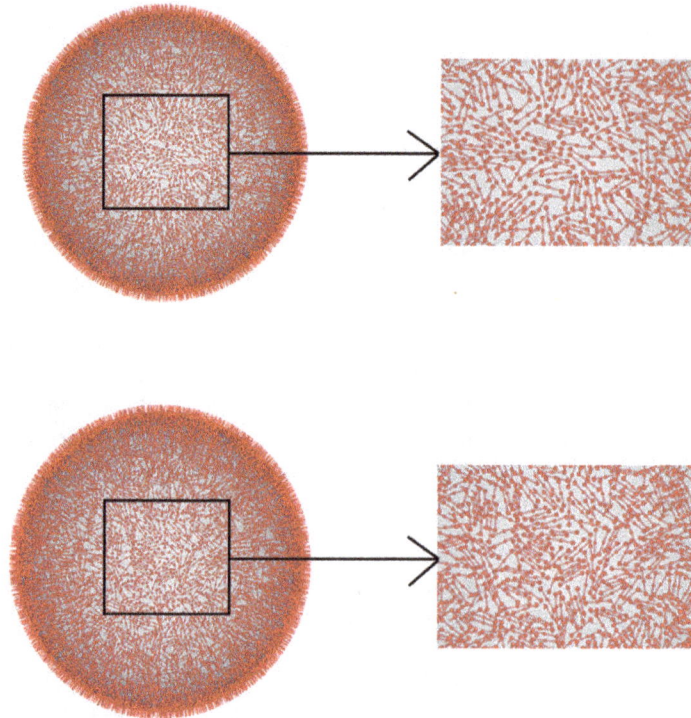

Figure 6. (Color online) The topology of fibrinogen molecules (approximated by the Bead Model B) on the positive latex particles derived from the RSA simulations at pH = 3.5 and for different electrolyte concentrations: top: 10^{-2} M, 1.13 mg m^{-2} and bottom: 0.15 M, 1.22 mg m^{-2}.

7.4, the net charge of the fibrinogen molecule is strongly negative (see table 2). In the case of positively charged microspheres, this discrepancy can be explained by the appearance of a strong electrostatic attraction between the particle surface and fibrinogen molecules promoting their side-on orientation.

However, at pH = 9.7, where the fibrinogen molecule charge is considerably more negative, the maximum coverages significantly increase attaining 2.0, 2.3 and 3.4 mg m^{-2} for NaCl concentration of 10^{-3}, 10^{-2} and 0.15 M, respectively. As can be noticed, (see table 3), these results are similar to the previously reported for the fibrinogen adsorption at negatively charged latex at pH = 3.5 [30]. They were interpreted in terms of the RSA model assuming unoriented adsorption of fibrinogen, i.e., simultaneously occurring in the side-on and end-on orientations that prevails for long adsorption times. Additionally, in these calculations the lateral electrostatic repulsion among the adsorbed fibrinogen molecules was considered to be responsible for the decrease of the maximum coverage for a lower NaCl concentration (see the last column in table 3). Therefore, the agreement of our results obtained for positive latex with these theoretical predictions confirms that at pH = 9.7, fibrinogen also adsorbs at positively charged latex in the end-on orientation.

3.3. Stability of fibrinogen monolayers on latex particles

In these series of experiments, the stability of fibrinogen monolayers on microspheres was checked by measuring their hydrodynamic diameter as function of time and in pH cycling experiments. The experimental results are shown in figure 7 as the dependence of d_i / d_0 (d_0 is the initial hydrodynamic diameter of fibrinogen-covered latex) on the storage time for NaCl concentration equal to 10^{-2} M and pH = 3.5.

As can be seen in figure 7, there were no significant changes in the normalized hydrodynamic diameter for the time period up to 35 h.

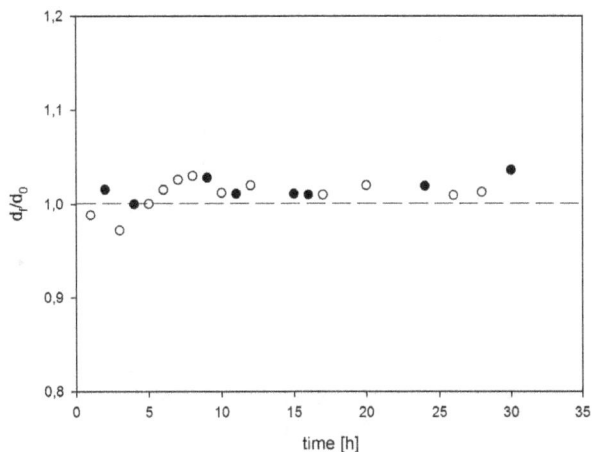

Figure 7. The dependence of normalized hydrodynamic diameter of fibrinogen d_f/d_0 on time for $I = 10^{-2}$ M NaCl, pH = 3.5, $\Gamma_f = 0.9$ mg m^{-2} and $\Gamma_f = 2.0$ mg m^{-2} for white and black dots, respectively.

Additionally, the stability of fibrinogen monolayers was checked in pH cycles. This experiment consisted of the following steps:

(i) a fibrinogen monolayer of a well-defined coverage was adsorbed on latex particles at pH = 3.5 and NaCl concentration of 0.15 M,

(ii) pH was stepwise changed by the addition of HCl (pH decrease) or NaOH (pH increase) by keeping the changes in the initial ionic strength negligible

(iii) after stabilization of pH, the electrophoretic mobility of latex was measured

(iv) the entire process was repeated three times in the range of pH between 3.5 and 9.7.

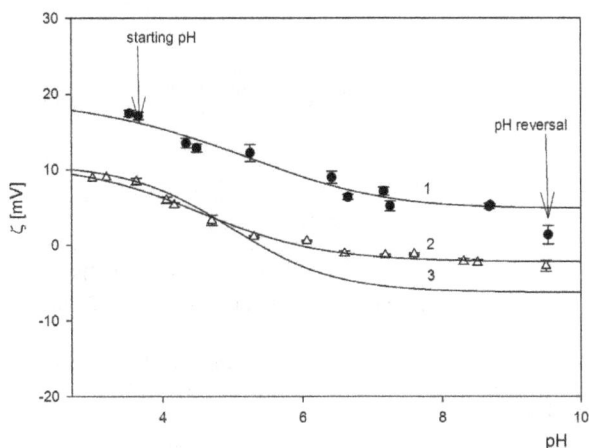

Figure 8. The dependence of fibrinogen monolayers adsorbed at pH = 3.5, $I = 0.15$ M NaCl, on pH cycling starting from 3.5 to 9.7 and back to 3.5. The points denote experimental results for: $\Gamma_f = 2$ mg m^{-2} (triangles), $\Gamma_f = 0.9$ mg m^{-2} (dots). Lines 1 and 2 are fits of experimental data, and line 3 denotes the reference results for fibrinogen zeta potential in the bulk.

Experimental results obtained in these measurements are shown in figure 8. It can be observed that under these conditions, no irreversible conformational changes in the protein monolayers were observed.

4. Conclusions

The AFM and micro-electrophoretic measurements complemented by extensive RSA modelling delivered reliable information about the mechanisms of fibrinogen adsorption on positively charged microspheres for a broad range of pHs and ionic strength. It was confirmed that fibrinogen adsorption was irreversible and the monolayer properties were stable in time.

Interesting results were obtained at pH = 3.5, where for NaCl concentrations of 10^{-2} and 0.15 M, the maximum fibrinogen coverage of fibrinogen on microspheres was quite significant amounting to 0.9 and 1.1 mg m^{-2}, respectively. This contradicts the predictions derived from the classical mean-field approach because both the fibrinogen molecules and the latex particles exhibit a positive zeta potential at this pH. This discrepancy was explained in terms of a heterogeneous charge distribution over the fibrinogen molecule with a negative charge located at its core part whereas the rest of the molecule remains positively charged. This anomalous adsorption of fibrinogen confirms that, in contrast to common views, a negative charge is present at the body of the fibrinogen molecule even at such low pH. This enables a side-on adsorption of the molecules interpreted by the random sequential adsorption modelling.

The increase in the fibrinogen coverage with NaCl concentration was interpreted as due to the electrostatic repulsion among the adsorbed protein molecules. On the other hand, for pH = 9.7, the maximum coverage of fibrinogen on latex was much larger, varying between 2.0 mg m^{-2} and 3.4 mg m^{-2} for 10^{-3} and 0.15 M NaCl, respectively, which almost fully matches the previous results obtained for negative latex at pH = 3.5. These results were in agreement with the theoretical RSA modelling carried out by assuming the side-on and end-on orientations prevailing at longer times. This confirmed a different adsorption mechanism of fibrinogen on latex at pH = 9.7, characterized by the presence of a considerable amount of end-on adsorbed molecules. The increase in the coverage with NaCl concentration was interpreted in terms of the diminishing range of the repulsive electrostatic interactions among the adsorbed molecules.

Besides a significance to basic science, our results have practical implications for developing a robust procedure of preparing stable fibrinogen monolayers at latex particles of a well-controlled coverage and orientation of molecules. Such latex/fibrinogen complexes can be exploited to study interactions with antibodies and other ligands.

Acknowledgements

This work was financially supported by the NCN Grants: PRELUDIUM-2013/09/N/ST4/00320 and UMO-2012/07/B/ST4/00559.

References

1. Hlady V., Buijs J., Curr. Opin. Biotechnol., 1996, 7, 72; doi:10.1016/S0958-1669(96)80098-X.
2. Adamczyk Z., Bratek-Skicki A., Żeliszewska P., Wasilewska M., Curr. Top. Med. Chem., 2014, 14, 702; doi:10.2174/1568026614666140118215158.
3. Hu Y., Jin J., Han Y., Yin J., Jiang W., Lian H., RSC Adv., 2014, 4, 7716; doi:10.1039/c3ra46934d.
4. Brash J.L., Uniyal S., Pusineri C., Schmitt A., J. Colloid Interface Sci., 1983, 95, 28; doi:10.1016/0021-9797(83)90068-1.
5. Elgersma A.V., Zsom R.L.J., Norde W., Lyklema J., J. Colloid Interface Sci., 1990, 138, 145; doi:10.1016/0021-9797(90)90190-Y.
6. Rezwan K., Meier L.P., Gauckler L.J., Biomaterials, 2005, 26, 4351; doi:10.1016/j.biomaterials.2004.11.017.
7. Cai K., Frant M., Bossert J., Hildebrand G., Liefeith K., Jandt K.D., Colloids Surf. B, 2006, 50, 1; doi:10.1016/j.colsurfb.2006.03.016.
8. Patil S., Sandberg A., Heckert E., Self W., Seal S., Biomaterials, 2007, 28, 4600; doi:10.1016/j.biomaterials.2007.07.029.

9. Baier G., Costa C., Zeller A., Baumann D., Sayer C., Araujo P.H.H., Mailander V., Musyanovych A., Landfester K., Macromol. Biosci., 2011, **11**, 628; doi:10.1002/mabi.201000395.
10. Rezwan K., Studart A.R., Vörös J., Gauckler L.J., J. Phys. Chem. B, 2005, **109**, 14469; doi:10.1021/jp050528w.
11. Henschen A.H., Lottspeich F., Kehl M., Southan C.D., Ann. N.Y. Acad. Sci., 1983, **408**, 28; doi:10.1111/j.1749-6632.1983.tb23232.x.
12. Doolittle R.F., Goldbaum D.L., Doolittle L.R., J. Mol. Biol., 1978, **120** (2), 311.
13. Hall C.E., Slayter H.S., J. Biophys. Biochem. Cytol., 1959, **5**, 11; doi:10.1083/jcb.5.1.11.
14. Folwer W.E., Erickson H.P., J. Mol. Biol., 1979, **134**, 241; doi:10.1016/0022-2836(79)90034-2.
15. Weisel J.W., Stauffacher C.V., Bullitt E., Cohen C., Science, 1985, **230**, 1388; doi:10.1126/science.4071058.
16. Verklich Y.J., Gorkun O.V., Medved L.V., Nieuwenhuizen W., Weisel S.W., J. Biol. Chem., 1993, **268**, 13577.
17. Ortega-Vinuesa J.L., Tengvall P., Lundstrom I., J. Colloid Interface Sci., 1998, **207**, 228; doi:10.1006/jcis.1998.5624.
18. Cacciafesta P., Humphris A.D.L., Jandt K.D., Miles M.J., Langmuir, 2000, **16**, 8167; doi:10.1021/la000362k.
19. Sit S.P., Marchant R.E., Surf. Sci., 2001, **491**, 421; doi:10.1016/S0039-6028(01)01308-5.
20. Tunc S., Maitz M.F., Steiner G., Vazquez L., Pham M.T., Salzer R., Colloids Surf. B, 2005, **42**, 219; doi:10.1016/j.colsurfb.2005.03.004.
21. Tsapikouni T.S., Missirlis Y.F., Colloids Surf. B, 2007, **57**, 89; doi:10.1016/j.colsurfb.2007.01.011.
22. Hassan N., Maldonado-Valderrama J., Gunning A.P., Morris V.J., Ruso J.M., J. Phys. Chem., 2011, **115**, 6304; doi:10.1021/jp200835j.
23. Wasilewska M., Adamczyk Z., Langmuir, 2011, **27**, 686; doi:10.1021/la102931a.
24. Adamczyk Z., Cichocki B., Ekiel-Jeżewska M.L., Słowicka A., Wajnryb E., Wasilewska M., J. Colloid Interface Sci., 2012, **385**, 244; doi:10.1016/j.jcis.2012.07.010.
25. Malmsten M., J. Colloid Interface Sci., 1994, **166**, 333; doi:10.1006/jcis.1994.1303.
26. Wertz C.F., Santore M.M., Langmuir, 2001, **17**, 3006; doi:10.1021/la0017781.
27. Santore M.M., Wertz C.F., Langmuir, 2005, **21**, 10172; doi:10.1021/la051059s.
28. Bai Z., Filiaggi M.J., Dahn J.R., Surf. Sci., 2009, **603**, 839; doi:10.1016/j.susc.2009.01.040.
29. Kalasin S., Santore M.M., Colloid Surf. B, 2009, **73**, 229; doi:10.1016/j.colsurfb.2009.05.028.
30. Bratek-Skicki A., Żeliszewska P., Adamczyk Z., Cieśla M., Langmuir, 2013, **29**, 3700; doi:10.1021/la400419y.
31. Żeliszewska P., Bratek-Skicki A., Adamczyk Z., Cieśla M., Langmuir, 2014, **30**, 11165; doi:10.1021/la5025668.
32. Bratek-Skicki A., Żeliszewska P., Adamczyk Z., Curr. Top. Med. Chem., 2014, **14**, 640; doi:10.2174/1568026614666140118212409.
33. Smith P.K., Krohn R.I., Hermanson G.T., Mallia A.K., Gartner F.H., Provenzano M.D., Fujimoto E.K., Goeke N.M., Olson B.J., Klenk D.C., Anal. Biochem., 1985, **150**, 76; doi:10.1016/0003-2697(85)90442-7.
34. Adamczyk Z., Bratek-Skicki A., Dąbrowska P., Nattich-Rak M., Langmuir, 2012, **28**, 474; doi:10.1021/la2038119.
35. Feder J., Giaever I., J. Colloid Interface Sci., 1980, **78**, 144; doi:10.1016/0021-9797(80)90502-0.
36. Hinrisen E.L., Feder J., Jossang T., J. Stat. Phys., 1986, **44**, 793.
37. Viot P., Tarjus G., Ricci S., Talbot J., J. Chem. Phys., 1992, **97**, 5212; doi:10.1063/1.463820.
38. Evans J.W., Rev. Mod. Phys., 1992, **65**, 1281; doi:10.1103/RevModPhys.65.1281.
39. Talbot J., Tarjus G., Van Tassel P.R., Viot P., Colloids Surf. A, 2000, **165**, 287; doi:10.1016/S0927-7757(99)00409-4.
40. Adamczyk Z., Barbasz J., Cieśla M., Langmuir, 2011, **27**, 6868; doi:10.1021/la200798d.
41. Wasilewska M., Adamczyk Z., Jachimska B., Langmuir, 2009, **25**, 3698; doi:10.1021/la803662a.
42. Adamczyk Z., Nattich M., Wasilewska M., Zaucha M., Adv. Colloid Interface Sci., 2011, **168**, 3; doi:10.1016/j.cis.2011.04.002.

12

Simulation study of a rectifying bipolar ion channel: Detailed model versus reduced model[*]

Z.Ható[1], D. Boda[1][†], D. Gillespie[2], J. Vrabec[3], G. Rutkai[3], T. Kristóf[1]

[1] Department of Physical Chemistry, University of Pannonia, P. O. Box 158, Veszprém, H-8201, Hungary

[2] Department Molecular Biophysics and Physiology, Rush University Medical Center, Chicago, IL 60612, USA

[3] University of Paderborn, Laboratory of Thermodynamics and Energy Technology, 100 Warburger St., 33098 Paderborn, Germany

We study a rectifying mutant of the OmpF porin ion channel using both all-atom and reduced models. The mutant was created by Miedema et al. [Nano Lett., 2007, **7**, 2886] on the basis of the N-P semiconductor diode, in which an N-P junction is formed. The mutant contains a pore region with positive amino acids on the left-hand side and negative amino acids on the right-hand side. Experiments show that this mutant rectifies. Although we do not know the structure of this mutant, we can build an all-atom model for it on the basis of the structure of the wild type channel. Interestingly, molecular dynamics simulations for this all-atom model do not produce rectification. A reduced model that contains only the important degrees of freedom (the positive and negative amino acids and free ions in an implicit solvent), on the other hand, exhibits rectification. Our calculations for the reduced model (using the Nernst-Planck equation coupled to Local Equilibrium Monte Carlo simulations) reveal a rectification mechanism that is different from that seen for semiconductor diodes. The basic reason is that the ions are different in nature from electrons and holes (they do not recombine). We provide explanations for the failure of the all-atom model including the effect of all the other atoms in the system as a noise that inhibits the response of ions (that would be necessary for rectification) to the polarizing external field.

Key words: *Monte Carlo, primitive model electrolytes, ion channel, selectivity*

1. Introduction

Rectification mechanisms in nanopores and ion channels are based on asymmetries in the structure of the pore [1, 2]. The asymmetry is either geometrical or electrostatic in nature. In the former, the shape of the pore is asymmetrical as in the case of conical nanopores [3, 4].

The latter case, when the charge distribution in the pore is asymmetrical [5], is the subject of this study. This phenomenon is well known in the case of semiconductor diodes [6, 7], where the charge asymmetry is achieved by doping different regions of the device differently thus forming an N-P diode, where the majority charge carriers are electrons and holes in the N and P regions, respectively. The N-P junction between these two regions forms a depletion zone for both electrons and holes. An external electric field in forward (ON) and reverse (OFF) bias acts differently on this region by making it even wider in the OFF state and thinner in the ON state. In the ON state, the majority carriers will conduct the current, while in the OFF state, the minority carriers will do the job; hence, the rectification.

In this paper, we consider devices where the charge carriers are ions solvated in a liquid solvent (usually water) that migrate through a pore that is embedded in a membrane. The two major classes of these pores are artificial nanopores and biological ion channels. Nanopores with an N-P charge distribution on their pore walls are called bipolar nanopores [8–19]. Nanopores are etched into plastic membranes

---

[*] With this paper we intend to honor the achievements of Stefan Sokołowski.

[†] Author for correspondence: boda@almos.vein.hu

[20–23]. The charge distribution on the wall of the pore can be controlled by chemical methods. They are wider than ion channels, although the technology of nanopore fabrication is advancing rapidly resulting in increasingly narrow pores. Nanopores are stable and easy to regulate which makes them potential building blocks of nanodevices [24–26] and sensors [14, 23, 27–30].

Ion channels, on the other hand, are natural pores in proteins produced by evolution for specific purposes according to their specific gating, selectivity, and conductance properties [31–33]. They are much narrower than synthetic nanopores. Also, their experimental study is more problematic. Changing their structure, for example, requires point mutations of amino acids and synthesizing the protein by cells. Moreover, the accurate three-dimensional (3D) structure of ion channels is rarely known because they are hard to crystallize.

The OmpF ion channel, a bacterial porin, is an exception, because its structure has been determined relatively early [34, 35]. This explains the fact that numerous experimental and simulation works used this channel as a case study [36–46]. The work of Miedema et al. [47] is especially important from the point of view of our study. They mutated the OmpF channel aiming to create an N-P junction in its pore and showed that this mutant (abbreviated as RREE) rectifies. The study of Miedema et al. [47] inspired us to perform all-atom molecular dynamics (MD) simulation for the wild type (WT) OmpF channel and its mutant. The model of the WT channel is based on experimental X-ray data that are available. In the case of the RREE mutant, on the other hand, the structure is unavailable so that the model is based on changing the amino acids in the WT structure and optimizing it with the VMD program package.

The model of the mutant, therefore, just as in the paper of Miedema et al. [47], is just a guess. Surprisingly, our all-atom simulations did not show rectification for the model of the RREE mutant. This paper will undertake the risky business of searching for the explanation of the discrepancy between the experimental and simulation results.

We hypothesize that the sign of voltage cannot exert a decisive effect on the ionic distribution in the pore because there is too much noise in the all-atom model. In order to get rid of the noise and to achieve a better understanding of the rectification mechanism in bipolar pores [8, 10–16, 18, 19], we also constructed a reduced model of the ion channel, where only the "important" amino acids were modelled explicitly. These amino acids are those that form the N and P regions by preferentially attracting the counterions into the respective region. In this paper, we follow the nomenclature of the field of semiconductor devices and call the region where anions dominate the N region (and P region, where the cations dominate).

We study the reduced model with the Nernst-Planck (NP) equation that we couple to a simulation procedure (Local Equilibrium Monte Carlo, LEMC) that establishes the relation of the concentration profiles to the electrochemical potential profile [48–51]. This simulation method is an adaptation of the Grand Canonical Monte Carlo (GCMC) method to a non-equilibrium situation by using a spatially non-homogeneous electrochemical potential as the input variable of the simulation and yielding the concentration profile as an output. The resulting NP+LEMC method efficiently computes current-voltage (IV) profiles for the reduced model using modest computer time compared to the massive computational load needed to get a close-to-reasonable statistics for the all-atom model.

Since the reduced model has been constructed by building only those degrees of freedom into the model that are essential to produce rectification, it is not a surprise that rectification has been found in this case. These calculations are useful because they provide an understanding of the phenomenon under study. Since the bipolar ion channel created by Miedema et al. [47] is known to rectify, we encounter an example where a reduced model describes the reality better than a detailed model. This does not mean that detailed models are not useful. It just means that there are situations where "less is more", especially when long-range effects (electric field, polarization) are responsible for the phenomenon. In such cases, details do not necessarily serve understanding, because the effect is hidden in the noise and we just cannot see the wood for the trees.

2. A rectifying mutant of the OmpF ion channel

In this section, we present the experimental facts for the RREE mutant as obtained by Miedema et al. [47], the all-atom model that we constructed for the channel, details of the simulations, and the results given by the simulations.

2.1. Experimental facts for the RREE mutant

In the experimental work of Miedema et al. two filters have been identified inside the pore (see table 1 of reference [47]). In the first filter, the negative amino acids D113 and E117 have been mutated into positive arginines, R113 and R117. In the second filter, the positive arginines, R167 and R168, have been mutated into negative glutamates, E167 and E168. This way, the first filter has been positively doped, while the second filter has been negatively doped, at least, in theory (see figure 1). The point mutations aiming the N-P junction are hard facts, but we do not know whether the protein is folded in the way we want it to fold: crystal structure data are not available for the mutant.

Using 0.1 M NaCl and ± 100 mV voltage, the authors found a rectification 0.22 ± 0.02 for the RREE mutant as opposed to the value 1.14 ± 0.03 in the case of the WT channel. Rectification, which is a voltage-dependent quantity, is defined as

$$r(U) = \left| \frac{I(U)}{I(-U)} \right|. \tag{1}$$

In the case of a 1M NaCl electrolyte, the rectification values are 0.65 ± 0.06 and 0.99 ± 0.01 for the RREE and WT channels, respectively. Rectification, therefore, decreases as concentration increases.

The authors hypothesize in a cartoon (figure 5 (b) in their paper [47]) about the rectification mechanism that is adapted from the case of the semiconductor N-P diodes. The supposed mechanism is that a depletion zone is formed at the junction of the N and P regions that becomes wider and more depleted at the OFF sign of the voltage. It seems to be a widespread assumption that the rectification mechanism is the same in bipolar pores (where ions are the charge carriers) and in semiconductor diodes (where electrons and holes are the charge carriers). In this paper, we show that the mechanism of rectification is different, or, at least, that it can be different in narrow nanopores and ion channels.

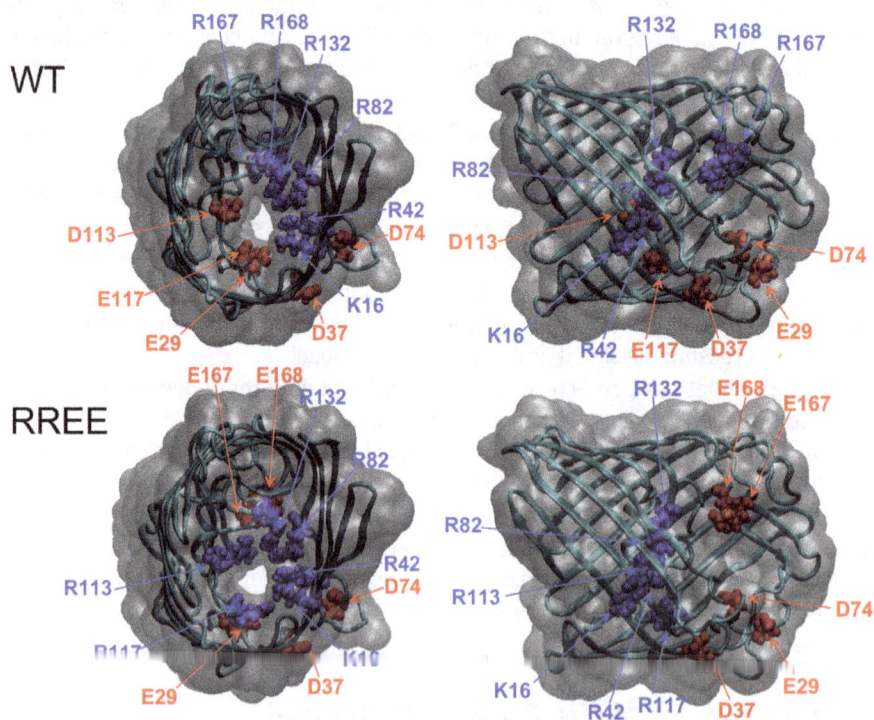

Figure 1. (Color online) The WT OmpF ion channel (top row) on the basis of the 2OMF structure [34, 35] and its RREE mutant (bottom row) [47] made by the VMD package [52] after changing the indicated amino acids: D113→R113, E117→R117, R167→E167, and R168→E168.

2.2. All-atom model and molecular dynamics simulations

The OmpF channel has been simulated in numerous studies [36–46, 53]. The simulations identified two distinct pathways for cations and anions with a slight cation selectivity. Several mutations of the WT OmpF have also been studied [37, 42].

The structure of the OmpF trimer [34, 35] was constructed according to the ProteinDataBank database (identifier: 2OMF). The protein/membrane complex was generated with the help of CHARMM-GUI [54], embedding the protein into a DMPC lipid bilayer. We used the VMD program package [52] to mutate the WT channel into the RREE mutant (see figure 1).

We performed all-atom MD simulations with the GROMACS program suite [55, 56] using the leap frog integrator with a 2 fs timestep. The system temperature was set with the Nose-Hoover thermostat [57]. Simulations in the NpT ensemble were conducted with a Parrinello-Rahman barostat [58]. We used CHARMM27 force-field based flexible models together with position restraints for the backbone atoms of the protein [59]. The bonds of hydrogen atoms were considered rigid; this allows us to use a slightly larger timestep (larger than that required for an accurate simulation of bond vibration with hydrogen atoms). In simulations with electric fields we applied a ± 200 mV potential (with the ground at the left-hand side). Periodic boundary conditions were present in all spatial directions.

Most of our simulations were performed in a simulation cell with the size of $105.6 \times 105.6 \times 114.5$ Å^3 in x, y, and z dimensions with z being the transport direction. The solvent phase was constructed of 561 Na$^+$, 528 Cl$^-$, and 29 317 TIP3P water molecules resulting in $\approx 132\,000$ atoms including $\approx 15\,000$ from the protein trimer, and $\approx 28\,000$ from the DMPC lipid layers.

To check for a system size dependence, we performed two simulations (for 200 and -200 mV) for a larger simulation volume with approximate dimensions $220 \times 220 \times 1\,130$ Å^3 containing four RREE trimers and $\approx 5\,000\,000$ atoms. The protein and lipid membrane were constructed using four times the smaller simulation volume that was elongated in the direction of the transfer (along axis z) and filled with water and ions.

We followed the simulation procedure of Faraudo et al. [44]. In five preliminary equilibration runs we did not apply an external electric field. We started with an energy minimization run after the construction of the simulation cell. This was followed by a 100 ps NVT run at 100 K and another 100 ps NVT simulation at 296 K. After these steps we turned the barostat on and performed a 1 ns NpT calculation at 296 K and 1 bar with isotropic pressure coupling. The last preliminary equilibration step was to perform a 3 ns NpT simulation at 296 K and 1 bar pressure with semi-isotropic pressure coupling (independent coupling in the direction of transfer).

After we let the system relax, we started the simulations with an applied external field. To achieve a stationary state we did a 10 ns long NVT run at 296 K and with an external electric field corresponding to a 200 mV potential difference across the simulation cell in the z direction. Next, we did the actual production run in which we counted the diffusing particles through the membrane. We have monitored the number of ions that completely crossed the protein by following the individual trajectories of each ion. An ion was considered to cross the channel if it is initially at one side of the membrane, and then ends at the opposite side of the membrane after propagating through the protein channel (some ions enter the channel but instead of crossing it, they return to the bulk where they started).

2.3. Results for the all-atom model

The number of counted ion-crossings as a function time (in the final production run) is plotted in figure 2 for the WT channel. We found the channel slightly selective for Cl$^-$ at 200 mV. The real channel is known to be slightly cation selective. We also have simulations for KCl, but with much shorter runs and weaker statistics. In this case, we found K$^+$ selectivity for 200 mV. No significant rectification was observed.

When the relevant amino acids are mutated (see figure 1), the channel becomes perfectly anion selective, so we plot only the Cl$^-$ currents in figure 3. The lack of cation current is probably due to the mutations made in the left-hand side filter; the ring formed by positive amino acids has a very narrow opening that repulses the cations effectively. The negative ring on the other side has a much larger hole in the middle that makes the passage of anions possible.

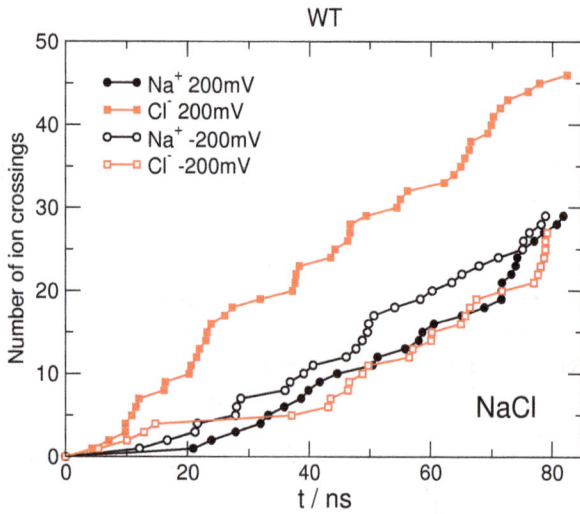

Figure 2. (Color online) The number of ion crossings as a function of simulation time for the WT OmpF porin at ±200 mV using 0.1 M symmetric NaCl.

It is more important that we have not found rectification for this model of the RREE mutant. The Cl^- current is practically the same for 200 mV and −200 mV within the statistical error of the simulation. These statistical errors can be estimated on the basis of the standard deviations of block averages; we obtained a large number for the error (±50 pA). Even if this large error indicates a weak statistics for the simulations, one thing can be concluded from figure 3 safely: rectification cannot be observed within the applied simulation lengths. From shorter runs for KCl we can draw the same conclusion.

If we want to find an explanation for the discrepancy between experiment and simulations, or, at least, we want to get closer to the explanation, we can look at the concentration profiles. Figure 4 shows the concentration profiles, $n_i(z)$, which are defined as the average number of ions in a slab divided

by the volume of the whole slab (the simulation cell is divided into slabs with a thickness of 2.5 Å in the z direction). An alternative way to plot the results is to show an effective local concentration, $c_i(z)$, where the average number of ions is divided by an effective volume. The effective volume is defined as the part of the whole slab, where the ions do not overlap with the body of the protein and the membrane — practically, the region of electrolyte. We will show results for the concentration profiles (in mol/dm^3) in the case of the all-atom model, because the effective volume is not a well-defined quantity due to the flexibility of the protein/membrane system.

In figure 4, one of the relevant observations is that the Na$^+$ ions are depleted inside the pore (note the

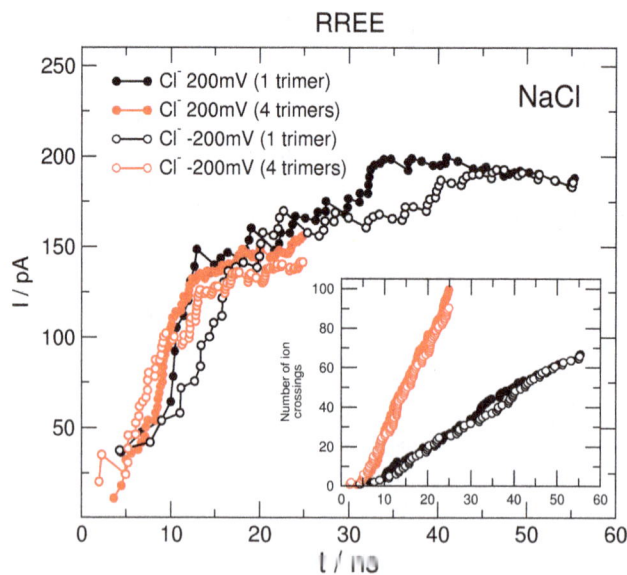

Figure 3. (Color online) Cumulative electrical currents carried by Na$^+$ and Cl$^-$ ions as a function of time for the RREE mutant. The red symbols refer to the simulations for the large cell with the four trimers. In this case, the current is divided by four, so the figure shows the current flowing through one trimer. In the inset, the number of ion crossings as a function of time is shown. Here, the number of crossings for the large simulation cell (four trimers) is not divided by four.

RREE

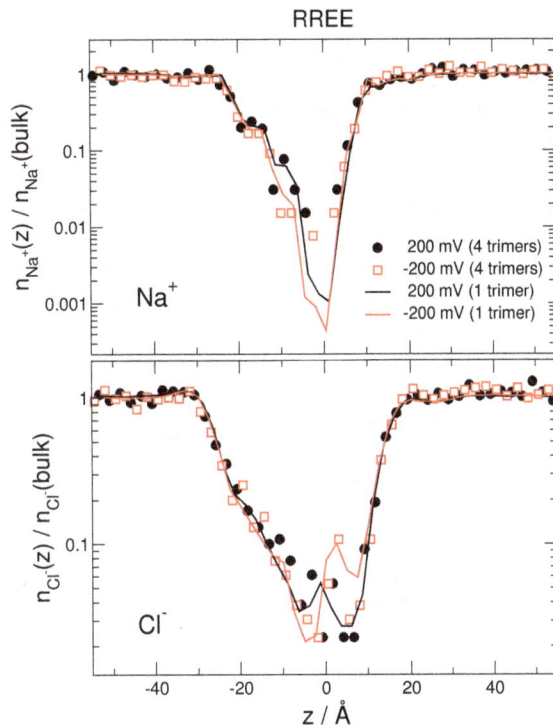

Figure 4. (Color online) Normalized concentration profiles (normalized with the bulk value, 1 M) for Na^+ and Cl^- ions for 200 and −200 mV from the all-atom simulations performed in the small simulation cell (one trimer, lines) and in the large simulation cell (four trimers, symbols) for the RREE mutant.

logarithmic scale of the concentration axis). This depletion zone acts as a high-resistance segment of the pore that effectively cuts the current of Na^+.

The other observation is that changing the sign of the voltage has little effect on the concentration profiles of Cl^- (the ion that conducts). The effect is that a depletion zone is formed at ≈ -5 Å for −200 mV, while for 200 mV the depletion zone is formed at ≈ 5 Å. Rectification would happen if the depletion zone were deeper at one voltage than at the opposite sign voltage. Here, the depletion zone is just shifted. From the point of view of conductance, the two profiles do not make a difference, therefore, the currents are the same for the two opposite signs of the voltage.

Third, the profiles obtained from the simulation for the large system (four trimers) and the small system (one trimer) agree. This justifies the use of the smaller simulation volume and indicates that the results obtained from it can be the basis of analysis.

3. Reduced model for a bipolar ion channel

The other way of figuring out what is going on in this system is to create a reduced model that takes into account only the "important" degrees of freedom and ignores the noise of the "unimportant" degrees of freedom. The "important" degrees of freedom are those that Miedema et al. [47] manipulated when they created their mutant in order to achieve a rectifying N-P junction in the ion channel. They are the amino acids that form an N-P junction inside the pore as shown in figure 1. To build a reduced model that is appropriate for our purpose, we choose the ion channel model that we used in our previous papers for the L-type calcium channel [50, 60–68], the Ryanodine Receptor calcium channel [51, 69–72], and the neuronal sodium channel [73–75]. These reduced models were able to capture the essential features of these channels and reproduce various anomalous selectivity behaviors.

3.1. Reduced model

In this model, we work with a reduced representation of the electrolyte, the protein, and the membrane. The ions are charged hard spheres immersed in a dielectric continuum that models the solvent implicitly. The ionic radii are 2 Å for both the cation and the anion (we work with a 1:1 electrolyte), the dielectric constant is $\epsilon = 78.5$, the temperature is 298.15 K. The ions electrostatically interact through the screened Coulomb potential if they do not overlap (which is forbidden). The membrane is confined between two hard walls (their distance is 30 Å), with which the ions cannot overlap.

A pore of radius 4 Å penetrates the membrane. The pore has hard walls with which the ions cannot overlap. The central cylindrical portion (of length 20 Å) represents the selectivity filter.

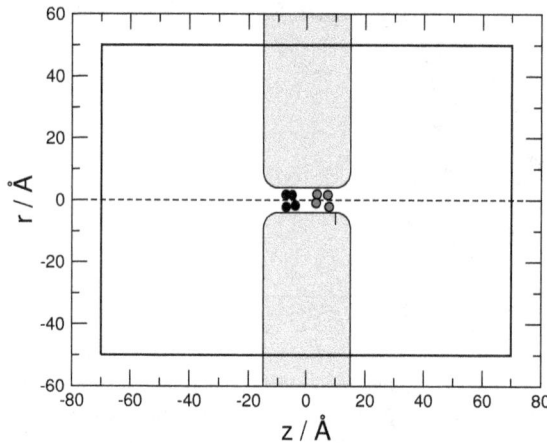

The amino acid side chains are represented with charged hard spheres with radius 1.4 Å. Four positive hard spheres ($0.5e$ charge) are confined in the $(-8 \text{ Å}, -2 \text{ Å})$ region, while four negative hard spheres ($-0.5e$ charge) are confined in the $(2 \text{ Å}, 8 \text{ Å})$ region. These structural ions are confined using a smooth potential described by Malasics et al. [their equation (1)] [76].

The diffusion coefficient of both ionic species was $D_i^{\text{bulk}} = 1.334 \times 10^{-9} \text{ m}^2\text{s}^{-1}$ in the bulk, while it is smaller in the selectivity filter (D_i^{filter}; it is a parameter we can change). In the vestibules the diffusion coefficient is interpolated between these two values in a way described by Boda [51].

The simulation cell is a finite cylinder with hard walls (the 3D cell is obtained by rotating figure 5 around the z-axis). The two cylindrical compartments on the two sides of the membrane represent the two bulk regions between which the ion transport flows. Such a bulk compartment has two parts: one is a transport region that is in non-equilibrium (indicated by a blue line), and the other is an equilibrium bulk region that surrounds it (outside of the blue line). The NP transport equation is solved for the transport region and the boundary conditions are specified on the outer surfaces of the transport regions (two half cylinders).

Figure 5. (Color online) Reduced model of a bipolar ion channel.

3.2. NP+LEMC method

The ion transport is described by the NP transport equation:

$$-k_B T \mathbf{j}_i(\mathbf{r}) = D_i(\mathbf{r})c_i(\mathbf{r})\nabla\mu_i(\mathbf{r}), \tag{2}$$

where $\mathbf{j}_i(\mathbf{r})$ is the particle flux density, k_B is Boltzmann's constant, $D_i(\mathbf{r})$ is the diffusion coefficient profile, $c_i(\mathbf{r})$ is the concentration profile, and

$$\mu_i(\mathbf{r}) = \mu_i^{\text{ch}}(\mathbf{r}) + z_i e \Phi(\mathbf{r}) \tag{3}$$

is the electrochemical potential profile that is the sum of the chemical potential

$$\mu_i^{\text{ch}}(\mathbf{r}) = \mu_i^0 + k_B T \ln c_i(\mathbf{r}) + \mu_i^{\text{ex}}(\mathbf{r}) \tag{4}$$

and the interaction with the mean electric potential, $\Phi(\mathbf{r})$. In these equations, z_i is the ionic valence, e is the elementary charge, μ_i^0 is a standard chemical potential, and $\mu_i^{\text{ex}}(\mathbf{r})$ is the excess chemical potential profile. The transport is driven by the gradient of the electrochemical potential, $\nabla\mu_i(\mathbf{r})$.

To solve the NP equation, we need a closure between $c_i(\mathbf{r})$ and $\mu_i(\mathbf{r})$. In the Poisson-Nernst-Planck (PNP) theory [8–10, 12–14, 19, 24, 77–83], this closure is provided by the Poisson-Boltzmann theory. For the hard sphere ions studied here, this theory cannot be applied, because it is a mean field approach for point charges. To handle the hard sphere ions, a more developed statistical mechanical theory is needed, for example, the Density Functional Theory of Gillespie et al. [84, 85].

Here, we use the LEMC method that is an adaptation of the GCMC method for a non-equilibrium situation [48–51]. The system is divided into small elementary cells, $\mathcal{D}_k$, in which different electrochemical potentials can be assumed [$\mu_i(\mathbf{r}_k)$, where $\mathbf{r}_k$ is the center of $\mathcal{D}_k$]. Such an elementary cell is assumed to be in local equilibrium that makes it possible to perform particle insertions and deletions with the acceptance criterion of GCMC simulations, but using the particle number in the given cell, N_k, its volume, V_k, and the electrochemical potential assigned to the cell, $\mu_i(\mathbf{r}_k)$. The energy of the ion insertion/deletion contains the interaction with all the ions in the whole simulation cell and the interaction with the applied field, $\Phi^{\text{app}}(\mathbf{r})$.

The applied field is computed by solving Laplace's equation for the empty solvation domain (all the charges removed) with the Dirichlet boundary condition that the potential is zero at the half cylinder on the left-hand side and the value of the voltage, U, at the half cylinder on the right-hand side. These surfaces are indicated with a blue line in figure 5. The NP equation is solved inside this surface.

The LEMC simulation provides the concentration profiles as an output, $c_i(\mathbf{r}_k)$, given an electrochemical potential profile, $\mu_i(\mathbf{r}_k)$. An iteration procedure is used to obtain a self-consistent system in which the flux satisfies the continuity equation, $\nabla \cdot \mathbf{j}_i(\mathbf{r}) = 0$, namely, the conservation of mass. The heart of the iteration can be summarized as

$$\mu_i[n] \xrightarrow{\text{LEMC}} c_i[n] \xrightarrow{\nabla \cdot \mathbf{j}^\alpha = 0} \mu_i[n+1]. \tag{5}$$

Starting from an electrochemical potential profile in iteration n, the concentration profile for that iteration, $c_i[n]$, is obtained from LEMC. The electrochemical potential profile for the next iteration is obtained from writing the integral form of the continuity equation for the elementary cell, $\mathcal{D}_k$, as

$$\oint_{\mathcal{D}_k} \mathbf{j}_i \cdot \mathbf{da} = 0 \tag{6}$$

and substituting the NP equation for $\mathbf{j}_i$:

$$\oint_{\mathcal{D}_k} D_i c_i[n] \nabla \mu_i[n+1] \cdot \mathbf{da} = 0. \tag{7}$$

The electrochemical potential for the next iteration, $\mu_i[n+1]$, satisfies conservation of mass together with the concentration in the previous iteration, $c_i[n]$. The iteration provides the $c_i(\mathbf{r})$ and $\mu_i(\mathbf{r})$ profiles fluctuating around their limiting distributions. The final results are obtained as running averages.

3.3. Results for the reduced model

The electrical current flowing through the pore is obtained from

$$I = -\sum_i z_i e \int_A \mathbf{j}_i \cdot \mathbf{da}, \tag{8}$$

where A is the cross section of the pore. The negative sign makes the current positive for positive voltage.

The current-voltage curves for different values of the filter diffusion constant are shown in figure 6. Rectification is clearly observed; the current is larger in magnitude at positive voltages than at negative voltages [see the rectification curves in the inset; rectification is defined in equation (1)]. Decreasing D_i^{filter} decreases the net current, but it has no effect on rectification. If we increase the number of positive/negative structural charges in the filters, rectification improves (results not shown). The fact that rectification is not sensitive to the diffusion constant indicates that rectification is rather determined by another factor in the NP equation: concentration, $c_i(\mathbf{r})$.

The effective local concentration profiles are shown in figure 7. The figure illuminates the rectification mechanism observed in the reduced model of a bipolar ion channel. It can be summarized as follows:

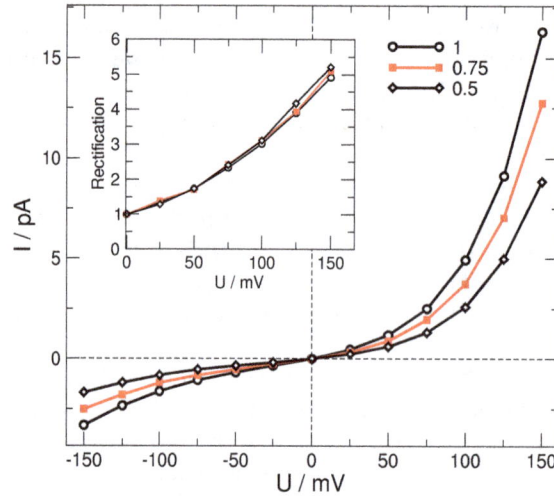

Figure 6. (Color online) Current-voltage curves for different values of the ratio $D_i^{\text{filter}}/D_i^{\text{bulk}}$.

(1) Both ions have depletion zones in the zones whose structural ions repulse them. Anion profiles are depleted on the right-hand side (top panel), while cation profiles are depleted on the left-hand side (bottom panel). (2) The profiles are more depleted in the OFF state (red curves with open symbols).

The rectification mechanism is similar to that observed in semiconductor diodes from the point of view that enhanced depletion in the OFF state produces rectification, but the list of similarities ends here. In the case of semiconductors, the width of the N-P junction is modulated by the voltage. When electrons get into the P zone, they produce a net current even if they are not the majority charge carriers there. The reason is that they recombine with holes arriving from the other direction.

In the case of the ion channel, ions are the charge carriers that cannot recombine. Therefore, the anions, for example, must conduct current in their own depletion zone (the P zone) if we want a net current. The same is true for the cations. The total current is determined by the depletion zone, because that is the largest resistance element of the resistors connected in series if we imagine the slabs as resistors in an equivalent circuit.

The OFF state of the voltage makes its own depletion zone of an ionic species even deeper. The important zone from the point of view of depletion is not the junction zone between the N and P regions, but the N and P regions themselves.

This finding contradicts the usual assumption in the ion channel and nanopore literature where authors assume that the rectification mechanism is the same in semiconductor and ionic devices. We showed here that this is not necessarily true. A deeper discussion of the rectification mechanism observed for the ionic diode follows.

4. Discussion

In the following, we analyze how the concentration profiles become more depleted in the OFF state. First, we must realize that electrical double layers are formed at the two sides of the membrane. For example, in figure 7 at 100 mV (black curves with full circles) there are more anions at the left-hand side than cations, and vice versa on the right hand side. The important thing is that double layers of the opposite sign are formed in the case of -100 mV.

To understand why these oppositely charged double layers are formed, we need to look at the potential profiles (figure 8). The average electrostatic potential can be computed during simulation by inserting test charges into the system, sampling the potential with them, and averaging. The total electrostatic potential has two components:

$$\Phi^{\text{reduced}}(\mathbf{r}) = \Phi^{\text{app}}(\mathbf{r}) + \Phi^{\text{ion}}(\mathbf{r}). \tag{9}$$

The applied potential created by the electrode charges, $\Phi^{\text{app}}(\mathbf{r})$, is obtained by solving Laplace's equation

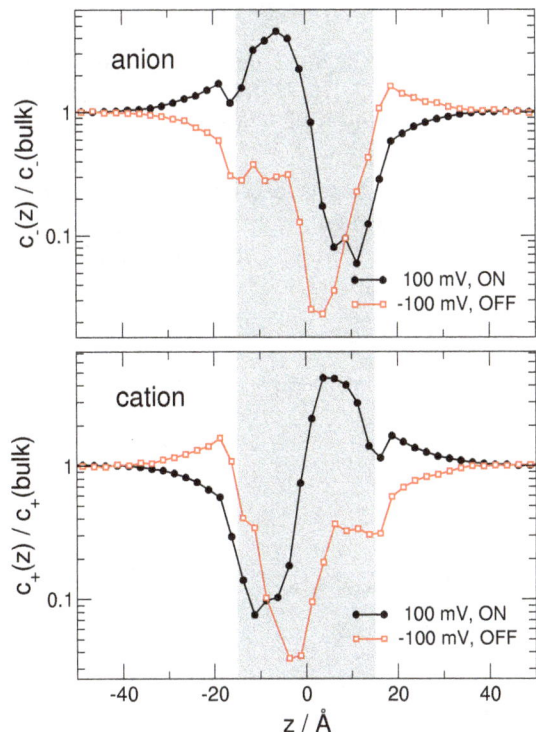

Figure 7. (Color online) Normalized local concentration profiles (normalized with the bulk value, 0.1 M) for Na^+ and Cl^- ions for 100 and -100 mV. The gray area indicates the membrane region.

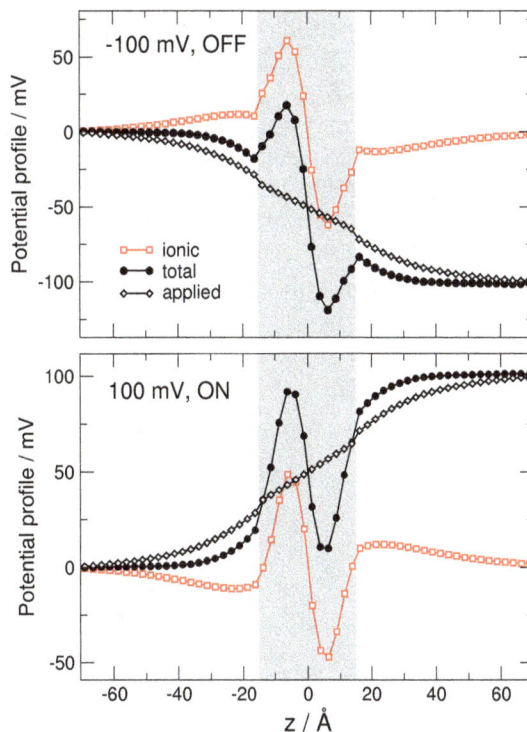

Figure 8. (Color online) Potential profiles and its two components (applied and ionic) for 100 and -100 mV. The z-dependent profiles are obtained by averaging the potentials in equation (9) over the cross section.

for the ion-free system with the prescribed Dirichlet boundary conditions. The other term is the potential produced by the ions, $\Phi^{ion}(\mathbf{r})$, that is related to the ionic charge distribution (including the structural ions) through Poisson's equation.

The slope of the total potential profile is supposed to be small in the bulk solutions because the resistance of the bulk electrolytes is small compared to the ion channel. The potential drop across the membrane region dominates over the drops in the bulk regions (solid curves with full circles). To achieve this, the ions have to arrange into a distribution that imposes an appropriate counterfield (red curves with open squares) against the applied potential. This is the $\Phi^{ion}(\mathbf{r})$ term that is zero at the boundaries of the system, as it should be, if we expect it from the total potential to satisfy the boundary conditions.

The $\Phi^{ion}(\mathbf{r})$ profile is decisively influenced by the double layers shown in figure 7. For example, in the OFF state we have a positive double layer on the left-hand side. That produces the positive ionic potential on the left-hand side in the OFF state (top panel of figure 8).

Now, let us return to figure 7 and analyse the concentration profiles further. We have a more depleted cation profile on the left-hand side of the pore, in the N region at -100 mV (red curves with open symbols, bottom panel). This seems to contradict the observation that we have more cations in the neighboring double layer on the left-hand side.

The contradiction can be resolved if we realize that the change of the sign of the double layer has a direct effect on the other ion, the majority carriers. For example, changing the voltage from 100 mV to -100 mV, the concentration of anions severely drops in the left-hand side double layer and also in the left-hand side N region (note the logarithmic scale). The drop of the cation profile is a consequence of the drop of the anion profile. The mere reason that there are cations in the N region is because the anions drag them along. If there is less anion, there is less cation.

This is an important distinction compared to the electron/hole charge carriers. Cations and anions do not recombine, but they are trying to stay close to each other and screen each other's electric field. If the amount of cations in the N zone is already small, it becomes even smaller if the amount of anions (that are eventually responsible for bringing the cations in) decreases.

Therefore, the change in the voltage sign has an indirect effect on the depletion zones of the minority charge carriers in a given zone. The OFF-state voltage creates double layers that deplete the majority charge carriers in the given zone. The further depletion of the minority charge carriers is a consequence of the depletion of the majority charge carriers.

The question arises why the all-atom model does not show the expected behavior. We do not see significant double layers in figure 4 and, what is more important, we do not see a significant effect of the sign change of the voltage. This can be seen even more clearly if we plot the charge profiles (the difference of cation and anion profiles). Figure 9 shows the profiles for both models.

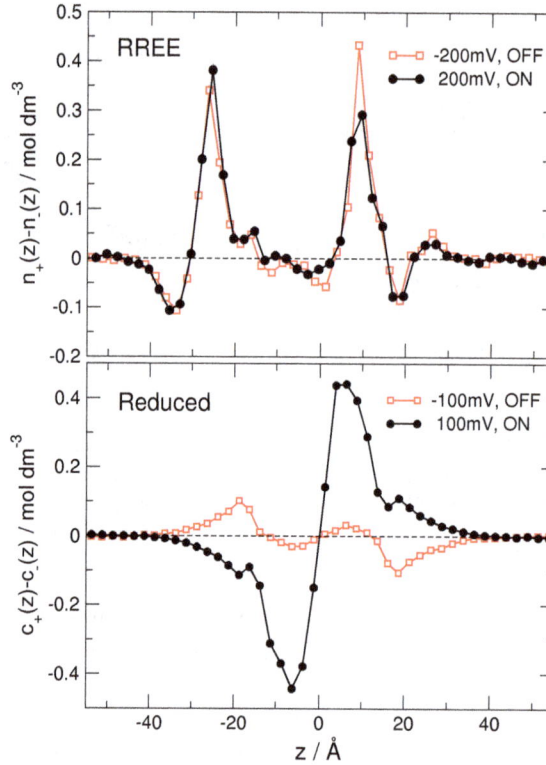

Figure 9. (Color online) Difference of the cation and anion distribution profiles for the RREE all-atom model (top panel, the difference of concentration profiles, $n_i(z)$, is shown) and for the reduced model (bottom panel, the difference of the local concentration profiles, $c_i(z)$, is shown) for positive and negative voltages.

While the charge profiles for the reduced model clearly exhibit the change in the sign of the double layers as a consequence of the change in sign of the voltage (bottom panel), we do not see such an effect in the case of the all-atom model (top panel). The oppositely charged double layers that are so distinct and important for rectification in the reduced model are also absent in the all-atom model.

To understand the absence of double layers, let us investigate the potential profiles obtained for the all-atom model of the RREE mutant (figure 10). Now there are more players in the simulation cell, so the potential has more components. In addition to the $\Phi^{ion}(\mathbf{r})$ term that was produced solely by the ions in the reduced model, now we have components due to the partial charges in the protein, the membrane, and water.

$$\Phi^{all\text{-}atom}(\mathbf{r}) = \Phi^{app}(\mathbf{r}) + \Phi^{ion}(\mathbf{r}) + \Phi^{protein}(\mathbf{r}) + \Phi^{membrane}(\mathbf{r}) + \Phi^{water}(\mathbf{r}). \qquad (10)$$

Figure 10 shows these four terms in the left-hand panels for voltages ±200 mV. The right-hand panels show the total potential with (top) and without (bottom) the applied potential. Our statistics, unfortunately, are quite poor, but a few major conclusions can be drawn nevertheless.

First, the qualitative statement can be laid down that the external field polarizes the system. In this

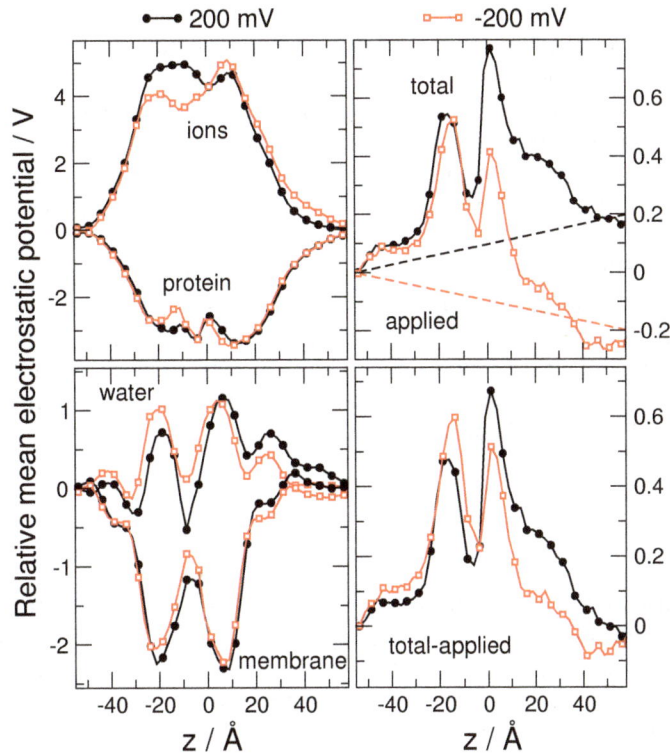

Figure 10. (Color online) Relative electrostatic potential profiles and its components [see equation (10)] for 200 and −200 mV for the all-atom model of the RREE mutant. The data are plotted relative to the left-hand side of the cell. The results have been obtained by inserting test charges into $2.5 \times 2.5 \times 2.5$ Å^3 cubes in that part of the system that is attainable for ions (practically, the electrolyte) and averaging over configurations.

case, however, it polarizes not only the ionic distribution, but also everything else in the system that carries partial charges. The external field must exert work to polarize the system. Only a portion of this work is spent on the ions, most of it is "wasted" on the protein, the membrane, and the water molecules. The ionic profiles, therefore, do not respond so sensitively to the polarizing field as in the case of the reduced model.

Although the poor statistics prevent us from drawing accurate quantitative conclusions, it seems that water is the component that is chiefly responsible for creating the counterfield to the applied field (bottom-left-hand panel). The ionic profiles also respond to some degree if we look at the potential (top-left-hand panel), but this does not manifest in the change of the ionic distribution that would be sufficient to produce rectification on the basis of the mechanism seen in the reduced model.

The slope of the total potential in the right hand bulk is close to zero (top-right-hand panel) as a result of the counterfield (bottom-right-hand panel) that is added to the applied field (dashed lines in the top-right-hand panel). This is clearly seen despite the poor statistics and the small size of the simulation cell.

As a matter of fact, this was the reason that we performed the simulations for the large (four trimer, 5 million atoms) simulation cell. We hoped that in a larger bulk we had more space for the double layers. Potential profiles have not been calculated for the large cell, but for the concentration profiles and the current we obtained the same results as in the small cell.

Summarizing, the presence of all the other atoms and charges in the all-atom system screen the small N-P region inside the pore so effectively that the external field has no observable effect on the ionic profiles, and, thus, its sign change does not produce any observable rectification.

Another possible explanation of the lack of rectification can be deduced from the top panel of figure 9. The charge profile looks like a P-N-P charge distribution rather than an N-P one. It is quite symmetric, albeit rectification requires asymmetry in the charge distribution.

Although these explanations of the failure of the all-atom model make sense, they do not explain why the mutated ion channel of Miedema et al. [47] does rectify. The answer can be some local structural effect that our model cannot capture.

It is strange that (1) Miedema et al. [47] assumed a mechanism of rectification (the N-P junction), (2) created an ion channel based on that assumption with point mutation, (3) showed that the channel really rectifies, but (4) the all atom model of this mutant does not rectify. (5) In the meantime, the reduced model — based on the same assumption Miedema et al. [47] started with — does rectify.

We do not really know where is the flaw in this chain. It can be that the mutant of Miedema et al. [47] folds in a way that has nothing to do with how we imagine its folding. It can be that the all-atom model is not accurate enough due to force field problems. By all accounts, there are several problems with classical force fields. They seem to be more appropriate to study local effects rather than long-range phenomena, including an applied field, screening double layers, and so on. Force fields that handle polarization more realistically are definitely needed. Finally, it can be some kind of problem with the MD methodology, although we think that this is improbable.

5. Summary

In this work we presented a system, in which a powerful experimental fact (rectification) is studied with all-atom and reduced models. We found the puzzling result that the all-atom model, that is supposed to be more "accurate", cannot reproduce rectification, while the reduced model, that is admittedly simplistic, can. The results show that there are cases when reduced representations can serve our purpose better than detailed representations if our purpose is to understand a given phenomenon.

Rectification is a result of the balance of long-range effects, such as the applied field and the counterfield of the ionic distributions. If we concentrate on these effects in our reduced model, we can better focus on the phenomenon at hand. Building efficient reduced models is far from being trivial. Our earlier works for ion channels [50, 51, 60–75] showed that such models can capture an essential portion of reality that is necessary, and in some cases sufficient, to explain a well-specified phenomenon. All-atom models, however, can guide us in creating these models.

The other main message of this paper is that rectification mechanism in bipolar ionic diodes (biological ion channels or narrow synthetic nanopores) is different from the mechanism in semiconductor N-P diodes. One of the reasons is that ions are different in nature from electrons and holes. Cations and anions do not recombine, so an ion must go through the whole pore all the way, including its own depletion zone. That depletion zone is more depleted at the OFF sign of the voltage than at the ON sign.

The explanation is the effect of the double layers formed at the entrances of the channel as detailed above. These double layers are everywhere. They form at the wall of the nanopore too. If the nanopore is too wide compared to the Debye length, a bulk electrolyte is formed in the center of the pore. In this case, depletion zones do not form and the rectification mechanism described in this paper does not work efficiently. The interesting and efficient pores, therefore, are those whose radius is smaller than the Debye length. Ion channels obviously belong to this category.

The other difference between bipolar ionic and semiconductor diodes, therefore, is that narrow pores are needed in the case of ions as carriers in order to make the formation of depletion zones possible. There is no such a requirement in the case of semiconductors. Furthermore, while the junction region between the N and P regions is important in the case of semiconductors, it is the junction region at the entrances of the pore that has a large impact on the behavior of the system. The double layers extend into the N and P zones and deplete the majority carriers that, in turn, deplete minority carriers further in the OFF state.

Rectification mechanism in long nanopores can be different from that in short nanopores because the resistance of the pore itself dominates over the access resistances at the pore entrances in the case of long pores. This question has been thoroughly discussed by Vlassiouk et al. [14] using both numerical and analytical solutions of PNP. Interestingly, their concentration profiles (figure 2 in reference [14]) do not seem very different from ours (figure 7): in the OFF state, both ions become depleted compared to the ON state. Furthermore, the ions become depleted not only at the junction in the middle (that Vlassiouk et al. call a depletion zone), but also in the entire half zones in the pore (these are the real depletion zones, in our view).

For us, the profiles of Vlassiouk et al. [14] imply a similarity with the mechanism described in the present paper. Although Vlassiouk et al. emphasize that they found "a striking similarity to the corresponding solid-state devices", we suspect that the similarity is limited. It will be fun to sort out these uncertainties in future studies for nanopores. We expect that our NP+LEMC method will provide an additional insight compared to PNP studies due to its improved capabilities to handle ionic correlations in the nanopore and in the ionic double layer.

Acknowledgements

We gratefully acknowledge the financial support of the Hungarian National Research Fund (OTKA NN113527) in the framework of ERA Chemistry and the computational support of the Paderborn Center for Parallel Computing (PC^2) for providing access to the OCuLUS cluster. The financial and infrastructural support of the State of Hungary and the European Union in the frame of the TÁMOP-4.2.2.B-15/1/KONV-2015-0004 and TÁMOP-4.2.1.D-15/1/KONV-2015-0006 projects is gratefully acknowledged. We thank Peter Gurin and Szabolcs Varga for the helpful discussion.

References

1. Siwy Z.S., Adv. Funct. Mater., 2006, **16**, No. 6, 735; doi:10.1002/adfm.200500471.
2. Zhang H., Tian Y., Jiang L., Chem. Commun., 2013, **49**, 10048; doi:10.1039/c3cc45526b.
3. Apel P.Y., Blonskaya I.V., Orelovitch O.L., Ramirez P., Sartowska B.A., Nanotechnology, 2011, **22**, No. 17, 175302; doi:10.1088/0957-4484/22/17/175302.
4. Kubeil C., Bund A., J. Phys. Chem. C, 2011, **115**, No. 16, 7866; doi:10.1021/jp111377h.
5. Stein D., Kruithof M., Dekker C., Phys. Rev. Lett., 2004, **93**, 035901; doi:10.1103/PhysRevLett.93.035901.
6. Shockley W., Electrons and Holes in Semiconductors to Applications in Transistor Electronics, Van Nostrand, New York, 1950.
7. Shur M., Physics of Semiconductor Devices, Prentice Hall, New York, 1990.
8. Daiguji H., Oka Y., Shirono K., Nano Lett., 2005, **5**, No. 11, 2274; doi:10.1021/nl051646y.
9. Karnik R., Duan C., Castelino K., Daiguji H., Majumdar A., Nano Lett., 2007, **7**, No. 3, 547; doi:10.1021/nl062806o.
10. Constantin D., Siwy Z.S., Phys. Rev. E, 2007, **76**, No. 4, 041202; doi:10.1103/PhysRevE.76.041202.
11. Gracheva M.E., Vidal J., Leburton J.P., Nano Lett., 2007, **7**, No. 6, 1717; doi:10.1021/nl0707104.
12. Vlassiouk I., Siwy Z.S., Nano Lett., 2007, **7**, No. 3, 552; doi:10.1021/nl062924b.
13. Kalman E.B., Vlassiouk I., Siwy Z.S., Adv. Mater., 2008, **20**, No. 2, 293; doi:10.1002/adma.200701867.
14. Vlassiouk I., Smirnov S., Siwy Z., ACS Nano, 2008, **2**, No. 8, 1589; doi:10.1021/nn800306u.
15. Yan R., Liang W., Fan R., Yang P., Nano Lett., 2009, **9**, No. 11, 3820; doi:10.1021/nl9020123.
16. Cheng L.J., Guo L.J., ACS Nano, 2009, **3**, No. 3, 575; doi:10.1021/nn8007542.
17. Cheng L.J., Guo L.J., Chem. Soc. Rev., 2010, **39**, 923; doi:10.1039/B822554K.
18. Nguyen G., Vlassiouk I., Siwy Z.S., Nanotechnology, 2010, **21**, No. 26, 265301; doi:10.1088/0957-4484/21/26/265301.
19. Van Oeffelen L., Roy W.V., Idrissi H., Charlier D., Lagae L., Borghs G., PLOS ONE, 2015, **10**, No. 5, e0124171; doi:10.1371/journal.pone.0124171.
20. Siwy Z., Fulinski A., Phys. Rev. Lett., 2002, **89**, No. 19, 198103; doi:10.1103/PhysRevLett.89.198103.
21. Siwy Z., Apel P., Baur D., Dobrev D.D., Korchev Y.E., Neumann R., Spohr R., Trautmann C., Voss K.O., Surf. Sci., 2003, **532**, 1061; doi:10.1016/S0039-6028(03)00448-5.
22. Siwy Z., Apel P., Dobrev D., Neumann R., Spohr R., Trautmann C., Voss K., Nucl. Instrum. Methods Phys. Res., Sect. B, 2003, **208**, 143; doi:10.1016/S0168-583X(03)00884-X.
23. Howorka S., Siwy Z., In: Handbook of Single-Molecule Biophysics, Advances in Chemical Physics, Hinterdorfer P., van Oijen A. (Eds.), Springer, New York, 2009, Chapter 11, 293–339; doi:10.1007/978-0-387-76497-9_11.
24. Cervera J., Ramírez P., Mafe S., Stroeve P., Electrochim. Acta, 2011, **56**, No. 12, 4504; doi:10.1016/j.electacta.2011.02.056.
25. Tybrandt K., Forchheimer R., Berggren M., Nat. Commun., 2012, **3**, 871; doi:10.1038/ncomms1869.
26. Guan W., Li S.X., Reed M.A., Nanotechnology, 2014, **25**, No. 12, 122001; doi:10.1088/0957-4484/25/12/122001.
27. Sexton L.T., Horne L.P., Martin C.R., Mol. Biosyst., 2007, **3**, 667; doi:10.1039/b708725j.
28. Howorka S., Siwy Z., Chem. Soc. Rev., 2009, **38**, No. 8, 2360; doi:10.1039/b813796j.

29. Vlassiouk I., Kozel T.R., Siwy Z.S., J. Am. Chem. Soc., 2009, **131**, No. 23, 8211; doi:10.1021/ja901120f.

30. Piruska A., Gong M., Sweedler J.V., Chem. Soc. Rev., 2010, **39**, 1060; doi:10.1039/B900409M.

31. Hille B., Ion Channels of Excitable Membranes, 3rd Edn., Sinauer Associates, Sunderland, 2001.

32. Biological Membrane Ion Channels; Dynamics, Structure, and Applications, Chung S.H., Andersen O.S., Krishna-murthy V. (Eds.), Springer, New York, 2007; doi:10.1007/0-387-68919-2.

33. Eisenberg R.S., In: Recent Developments in Theoretical Studies of Proteins, Advanced Series in Physical Chem-istry, Vol. 7, Elber R. (Ed.), World Scientific, Philadelphia, 1996, 269–357; doi:10.1142/9789814261418_0005.

34. Pauptit R.A., Zhang H., Rummel G., Tilman S., Jansonius J.N., Rosenbusch J.P., J. Mol. Biol., 1991, **218**, No. 3, 505; doi:10.1016/0022-2836(91)90696-4.

35. Cowan S.W., Schirmer T., Rummel G., Steiert M., Ghosh R., Pauptit R.A., Jansonius J.N., Rosenbusch J.P., Nature, 1992, **358**, No. 6389, 727; doi:10.1038/358727a0.

36. Tieleman D.P., Berendsen H.J.C., Biophys. J., 1998, **74**, No. 6, 2786; doi:10.1016/S0006-3495(98)77986-X.

37. Phale P.S., Philippsen A., Widmer C., Phale V.P., Rosenbusch J.P., Schirmer T., Biochem., 2001, **40**, No. 21, 6319; doi:10.1021/bi010046k.

38. Robertson K.M., Tieleman D.P., FEBS Lett., 2002, **528**, No. 1–3, 53; doi:10.1016/S0014-5793(02)03173-3.

39. Danelon C., Suenaga A., Winterhalter M., Yamato I., Biophys. Chem., 2003, **104**, No. 3, 591; doi:10.1016/S0301-4622(03)00062-0.

40. Baaden M., Sansom M.S., Biophys. J., 2004, **87**, No. 5, 2942; doi:10.1529/biophysj.104.046987.

41. Varma S., Chiu S.W., Jakobsson E., Biophys. J., 2006, **90**, No. 1, 112; doi:10.1529/biophysj.105.059329.

42. Aguilella-Arzo M., García-Celma J.J., Cervera J., Alcaraz A., Aguilella V.M., Bioelectrochem., 2007, **70**, No. 2, 320; doi:10.1016/j.bioelechem.2006.04.005.

43. Biró I., Pezeshki S., Weingart H., Winterhalter M., Kleinekathöfer U., Biophys. J., 2010, **98**, No. 9, 1830; doi:10.1016/j.bpj.2010.01.026.

44. Faraudo J., Calero C., Aguilella-Arzo M., Biophys. J., 2010, **99**, No. 7, 2107; doi:10.1016/j.bpj.2010.07.058.

45. Alcaraz A., Queralt-Martín M., García-Giménez E., Aguilella V.M., Biochim. Biophys. Acta, Biomembr., 2012, **1818**, No. 11, 2777; doi:10.1016/j.bbamem.2012.07.001.

46. Matsuura Y., Yamato I., Ando T., Suenaga A., J. Comput. Chem. Jpn., 2014, **13**, No. 5, 278; doi:10.2477/jccj.2014-00.

47. Miedema H., Vrouenraets M., Wierenga J., Meijberg W., Robillard G., Eisenberg B., Nano Lett., 2007, 7, No. 9, 2886; doi:10.1021/nl0716808.

48. Boda D., Gillespie D., J. Chem. Theor. Comput., 2012, **8**, No. 3, 824; doi:10.1021/ct2007988.

49.Ható Z., Boda D., Kristóf T., J. Chem. Phys., 2012, **137**, No. 5, 054109; doi:10.1063/1.4739255.

50. Boda D., Kovács R., Gillespie D., Kristóf T., J. Mol. Liq., 2014, **189**, 100; doi:10.1016/j.molliq.2013.03.015.

51. Boda D., Ann. Rep. Comp. Chem., 2014, **10**, 127; doi:10.1016/B978-0-444-63378-1.00005-7.

52. Humphrey W., Dalke A., Schulten K., J. Mol. Graphics, 1996, **14**, 33; doi:10.1016/0263-7855(96)00018-5.

53. Im W., Roux B., J. Mol. Biol., 2002, **319**, No. 5, 1177; doi:10.1016/S0022-2836(02)00380-7.

54. Jo S., Kim T., Iyer V.G., Im W., J. Comp. Chem., 2008, **29**, No. 11, 1859; doi:10.1002/jcc.20945.

55. Berendsen H.J.C., van der Spoel D., van Drunen R., Comput. Phys. Commun., 1995, **91**, No. 1–3, 43; doi:10.1016/0010-4655(95)00042-E.

56. Pronk S., Páll S., Schulz R., Larsson P., Bjelkmar P., Apostolov R., Shirts M.R., Smith J.C., Kasson P.M., van der Spoel D., Hess B., Lindahl E., Bioinformatics, 2013, **29**, No. 7, 845; doi:10.1093/bioinformatics/btt055.

57. Nosé S., Klein M.L., Mol. Phys., 1983, **50**, No. 5, 1055; doi:10.1080/00268978300102851.

58. Parrinello M., Rahman A., J. Appl. Phys., 1981, **52**, No. 12, 7182; doi:10.1063/1.328693.

59. Bjelkmar P., Larsson P., Cuendet M.A., Hess B., Lindahl E., J. Chem. Theory Comput., 2010, **6**, No. 2, 459; doi:10.1021/ct900549r.

60. Nonner W., Catacuzzeno L., Eisenberg B., Biophys. J., 2000, **79**, No. 4, 1976; doi:10.1016/S0006-3495(00)76446-0.

61. Nonner W., Gillespie D., Henderson D., Eisenberg B., J. Phys. Chem. B, 2001, **105**, No. 27, 6427; doi:10.1021/jp010562k.

62. Boda D., Valiskó M., Eisenberg B., Nonner W., Henderson D., Gillespie D., J. Chem. Phys., 2006, **125**, No. 3, 034901; doi:10.1063/1.2212423.

63. Boda D., Valiskó M., Eisenberg B., Nonner W., Henderson D., Gillespie D., Phys. Rev. Lett., 2007, **98**, No. 16, 168102; doi:10.1103/PhysRevLett.98.168102.

64. Gillespie D., Boda D., Biophys. J., 2008, **95**, No. 6, 2658; doi:10.1529/biophysj.107.127977.

65. Boda D., Valiskó M., Henderson D., Eisenberg B., Gillespie D., Nonner W., J. Gen. Physiol., 2009, **133**, No. 5, 107; doi:10.1085/jgp.200910211.

66. Malasics M., Boda D., Valiskó M., Henderson D., Gillespie D., Biochim. Biophys. Acta, Biomembr., 2010, **1798**, No. 11, 2013; doi:10.1016/j.bbamem.2010.08.001.

67. Boda D., Giri J., Henderson D., Eisenberg B., Gillespie D., J. Chem. Phys., 2011, **134**, No. 5, 055102; doi:10.1063/1.3532937.

68. Boda D., Henderson D., Gillespie D., J. Chem. Phys., 2013, **139**, No. 5, 055103; doi:10.1063/1.4817205.

69. Gillespie D., Xu L., Wang Y., Meissner G., J. Phys. Chem. B, 2005, **109**, No. 32, 15598; doi:10.1021/jp052471j.

70. Gillespie D., Biophys. J., 2008, **94**, No. 4, 1169; doi:10.1529/biophysj.107.116798.

71. Gillespie D., Fill M., Biophys. J., 2008, **95**, No. 8, 3706; doi:10.1529/biophysj.108.131987.

72. Gillespie D., Giri J., Fill M., Biophys. J., 2009, **97**, No. 8, 2212; doi:10.1016/j.bpj.2009.08.009.

73. Boda D., Henderson D., Busath D.D., Mol. Phys., 2002, **100**, No. 14, 2361; doi:10.1080/00268970210125304.

74. Boda D., Nonner W., Valiskó M., Henderson D., Eisenberg B., Gillespie D., Biophys. J., 2007, **93**, No. 6, 1960; doi:10.1529/biophysj.107.105478.

75. Boda D., Leaf G., Fonseca J., Eisenberg B., Condens. Matter Phys., 2015, **18**, No. 1, 13601; doi:10.5488/CMP.18.13601.

76. Malasics A., Gillespie D., Nonner W., Henderson D., Eisenberg B., Boda D., Biochim. Biophys. Acta, Biomembr., 2009, **1788**, No. 12, 2471; doi:10.1016/j.bbamem.2009.09.022.

77. Nonner W., Chen D.P., Eisenberg B., Biophys. J., 1998, **74**, No. 5, 2327; doi:10.1016/S0006-3495(98)77942-1.

78. Nonner W., Eisenberg B., Biophys. J., 1998, **75**, 1287; doi:10.1016/S0006-3495(98)74048-2.

79. Aguilella-Arzo M., Aguilella V.M., Eisenberg R.S., Eur. Biophys. J., 2005, **34**, No. 4, 314; doi:10.1007/s00249-004-0452-x.

80. Daiguji H., Yang P., Majumdar A., Nano Lett., 2004, **4**, No. 1, 137; doi:10.1021/nl0348185.

81. Cervera J., Schiedt B., Neumann R., Mafe S., Ramirez P., J. Chem. Phys., 2006, **124**, No. 10, 104706; doi:10.1063/1.2179797.

82. Wolfram M.T., Burger M., Siwy Z.S., J. Phys.: Condens. Matter, 2010, **22**, No. 45, 454101; doi:10.1088/0953-8984/22/45/454101.

83. Pietschmann J.F., Wolfram M.T., Burger M., Trautmann C., Nguyen G., Pevarnik M., Bayer V., Siwy Z., Phys. Chem. Chem. Phys., 2013, **15**, 16917; doi:10.1039/c3cp53105h.

84. Gillespie D., Nonner W., Eisenberg R.S., J. Phys.: Condens. Matter, 2002, **14**, No. 46, 12129; doi:10.1088/0953-8984/14/46/317.

85. Gillespie D., Nonner W., Eisenberg R.S., Phys. Rev. E, 2003, **68**, No. 3, 031503; doi:10.1103/PhysRevE.68.031503.

13

Proteins in solution: Fractal surfaces in solutions

R. Tscheliessnig[1], L. Pusztai[2]

[1] Austrian Centre for Industrial Biotechnology (ACIB), Muthgasse 11, A-1190 Wien, Austria

[2] Wigner Research Centre for Physics, Hungarian Academy of Sciences,
Konkoly Thege út. 29-33, H-1121, Budapest, Hungary

The concept of the surface of a protein in solution, as well of the interface between protein and 'bulk solution', is introduced. The experimental technique of small angle X-ray and neutron scattering is introduced and described briefly. Molecular dynamics simulation, as an appropriate computational tool for studying the hydration shell of proteins, is also discussed. The concept of protein surfaces with fractal dimensions is elaborated. We finish by exposing an experimental (using small angle X-ray scattering) and a computer simulation case study, which are meant as demonstrations of the possibilities we have at hand for investigating the delicate interfaces that connect (and divide) protein molecules and the neighboring electrolyte solution.

Key words: *protein solution, protein hydration, protein surface, small angle scattering*

1. Introduction

The appearance of aqueous solutions of even large proteins is, in many cases, similar to that of dilute solutions of simple salts: the liquid may be completely transparent, even though the size of solute molecules may be two orders of magnitude larger than that of the solvent particles (i.e., dozens of nanometers). This is made possible by strong interactions between the charged 'surface' of a protein and the dipolar solvent molecules that surround a large particle; sometimes even tiny changes of the conditions (of e.g., composition, temperature) can alter the situation completely and make protein molecules aggregate and precipitate (see, e.g., reference [1, 2]).

The 'surface' of protein molecules in aqueous solutions may be considered as being defined by the hydration sphere of a macromolecule. The natural tool for studying the hydration structure, within distances of a few Å, would be wide angle X-ray (and/or neutron) scattering — just as it is routinely done for solutions of simple salts (see, e.g., reference [3]). However, due to a large number of components in a solution (water, protein, stabilizers), as well as due to the complicated internal structure and relatively low molar concentration of the protein, this route has not been very frequently chosen; examples of such studies are references [4, 5].

Perhaps surprisingly, it is the microscopic dynamics of the hydration sphere that has been more extensively studied than the static structure: this can be readily understood by considering that most of the dynamical studies are based on examining the dynamics of water molecules only. NMR spectroscopy [6, 7], dielectric relaxation spectroscopy [8, 9], as well inelastic neutron scattering [10, 11] have all been applied for the purpose. More recently, terahertz (THz) spectroscopy has been used for tracking changes of the broadly defined hydration layer, up to a thickness of about 1 nm [12, 13].

In the pursuit of revealing the surface of a protein molecule in solutions, small angle scattering (SAS) [14–16] is our chosen experimental method for the present report. SAS provides a (or arguably, the only) viable experimental possibility for studying the shape of a biomolecule in solution, as it has been exemplified in references [16–18]. Unfortunately, the interpretation of SAS data is far from being straightforward: this issue is considered in detail later in this work (see below).

In any case, to make the surface of a biomolecule 'visible', one needs to possess a high (most preferably, atomic) resolution picture of the molecule in solution. No experimental technique is capable of providing such pictures so far: for this reason, we must turn to computer simulations, such as the molecular dynamics (MD) method [19]. Proteins and their solutions have been targeted by MD for quite some time, due to the pioneering works of Karplus and co-workers (see, e.g., reference [20]). The MD methodology will also be made use of extensively in the present work; more details will be provided in due course.

One way of defining the surface of a protein is to evaluate the 'solvent inaccessible' volume of the biomolecule. In the cube method [21], the biomolecule is placed in a parallelepiped-shape box which is subdivided into small cubes with edges of 0.5–1.5 Å. The boundary of the biomolecule is determined by examining whether each cube belongs to the biomolecule or to the solvent (see, e.g., reference [22]). A more complicated method is to calculate the 'electron envelope' of the macromolecule: an algorithm for this is implemented in the program CRYSOL17 [23].

In general, due to a large variety of the ways the beta sheets and alpha helices are put into sequences in protein molecules, the surface of such molecules is rather complicated. In the present contribution, we consider that in general (or at least, in a large number of cases) the boundary of a protein molecule may have a fractal dimension. We pursue this idea by presenting theoretical and experimental arguments; we finish with providing computer simulation results based on simple concepts.

2. Scattering from fractal surfaces

What is a surface in terms of scattering theories? Small angle scattering may provide information on surface areas that are larger and more uniform than that of a biomolecule (see references [14–16]); we must, therefore, take a more indirect way. In fact, connections between the measured intensity and fractal (or 'rough') surfaces have already been sought for [24–26]; note that these investigations have not considered protein surfaces directly.

While the 'reaction coordinate', i.e., the location of a site of importance (e.g., of a scattering site) within the investigated volume in a slit pore seems obvious, as it follows from the symmetry of the pore, it is a complex task to determine if we deal with soft matter, e.g., proteins. Let us take a rather simple protein: it will be formed by alpha helical domains (a typical one is indicated orange in figure 1) and joined by random coils. For the present considerations, we have chosen a well-known globular protein, selected out of thousands of possibilities: Bovine serum albumin (BSA) (for its crystalline structure see reference [27]). We determine its point of reference (in other words, the 'origin' of the system). This is a crucial step because it will mathematically determine what we term a *fractal* surface.

We assume scattering sites in the vicinity of, or indeed, within amino acids. We compute their centroid and from their relative distance we compute the pair densities. The chosen system lacks any symmetry

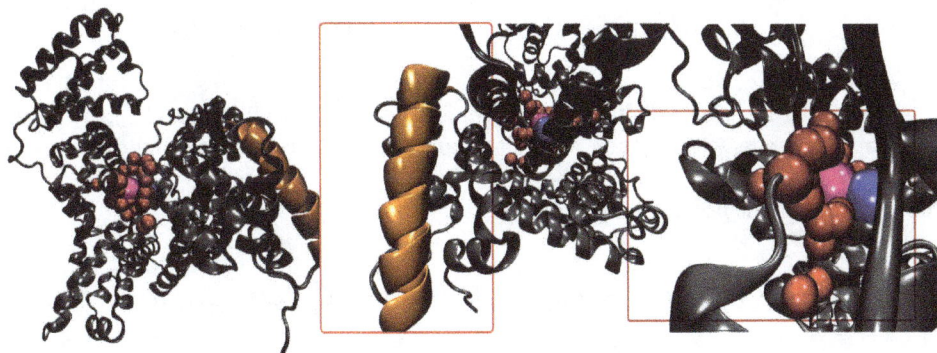

Figure 1. (Color online) The point of reference. Left-hand panel: Any protein is a complex structure of scattering sites. The issue to decide upon: which is the one we refer to? What are then reaction coordinates? Red spheres mark the sites for which the Fourier transform of pair and radial distributions provides comparable results. Their centroid is colored magenta, while blue spheres mark the centroid of all of the protein sites. (From left to right: the important part considered is gradually enlarged.)

and that is why we use equation (2.2) to determine the point of reference and compute, with respect to it, the radial density function.

First, we sketch a mathematical methodology to access structural information from small angle and neutron scattering data; this information will be related to the issue of the surface of a protein. We link the distribution of scattering sites to the definition of α stable distributions [28, 29].

We assume that $\{X_i\}$ and $\{X_j\}$ are random variables. Here, they are distances of scattering sites, with respect to sites i or site j of those variables, and they are distributed according to a particular probability density $\phi(\zeta)$. The distribution is called stable if the probability density $p(\zeta)$ of any linear combination $Y = \lambda_1 X_1 + \lambda_2 X_2$ then,

$$\gamma(\zeta) = \lambda_1 \lambda_2 \int_0^\infty d\zeta' \phi\big((\zeta - \zeta')\lambda_1\big)\phi(\zeta'\lambda_2). \tag{2.1}$$

The distributions coincide subject to rescaling, i.e.,

$$\langle \mathscr{F}(\gamma(\zeta))[Q] \rangle = \langle |\mathscr{F}(\phi(\lambda\zeta))[Q]|^2 \rangle. \tag{2.2}$$

It is a significant extension to the formulation of Kotlarchyk's work [30], as it includes scaling to relate the pair density of scattering sites of a protein with their radial distribution.

Kotlarchyk relates the scattering intensity, $I(Q)$, to the protein form factor $P(Q)$ and the protein-protein structure factor $S(Q)$ by

$$I(Q) = P(Q)\big\{1 + \beta(Q)[S(Q) - 1]\big\}, \tag{2.3}$$

wherein the term $\beta(Q) = P^\star(Q)/P(Q)$ takes into account the possible anisotropic form of a protein. In the previous paper [31] we introduced the fractal pendant to Debye's formula:

$$\mathscr{J}_D[Q\zeta_b] = \frac{J_{D/2-1}(Q\zeta_b)}{(Q\zeta_b)^{D/2-1}}. \tag{2.4}$$

We did give clear evidence [31] that the fractal dimension, D, may be related to the Debye screening length, and that it is not necessarily $D = 3$.

We rewrite the scattering intensity:

$$I(Q) = \mathscr{F}_D(\gamma(\zeta_R))[Q] = \int_0^\infty d\zeta_R \zeta_R^{D-1} \gamma(\zeta_R)\mathscr{J}_D[Q\zeta_b], \tag{2.5}$$

as a function of the pair density of the protein scattering sites $\gamma(\zeta_R)$. The pair density is a function of the relative distances, ζ_R, between individual protein scattering sites. The protein form factor

$$P(Q) = \langle |F(Q)|^2 \rangle = \mathscr{F}_D^2(\phi(\zeta_b))[Q] = \int_0^\infty d\zeta_b \zeta_b^{D-1} \phi(\zeta_b)\mathscr{J}_D[Q\zeta_b]^2, \tag{2.6}$$

however, is the Fourier transform of the radial probability density. It is a function of ζ_b, the distance with respect to an arbitrary site, within the protein. Commonly, the center of mass of the protein centroid is chosen.

In order to compute the protein anisotropy, $\beta(Q)$, we need

$$P^\star(Q) = |\langle F(Q) \rangle|^2 = \mathscr{F}_D(\phi(\zeta_b))[Q]^2 = \left(\int_0^\infty d\zeta_b \zeta_b^{D-1} \phi(\zeta_b)\mathscr{J}_D[Q\zeta_b] \right)^2. \tag{2.7}$$

We shall not explore a detailed deduction but draft the essence, and provide motivation from the observation of scalability by wet lab and computer experiments

$$I(Q) = \mathscr{F}_D(\gamma(\zeta_R))[Q] = \int_0^\infty d\zeta_R \zeta_R^{D/\lambda-1} \gamma(\zeta_R)\mathscr{J}_{D/\lambda}[Q\zeta_b]. \tag{2.8}$$

Due to the scaling capability and alpha stability of the pair density and radial probability density, we are allowed to introduce $\zeta_b = \lambda\zeta_R$ and rewrite the scattering intensity in terms of the protein form factor:

$$P(Q) \quad \propto \quad \int_0^\infty \mathrm{d}\zeta_b \zeta_b^{D-1} \phi(\zeta_b/\lambda)\, \mathscr{J}_D[Q\zeta_b]^2 = \lambda^D \int_0^\infty \mathrm{d}\zeta_R \zeta_R^{D-1} \phi(\zeta_R)\, \mathscr{J}_D[Q\lambda\zeta_R]^2$$

$$\propto \quad \int_0^\infty \mathrm{d}\zeta_R \zeta_R^{D/\lambda-1} \phi(\zeta_R)\, \mathscr{J}_{D/\lambda}[Q\zeta_R]^2. \tag{2.9}$$

The above is a set of equations that we term as 'fractal scattering theory'.

3. Small angle neutron and X-ray scattering from biological soft matter

In this section we briefly discuss the origin of the fractal dimension D.

Small angle neutron and small angle X-ray scattering data were collected to obtain structural information for BSA (concentration: 5 mg/ml) in three different aqueous salt environments (i.e., in different electrolyte solutions). The data are displayed in figure 2. These three different environments contained zero ammonium sulfate (state 1), 0.7 mol/kg ammonium sulfate (state 2), and 1.2 mol/kg (state 3). The pH of the solutions is very close to neutral (just below 7) and these electrolyte concentrations are far below the salting-out limit of the protein. Detailed description of the experiments can be found in references [31, 32].

We argue that the parameter D is considered to be of electrostatic origin and proportional to the salt concentrations in bulk solutions [32]. In figure 2 (right-hand panel) the pair density function of the initial crystallographic model is shown as dashed blue line with grey marker. The fits (solid lines, left-hand panel) were obtained by calculating the pair density function (solid blue line, right-hand panel) and scattering intensity using equation (2.4), with $D = 3$. Clearly, irrespective of the ion concentrations, we do not see significant changes in terms of the electronic contrast.

The SANS measurements were complemented by SAXS measurements for identical solutions. SAXS data are presented in figure 3. Note the discrepancy between the results of the two experimental techniques. Though the systems are identical, their scattering intensities $I(Q)$ differ.

Typically, for small angle X-ray scattering, one is tempted to interpret the changes in $I(Q)$ by the changes in their individual pair density distributions, and then, consequently, argue the changes in the protein conformation. However, this line of arguments is not supported by small angle neutron scattering data.

In what follows, we interpret the data differently: we leave the pair correlation untouched and change the parameter D, which may be interpreted as a fractal Dimension. Note that the fits of experimental data

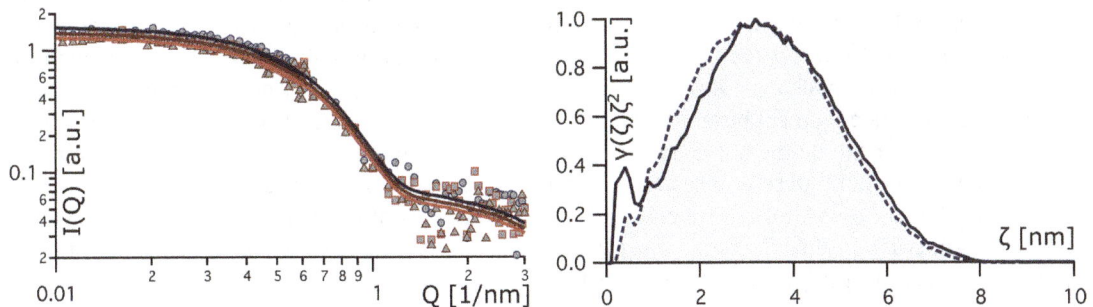

Figure 2. (Color online) SANS data of BSA at different salt concentrations. Left-hand panel: SANS data of BSA dissolved in three different environments which contained zero ammonium sulfate (blue squares), 0.7 mol/kg (i.e., per kg of solution) ammonium sulfate (red circles), and 1.2 mol/kg $(NH_4)_2SO_4$ (brown triangles). Solid lines give fits of experimental data. Right-hand panel: Dashed blue line with grey markers indicates the pair density computed from a crystallographic model of BSA, whereas solid blue line marks the pair density that corresponds to fits shown in the left-hand panel.

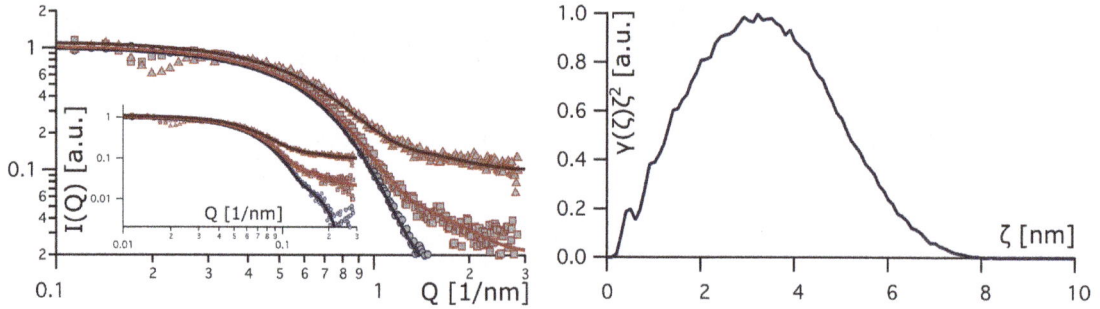

Figure 3. (Color online) SAXS data of BSA at different salt concentrations. Left-hand panel: SAXS data of BSA dissolved in three different environments which contained zero ammonium sulfate (blue squares), 0.7 mol/kg ammonium sulfate (red circles), and 1.2 mol/kg (brown triangles). Solid lines give fits of experimental data, for $D = 3.2$, $D = 2.9$ and $D = 2.5$. Right-hand panel: Solid blue line indicates the pair density computed from a crystallographic model of BSA. All data presented in the left-hand panel were computed from it for different fractal dimensions.

have been achieved by changing $D = 3.2$, $D = 2.9$ and $D = 2.5$ without changing the protein conformation. We left the density distribution of the protein untouched. The linear relation of D to the ionic strength of the particular solution is obvious. The higher the salt concentration the lower D should be put.

For a quick exploration of the effects of varying D, we use a computational approach, using molecular dynamics simulations (see next section).

4. The (fractal) surface of biomolecules: demonstration via computer simulation

Having defined three different quantities, i.e., the form factor, the structure factor and the anisotropic factor [$\beta(Q)$], it is time to explore these and put them in relation to computational approaches, such as density functional theory [33]. Therefore, we set up three systems. We discuss two of them qualitatively, whereas the third one we explore in detail. Since many theoretical systems, especially in the density functional theory, deal with slit pores [33], we shall start with these.

From a mathematical point of view it is difficult to compute the pair distribution of an infinite planar slit pore numerically, as one would need to compute the pairwise densities over all sites of a slit pore. The sum, or moments of the sum, would not necessarily converge: one might think of particle interactions that produce in plane pair densities that we may consider α stable. The common way out is to measure and compute density distributions perpendicular to the surface.

Let us switch to spherical coordinates: we do so for different reasons. They seem mathematically easier as well as they are very frequently applicable in soft matter as many a system investigated is of spherical symmetry. In fact, it may be the experiment as well that imposes spherical symmetry to the measured data, just as small angle X-ray and neutron scattering certainly do (see the previous section).

Let us rethink the planar slit pore to be an infinite spherical one. Then, we have to consider the point of reference, in order to define a reaction coordinate, ζ. For planar slit pores, its particular symmetry suggests to place the point of reference in the center of the slit pore. This may also be used for finite and infinite spherical slit pores. Now, a spherical slit pore will consist of two concentric spheres. The inner shell has a radius of r_∞ while the outer one, a radius of $r_\infty + \Delta$. We define a *radial density distribution* by exploiting the shift property of the Fourier transform:

$$\langle |\mathscr{F}(\phi(\lambda(r_\infty + \zeta)))[Q]|^2 \rangle = \langle |\mathscr{F}(\phi(\lambda\zeta))[Q]|^2 \rangle. \tag{4.1}$$

We hereby reinterpret the planar slit pore to be an infinite spherical slit pore. We shift the point of the origin next to one planar surface since it would be numerically cumbersome to compute the pair distribution of an infinite spherical slit pore, but by the use of equation (2.2). Next, we drop the inner spherical surface and replace it by a 'protein'. We use a simple Lennard Jones (LJ) model. We used the molecular dynamics package LAMMPS [34]. All parameters listed in the subsequent paragraphs are reduced to the wall LJ parameters. To save computational time, we rescale the protein by a factor of five.

We compute the centroids of each amino acid and replace these by LJ sites. 'Pair styles', i.e., specific parameters for the particular pairs of sites (for details, see the LAMMPS Manual [35]) between protein and liquid were put to $\epsilon = 0.1$ and $\sigma = 2.5$. The protein is positioned in its appropriate center, as computed from (2.2). For simplicity, we fix the protein "amino acids" by springs to the centroids. The spring constant that kept sites of the protein was put to $k = 10$. This value was chosen so that the liquid may slightly penetrate the protein. Interactions within the protein were turned off. The protein is dissolved in a LJ liquid. For liquid-liquid interactions, we constructed a hybrid potential by superposing two pair styles, a *lj/soft/cut* and a *gauss/cut*. LJ parameters for liquid-liquid interactions were set to $\epsilon = 0.05$ and $\sigma = 1.5$. A repelling Gaussian potential was added to the liquid-liquid interactions, whose amplitude was set to 0.05. A repelling distance of $\zeta = 1.0$ and a variance of 1 were used. The liquid comprised 3553 sites.

The construction of wall and liquid is enclosed in a spherical wall of a diameter $\zeta_d = 32$. It is a LJ wall of type *wall/lj93* (according to LAMMPS terminology) and parameterized as $\epsilon = 1.0$ and $\sigma = 1.0$.

We performed simple NVE simulations and initially gave all sites to a velocity of 3. After equilibrations of 500 steps, we performed simulations of 5000 steps. The system was reduced to a configuration as shown in figure 4. In the right-hand panel of figure 4 we find the radial distribution of the protein (solid blue line) and the radial distribution of the LJ liquid (solid red line with grey markers). Both were normalized to their maximum value. We rescaled these results to run them comparable to the experimental data. Arrows in figure 4 right-hand panel mark three regions. The reaction coordinate up to the blue arrow is termed protein. We attribute the linear regime (in-between blue and white arrow) to the protein surface, whereas the planar regime (in-between white and red arrow) is attributed to the LJ liquid bulk. The corrugation in the radial density of the LJ liquid around 8 nm proves the liquid-like state of it.

In figure 5, scattering profiles for protein plus protein surfaces of different thicknesses are displayed. All these complexes are in the linear regime shown by the insert of figure 4. While the blue line refers to the hypothetical scattering profile of a blank protein, the orange lines refer to scattering profiles and pair densities of the protein embedded in LJ liquid of different thickness. The pair densities are self-similar. The larger is the construct, then the corresponding scattering profile is found more to the left.

In the structure model in figure 5 we discriminate the protein (blue beads) from the protein surface (i.e., the 'hydration shell' of the protein, orange beads). The protein surface was determined as follows. For each amino acid we computed ten closest LJ sites. These form the protein surface. Clearly, we do see areas of low density of LJ sites in the protein surface surrounded by areas of high density of LJ sites.

It is evident that within the hydration shell, the local density of LJ sites differ. Their distribution is (though influenced by the parameters chosen) altogether a consequence of the protein morphology. It is a key difference from planar surfaces, where we expect a homogeneous distribution perpendicular to the

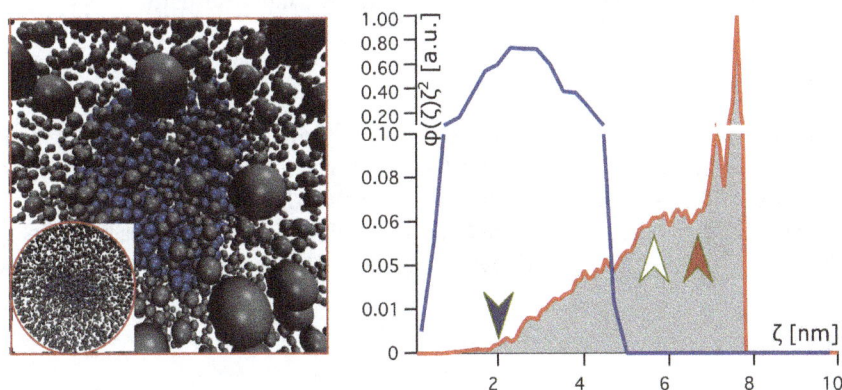

Figure 4. (Color online) MD simulation of a protein dissolved in a LJ liquid. Left-hand panel: a protein (blue beads) is dissolved in a LJ liquid (grey beads). The simulation box is not periodic: both types of particles are enclosed in a spherical wall. Right-hand panel: the radial density, $\phi(\zeta)$ is displayed for the protein (blue) and the LJ liquid (red line with grey markers). We distinguish the protein, the protein surface and the LJ liquid bulk.

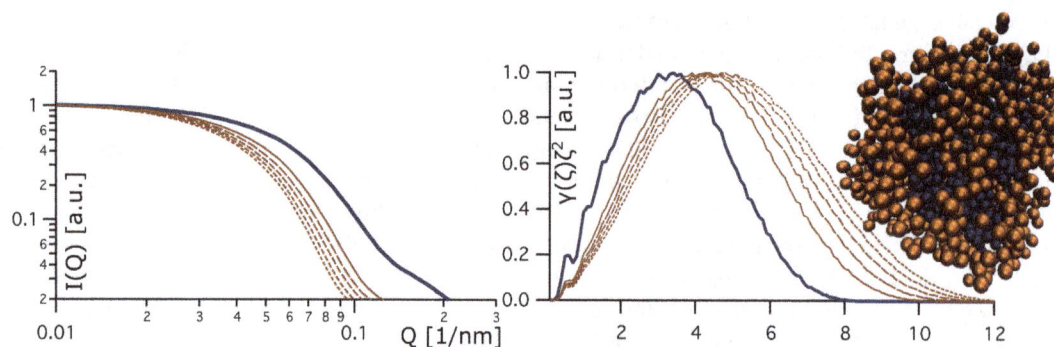

Figure 5. (Color online) Self similar SAXS signals. Left-hand panel: SAXS profiles computed from pair density distributions. The blue solid line mimics the scattering profile computed from the protein crystallographic model. There was no background added or subtracted. The orange (solid and dashed) lines give scattering profiles computed from the protein and from particles from the LJ liquid in the proximity of the protein. Right-hand panel: The blue solid line gives the pair density distribution for the LJ protein. The full and dashed orange lines indicate pair density distributions computed from the LJ protein and particles from the LJ liquid that are in the proximity of the protein. They scale invariantly.

surface.

Another difference is the linearity of the hydration shell, while the spherical surface already enforces a layered structure. This seems to suggest that protein fractal morphology extends the Henry regime to higher bulk densities — a conjecture that needs clarification in the future.

5. Summary and outlook

In this work we provide a (somewhat limited) collection of mathematical formulae that may be useful to link theoretical findings of classical density functional theory to experimental results derived from scattering techniques, such as small angle neutron and small angle X-ray scattering. We discuss the necessity of these and their fractal flavour. Though we lack a detailed mathematical discussion of the possible physical origin, we have experimental evidence that may be found in the electrostatics of the system investigated. We compare experimental data from small angle neutron scattering to the data of small angle X-ray scattering. While neutron scattering data do not change upon different salt concentrations, small angle X-ray data do. These changes in the scattering data can be explained by a fractal dimension, which is of electrostatic origin. We performed molecular dynamics simulation and presented a structure model. We distinguish protein from protein surface and find scale invariance for both.

Acknowledgements

ACIB is supported by the Federal Ministry of Economy, Family and Youth (BMWFJ), the Federal Ministry of Traffic, Innovation and Technology (BMVIT), the Styrian Business Promotion Agency SFG, the Standortagentur Tirol and ZIT-Technology Agency of the City of Vienna through the COMET-Funding Program managed by the Austrian Research Promotion Agency FFG. LP aknowledges financial support from the National Research, Development and Innovation Office of Hungary (NKFIH), grant no. SNN 116198.

References

1. Arakawa T., Timasheff S.N., Biochemistry, 1982, **21**, No. 25, 6545; doi:10.1021/bi00268a034.
2. Arakawa T., Timasheff S.N., Biochemistry, 1984, **23**, No. 25, 5912; doi:10.1021/bi00320a004.
3. Mile V., Pusztai L., Dominguez H., Pizio O., J. Phys. Chem. B, 2009, **113**, 10760; doi:10.1021/jp900092g.
4. Makowski L., J. Struct. Funct. Genomics, 2010, **11**, 9; doi:10.1007/s10969-009-9075-x.
5. Makowski L., Bardhan J., Gore D., Lal J., Mandava S., Park S., Rodi D.J., Ho N.T., Ho C., Fischetti R.F., J. Mol. Biol., 2011, **408**, No. 5, 909; doi:10.1016/j.jmb.2011.02.062.

6. Bax A., Protein Sci., 2003, **12**, No. 1, 1; doi:10.1110/ps.0233303.
7. Halle B., Philos. Trans. R. Soc. London, Ser. B, 2004, **359**, 1207; doi:10.1098/rstb.2004.1499.
8. Nandi N., Bhattacharyya K., Bagchi B., Chem. Rev., 2000, **100**, No. 6, 2013; doi:10.1021/cr980127v.
9. Murakra R.K., Head-Gordon T., J. Phys. Chem. B, 2008, **112**, 179; doi:10.1021/jp073440m.
10. Frölich A., Gabel F., Jasnin M., Lehnert U., Oesterhelt D., Stadler A.M., Tehei M., Weik M., Wood K., Zaccai G., Faraday Discuss., 2009, **141**, 117; doi:10.1039/B805506H.
11. General discussion, Faraday Discuss., 2009, **141**, 175; doi:10.1039/B818384H.
12. Dexheimer S., Terahertz Spectroscopy: Pprinciples and Applications, Taylor & Francis, London, 2007.
13. Leitner D.M., Havenith M., Gruebele M., Int. Rev. Phys. Chem., 2006, **25**, No. 4, 553; doi:10.1080/01442350600862117.
14. Glatter O., Kratky O., Small Angle X-ray Scattering, Academic Press, New York, 1982.
15. Feigin L.A., Svergun D.I., Structure Analysis by Small-angle X-ray and Neutron Scattering, Plenum Press, New York, 1987.
16. Putnam C.D., Hammel M., Hura G.L., Tainer J.A., Q. Rev. Biophys., 2007, **40**, 191; doi:10.1017/S0033583507004635.
17. Zhang F., Roth R., Wolf M., Roosen-Runge F., Skoda M.W.A., Jacobs R.M.J., Sztucki M., Schreiber F., Soft Matter, 2012, **8**, 1313; doi:10.1039/C2SM07008A.
18. Zhang F., Roosen-Runge F., Sauter A., Roth R., Skoda M.W.A., Jacobs R.M.J., Sztucki M., Schreiber F., Faraday Discuss., 2012, **159**, 313; doi:10.1039/c2fd20021j.
19. Allen M.P., Tildesley D.J., Computer Simulation of Liquids, Clarendon Press, Oxfords, 1987.
20. Karplus M., McCammon J.A., Nat. Struct. Biol., 2002, **9**, No. 9, 646; doi:10.1038/nsb0902-646.
21. Fedorov B.A., Denesyuk A.I., FEBS Lett., 1987, **88**, 114; doi:10.1016/0014-5793(78)80620-6.
22. Hubbard S.R., Small-angle x-ray scattering studies of calcium-binding proteins in solution, Ph.D. Thesis, Stanford University, 1987.
23. Svergun D., Barberato C., Koch M.H.J., J. Appl. Crystallogr., 1995, **28**, 768; doi:10.1107/S0021889895007047.
24. Wong P-Z., Bray A.J., J. Appl. Crystallogr., 1988, **21**, 786; doi:10.1107/S0021889888004686.
25. Schmidt P.W., J. Appl. Crystallogr., 1991, **24**, 414; doi:10.1107/S0021889891003400.
26. Foster T., Safran S.A., Sottmann T., Strey R., J. Chem. Phys., 2007, **127**, 204711; doi:10.1063/1.2748754.
27. Majorek K.A., Porebski P.J., Dayal A., Zimmerman M.D., Jablonska K., Stewart A.J., Chruszcz M., Minor W., Mol. Immunol., 2012, **52**, 174; doi:10.1016/j.molimm.2012.05.011.
28. Zolotarev V.M., One-dimensional Stable Distributions, Translations of Mathematical Monographs Series, Vol. 65, American Mathematical Society, Providence, 1986.
29. Mandelbrot B.B., The Fractal Geometry of Nature, Henry Holt and Company, New York, 1983.
30. Kotlarchyk M., Chen S.-H., J. Chem. Phys., 1983, **79**, 2461; doi:10.1063/1.446055.
31. Horejs C., Gollner H., Pum D., Sleytr U.B., Peterlik H., Jungbauer A., Tscheliessnig R., ACS Nano, 2011, **5**, 2288; doi:10.1021/nn1035729.
32. Tscheliessnig R., Sommer R., Überbacher R., Pusztai L., Székely N., Jungbauer A., Peterlik H., Soft Matter, 2016 (submitted).
33. Pizio O., Sokołowski S., Condens. Matter Phys., 2014, **17**, 23603; doi:10.5488/CMP.17.23603.
34. Plimpton S.J., J. Comput. Phys., 1995, **117**, 1; doi:10.1006/jcph.1995.1039.
35. http://lammps.sandia.gov/doc/Manual.html.

14

Influence of anisotropic ion shape, asymmetric valency, and electrolyte concentration on structural and thermodynamic properties of an electric double layer

M. Kaja[1], S. Lamperski[1]*, W. Silvestre-Alcantara[2], L.B. Bhuiyan[2], D. Henderson[3]

[1] Department of Physical Chemistry, Adam Mickiewicz University of Poznań, Umultowska 89b, 61-614 Poznań, Poland

[2] Laboratory of Theoretical Physics, Department of Physics, University of Puerto Rico, San Juan, 00931-3343, Puerto Rico

[3] Department of Chemistry and Biochemistry, Brigham Young University, Provo UT 84602-5700, USA

Grand canonical Monte Carlo simulation results are reported for an electric double layer modelled by a planar charged hard wall, anisotropic shape cations, and spherical anions at different electrolyte concentrations and asymmetric valencies. The cations consist of two tangentially tethered hard spheres of the same diameter, d. One sphere is charged while the other is neutral. Spherical anions are charged hard spheres of diameter d. The ion valency asymmetry 1:2 and 2:1 is considered, with the ions being immersed in a solvent mimicked by a continuum dielectric medium at standard temperature. The simulations are carried out for the following electrolyte concentrations: 0.1, 1.0 and 2.0 M. Profiles of the electrode-ion, electrode-neutral sphere singlet distributions, the average orientation of dimers, and the mean electrostatic potential are calculated for a given electrode surface charge, σ, while the contact electrode potential and the differential capacitance are presented for varying electrode charge. With an increasing electrolyte concentration, the shape of differential capacitance curve changes from that with a minimum surrounded by maxima into that of a distorted single maximum. For a 2:1 electrolyte, the maximum is located at a small negative σ value while for 1:2, at a small positive value.

Key words: *charged dimers, valency asymmetry, electrical double layer, grand-canonical Monte Carlo simulation*

This article is dedicated to Professor Stefan Sokołowski, the famous Polish scientist and our friend, on the occasion of his 65th birthday.

1. Introduction

Development of the electrical double layer (EDL) theory as well as the progress in the numerical technology have resulted in proposition and examination of an increasing number of EDL models. The Gouy-Chapman theory (GC) [1, 2] is based on solving the Poisson-Boltzmann equation and applies a mean field approximation to describe the electrostatic interactions. The GC theory can be applied to the simplest model of an electrolyte, which does not take into account the volume of ions. In this model, ions are represented by point electric charges, and the solvent is a dielectric continuum with the relative permittivity ϵ_r. It is worth noting that the Debye-Hückel theory uses the same model of an electrolyte. The

---

*E-mail: slamper@amu.edu.pl

GC theory describes well low density electrolytes at small electrode charges. It breaks down at higher concentrations and charges because the excluded volume and correlation effects are disregarded.

The hard-sphere and electrostatic correlation effects are considered in modern theories of EDL such as the Modified Poisson-Boltzmann Theory (MPB) [3, 4], the Mean Spherical Approximation (MSA) [5], the Hypernetted Chain Theory (HNC) [6, 7] or more recently the Density Functional Theory (DFT) [8]. They have been used to describe the primitive model of an electrolyte (PM) [9] and the restricted primitive model (RPM) [10]. In the PM model, ions are represented by hard spheres of different diameters with a point electric charge located at the centre. The charged spheres are immersed in a medium with the relative permittivity ϵ_r characteristic of a solvent. In the RPM model, the ion diameters are the same. For EDL composed of a planar electrode and a PM or RPM electrolyte, the modern theories predict the layering effect at high absolute values of the electrode charges [11, 12] and the transition of the capacitance minimum into a maximum at small electrode charges caused by an increasing electrolyte concentration [4]. The GC theory fails to describe these effects.

Due to correlation effects that are included, the MPB and DFT theories have been successfully applied to the solvent primitive model (SPM) electrolyte. SPM is the simplest model of an electrolyte which includes the volume of solvent molecules. In SPM, solvent molecules are modelled by neutral hard spheres whose diameter is the same as [13, 14] or different [15] from that of ions. Neutral spheres as well as cations and anions are immersed in the continuous dielectric medium characterised by ϵ_r. The presence of neutral spheres leads to the generation of density oscillations near the electrode surface and to an increase in the differential capacitance of EDL [16].

Of the formal statistical mechanical theories, the HNC has been applied to the non-primitive model (NP) [17] in which solvent molecules are modelled by hard spheres with a point permanent electric dipole moment located at the centre. However, it leads to lowered values of the relative permittivity.

The DFT theory gives fresh possibilities. Due to the free energy term of intra-molecular interactions, the DFT theory is applicable to ions of more complicated topology than spherical. Here, we must mention the work by Sokołowski [18] who has applied the modified DFT theory to the study of orientation ordering of electrostatic neutral hard dumbbells at the planar surface. Charged dumbbells, called dimers, have aroused great interest in investigation of the EDL properties [19, 20]. This advanced model of an electrolyte has been used for modelling systems of high densities like ionic liquids [21].

Torrie and Valleau [10] have introduced the computer simulation technique into the investigation of EDL. At that time computer simulation results were used to confirm the correctness of EDL theories. Now, the tremendous development of numerical technology has opened new area of application inaccessible for theory. Fedorov et al. [22] have studied models of ionic liquids made of one, two or three beads, assuming that one of the hard spheres in the chain had a positive charge. Breitsprecher et al. [23, 24] have conducted molecular dynamics simulations for ions with different size and valency, represented by a coarse-grained model. Silvestre-Alcantara et al. [25] have investigated the properties of ELD containing fused dimer electrolyte. Charged hard walls and hard spheres have been replaced by molecular electrodes and soft sphere ions [26–28]. Ions of topology characteristic of ionic liquids are investigated. The solvent is no longer a dielectric continuum but is modelled by explicit molecular models [29]. Thus, the present models of EDL have become more realistic.

Recently, we have intensively investigated the EDL containing a dimer electrolyte. Among other effects, we have analysed the influence of concentration of 1:1 electrolyte [21] and of ion valence asymmetries [30] of charged dimers on the properties of EDL. However, we did not consider the properties of EDL for asymmetric ion valencies at different electrolyte concentrations. In particular, we expect new shapes of the differential capacitance. Thus, in this paper, we discuss the influence of anisotropic ion shape on the structural and thermodynamic properties of ELD containing asymmeteric 2:1 and 1:2 dimer electrolytes of different concentrations.

2. Model and methods

The electric double layer is composed of a planar electrode and an electrolyte, which is a mixture of spherical anions and anisotropic cations in the shape of dimers. The dimer consists of two tangentially tethered hard spheres, one of which has a point electric charge immersed at the centre and the other is neutral. The diameters of the two spheres of a dimer and the sphere of a monomer anion are the same. The ion valencies are asymmetric with the ions being immersed in a homogeneous medium of the relative permittivity, ϵ_r. The electrode is modelled by a hard planar wall with a uniformly distributed charge of surface density, σ. The image effect is not considered which means that ϵ_r of the electrode material and solvent are the same. The ion-ion and electrode-ion interactions are given by

$$u_{ij}(r) = \begin{cases} \infty, & r < (d_i + d_j)/2, \\ \frac{1}{4\pi\epsilon_0\epsilon_r}\frac{e^2 Z_i Z_j}{r}, & r \geq (d_i + d_j)/2, \end{cases} \tag{1}$$

and

$$u_{wi}(x) = \begin{cases} \infty, & x < d/2, \\ -\frac{\sigma Z_s e x}{\epsilon_0\epsilon_r}, & x \geq d/2, \end{cases} \tag{2}$$

respectively. Here, e is the magnitude of the elementary charge, and Z_s is the valency of the particle of species s. Also, ϵ_0 is the vacuum permittivity, r is the separation between the centres of the two hard spheres, and x is the perpendicular distance from the electrode surface to the centre of a hard sphere.

The local number density $\rho_s(x)$ of the species s at a distance x is the first average quantity obtained from our simulations. The reduced local density or the singlet distribution function $g_s(x) = \rho_s(x)/\rho_s^0$, ($\rho_s^0$ is the corresponding bulk number density) is used to describe the structure of ELD. Also, the mean electrostatic potential $\psi(x)$ is defined in terms of $\rho_s(x)$

$$\psi(x) = \frac{e}{\epsilon_0\epsilon_r}\sum_s Z_s \int_x^\infty \rho_s(x')(x - x')dx'. \tag{3}$$

The differential capacitance C_d is the property which can be compared with the experimental results. The differential capacity is defined as

$$C_d = d\sigma/d\psi(0). \tag{4}$$

In practice, C_d was calculated from the interpolation polynomials method introduced by Lamperski and Zydor [31].

The local (volume) charge density $\rho_Q(x)$

$$\rho_Q(x) = \sum_s Z_s e g_s(x) \tag{5}$$

and the local net charge per unit area, $\sigma_\Sigma(x)$

$$\sigma_\Sigma(x) = \sigma + \int_0^x \rho_Q(x')dx' \tag{6}$$

are the properties that can be used for indicating the charge inversion (CI) and charge reversal (CR) phenomena. The CI effect takes place when the electrode charge and the charge density of the second layer and, less commonly, subsequent layers of ions next to the electrode are of the same sign. When the charge of the first layer of counter-ions overcomes the charge of an electrode, the electric field reverses its direction. The function $\sigma_\Sigma(x)$ has the sign opposite to that of σ. This effect is called the charge reversal [32, 33]. Essentially at some x, the integrated charge overcompensates or overscreens the electrode charge. Hence, the often used term *charge overscreening* occurs in the literature.

The second average quantity obtained from our simulations is the mean orientation function $\langle\cos\theta\rangle$. It depends on the distance x from the electrode surface. The function $\langle\cos\theta\rangle$ is the average value of $\cos\theta$, where θ is the angle between the normal to the electrode surface and the straight line joining the centres of hard spheres constituting a dimer. The origin is located at the centre of the charged sphere.

The GCMC technique was applied to calculate the local densities, $\rho_s(x)$, and orientation $\langle\cos\theta\rangle$ profiles. This technique is recommended for inhomogeneous systems [34] as it eliminates difficulties associated with the determination of the bulk concentration which appears when MC simulations are carried out in the canonical ensemble. The details of the GCMC technique and its implementation to our investigation have been described in the previous papers [4, 35]. The procedure FLIP3 [36] was used to rotate the dimer while the long-range electrostatic interactions were estimated with the method proposed by Torrie and Valleau [10]. The ionic activity coefficients required by the GCMC technique as input were obtained from the inverse GCMC method [37].

3. Results

Simulations were carried out for the asymmetric ion valencies 1:2 and 2:1 at three electrolyte concentrations 0.1, 1.0, and 2.0 M in the range of the electrode surface charges varying from -1.0 to $+1.0$ C m^{-2}. Diameters of positively charged and neutral spheres of dimers and of spherical anions are equal to $d = 425$ pm. The other physical parameters were temperature $T = 298.15$ K, and the relative permittivity of water, $\epsilon_r = 78.5$. Because of the anisotropic shape of cations, the simulations were expected to require significantly longer configurational sampling. We sampled 2 billion configurations to obtain high precision averages.

Figure 1 shows the singlet distribution profiles of charged dimers and spherical anions for three electrolyte concentrations 0.1, 1.0 and 2.0 M at $\sigma = -0.30$ C m^{-2}. At this σ, a strong adsorption of dimers is observed, while anions are repelled from the vicinity of the electrode surface. By contrast, at positive values of σ, the adsorption of spherical anions occurs and now the dimer cations are repelled from the electrode. As a result, the internal structure of the distant cation has little influence on the double layer structure with the dimer behaving like a spherical ion (see for example, reference [30]). The surface properties of spherical ions are well known, so we do not discuss them here. The upper panel of figure 1 shows the results for the 1:2 electrolyte, while the lower one for the 2:1 case. The density profiles of charged spheres are similar to those observed for spherical counter-ions. A sharp peak corresponds to the contact distance, $d/2$. As expected, its height decreases with an increasing electrolyte concentration. The peak is higher and thinner for di-valent dimers as we have observed earlier [30]. The influence of a neutral sphere is hardly visible. The neutral sphere density profiles have two maxima. The first corresponds to the contact distance while the second is at $x/d \approx 1.45$. The height of the second maximum relative to the height of the first one increases with a decreasing electrolyte concentration. The first maximum indicates that the large fraction of dimers take parallel orientation to the electrode surface. Assuming the perpendicular orientation of a dimer with charged sphere adjacent the electrode surface, the centre of the neutral sphere is located at $x/d = 1.5$. A good agreement of this prediction with the simulation results

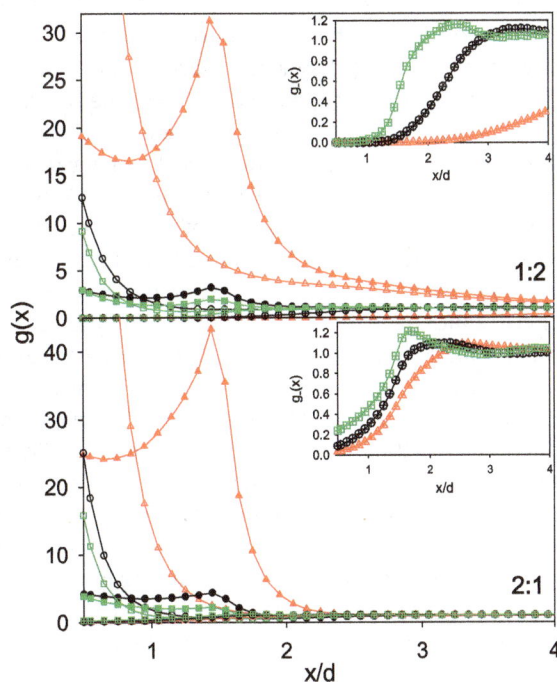

Figure 1. (Color online) Dependence of the singlet distribution function, $g(x)$, of charged dimers and spherical anions (see also the insets) on the distance, x/d, from the electrode at the surface charge $\sigma = -0.3$ C m^{-2} for electrolyte concentrations 0.10 M (triangles), 1.00 M (circles) and 2.00 M (squares) for ion valencies 1:2 (upper panel) and 2:1 (lower panel). Empty symbols show the results for charged spheres, filled for the neutral spheres of dimers and crossed for anions.

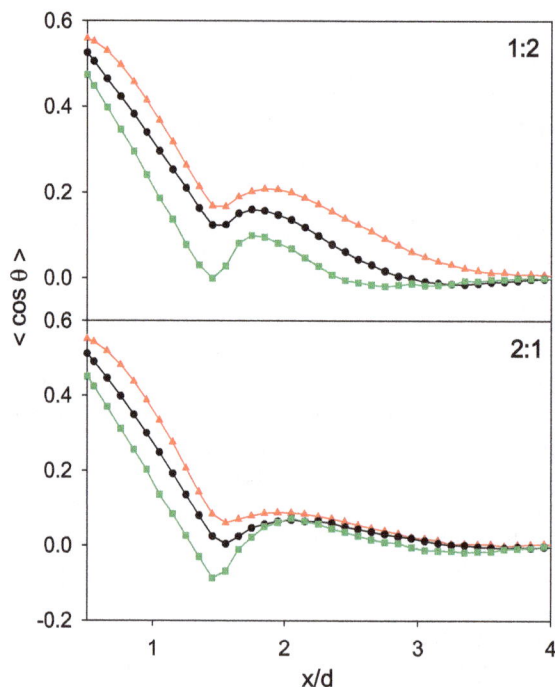

Figure 2. (Color online) Dependence of the angular orientation function, $\langle\cos\theta\rangle$, of dimers on the distance, x/d. Parameters and symbols have the same meaning as in figure 1.

indicates that the perpendicular orientation is also very probable and its probability increases with a decreasing electrolyte concentration. This conclusion confirms the mean orientation $\langle\cos\theta\rangle$ results.

The mean orientation $\langle\cos\theta\rangle$ profiles are shown in figure 2. The positive values of $\langle\cos\theta\rangle$ mean that dimers prefer the orientation with the charged sphere towards the electrode. The function $\langle\cos\theta\rangle$ has a nearly linear course from the contact distance to the position of the second maximum on the neutral sphere density profile. At this position, the $\langle\cos\theta\rangle$ function has a minimum. The minimum is negative for higher electrolyte concentrations and larger dimer valencies. At longer distances, the minimum transforms into a positive maximum. After leaving the maximum, the $\langle\cos\theta\rangle$ function tends to zero and the orientation randomisation takes place. Thus, the behaviour of $\langle\cos\theta\rangle$ suggests a generally perpendicular orientation of the dimers near the electrode with the charged head nearer to it than the neutral tail, a pattern that is consistent with our earlier studies. The contact values of $\langle\cos\theta\rangle$ are nearly independent of the dimer valency, but at some distance from the electrode surface they are lower for the divalent dimers. However, the course of curves remains similar. In the vicinity of the electrode surface, the rotation of the neutral sphere around the charged one is hindered mainly by the hard wall of the electrode. The wall effect is independent of electrolyte concentration. The value of $\langle\cos\theta\rangle$ varies from 0.5 at a contact distance to 0 at $x/d = 1.5$. The simulation contact results oscillate at around 0.5. They are greater than 0.5 for $c = 0.1$ M and lower than 0.5 for $c = 2.0$ M. It means that, as we have stated earlier, the low concentration electrolytes support the perpendicular orientation, while the electrolytes of high concentration enhance the parallel one. In this range of distances, the rotation of the neutral sphere around the charged one is hindered by steric and electrostatic interactions with neighbouring ions, only. The second maximum of $\langle\cos\theta\rangle$ shows that dimers form the second layer have their charged spheres oriented towards the electrode. It is worth noting that this layer is not visible at the charged sphere density distributions for $c = 1.0$ and 2.0 M.

Most of the anion distribution functions shown in figure 1 have a small maximum at $x/d > 1.5$ (see the insets). This maximum suggests the onset of the CI and CR phenomena. As a case in point, we have calculated the profiles of the local charge density, $\rho_Q(x)$, and the local net charge per unit area, $\sigma_\Sigma(x)$, which are presented in figure 3. It is seen that except the curve for the 1:2 electrolyte at concentration $c = 0.1$ M, which is monotonous, the remaining curves are all non-monotonous indicating the occurrence

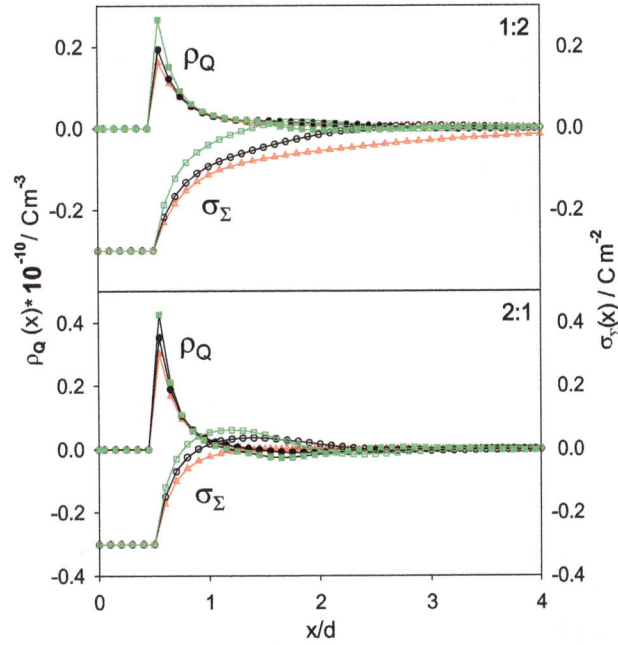

Figure 3. (Color online) Dependence of the local (volume) charge density $\rho_Q(x)$ (filled symbols) and the local net charge per unit area, $\sigma_\Sigma(x)$ (empty symbols) on the distance, x/d. Parameters and symbols have the same meaning as in figure 1.

of the CI and CR phenomena. The increase in electrolyte concentration or the presence of a higher valency counterion intensifies these features. Indeed, at the same electrolyte concentration, the effect is more pronounced for the 2:1 system than for the 1:2 system. The CI effect is observed closer to the electrode surface than to the CR one. The mechanism of CI and CR is not completely clear [33].

The mean electrostatic potential profiles calculated from equation (3) are presented in figure 4. For

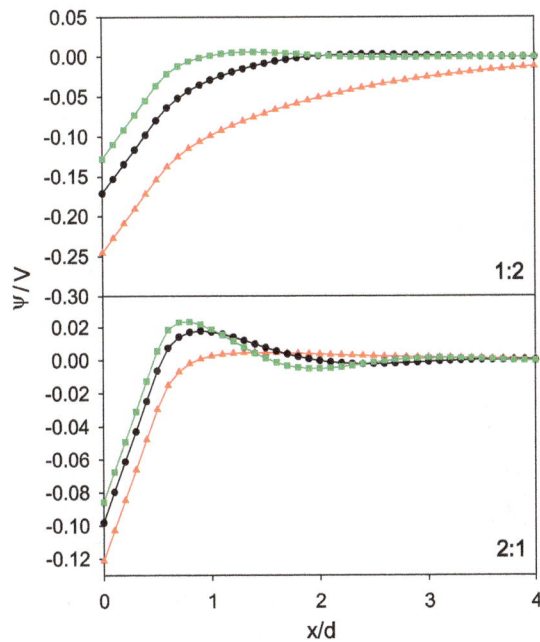

Figure 4. (Color online) Mean electrostatic potential ψ as a function of distance, x/d. Parameters and symbols have the same meaning as in figure 1.

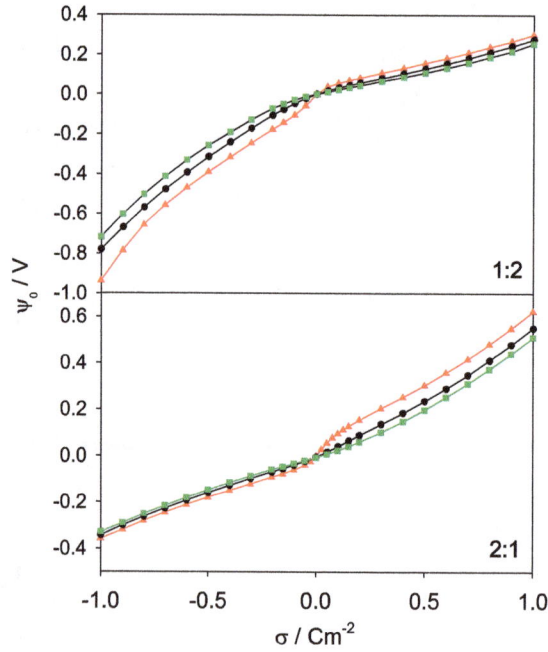

Figure 5. (Color online) Mean electrostatic potential of the electrode, ψ_0 as a function of the surface charge density, σ. Symbols have the same meaning as in figure 1.

the 1:2 electrolyte at $c = 0.1$ and 1.0 M, the potential is negative and the curves do not have any extrema. The remaining curves have a positive maximum characteristic of divalent electrolytes [38]. This maximum is related to the the CI effect. With an increasing electrolyte concentration and dimer valency, the absolute value of the mean potential and the width of EDL decrease.

The dependence of the mean electrostatic potential of the electrode, $\psi(0)$, on the electrode charge

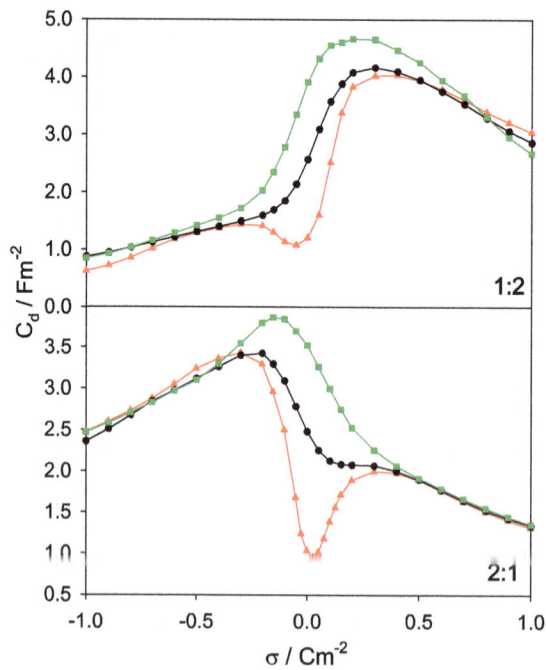

Figure 6. (Color online) Differential capacitance, C_d, of the electrical double layer as a function of the surface charge density, σ. Symbols have the same meaning as in figure 1.

density σ calculated for the investigated electrolyte concentrations and on ion valencies is shown in figure 5. The curves intersect in the vicinity of $\sigma = 0$. The slope of the curves in the range of adsorption of divalent ions is smaller than in the range of adsorption of monovalent ones. The $\psi(0)$ results are used for the calculation of the differential capacitance of EDL.

Figure 6 shows the dependence of the differential capacitance, C_d, of EDL on the electrode charge density calculated for different asymmetric valencies and electrolyte concentrations. For $c = 0.1$ M, the C_d curves have a minimum corresponding to $\sigma \approx 0$ and surrounded by two asymmetric maxima. An increase in the electrolyte concentration to $c = 2.0$ M transforms the minimum into a distorted maximum. For the 1:2 electrolyte, the maximum is located at a positive σ, while for the 2:1 electrolyte, at a negative σ value. The Gouy-Chapman theory [1, 2] predicts that the differential capacitance curve of EDL has the parabola-like shape with a minimum at $\sigma = 0$. Transition of the capacitance curve from that having a minimum to that having a maximum with an increasing electrolyte concentration has been explained for the RPM electrolyte by an increase in the thickness of EDL due to the formation of bi- or multi-layer structures of counterions [13, 39]. In the case of charged dimers, the increase in thickness of EDL is additionally caused by the neutral spheres of dimers which separate the two layers of charged spheres: adsorbed on the electrode surface and form the second layer. Finally, the increase in valency of counterions makes the EDL thinner. Explanation of the behaviour of $C_d - \sigma$ curve shown in figure 6 requires consideration of all these effects.

4. Conclusions

A charged dimer is a simple and useful model of molecules of charged surfactants and ionic liquids. Charged surfactants are composed of a charged head and a neutral, hydrophobic tail. In a charged dimer, the head is represented by a charged sphere while the tail by a neutral one. Surfactants are diluted in a solvent (water). Therefore, more advanced investigation of surfactants modelled by a charged dimer should include solvent molecules. The interaction of the charged head and the neutral tail with molecules of solvent depends strongly on the structural and polar properties of solvent molecules. Presumably they can influence the density and orientation profiles. Ionic liquids are solvent free molten crystals composed of large organic ions with the electric charge located off the centre of the ion. Charged dimers can model ionic liquids when their size is large and the concentration high. Current investigation of EDL composed of ionic liquids as an electrolyte is extremely interesting. However, the investigation of EDL formed by charged dimers by the method of molecular simulations breaks down because of high density of a system. The problem can be solved by replacing hard potentials with soft ones.

In this paper we have investigated the structural and thermodynamic properties of an electric double layer composed of a planar charged hard wall, charged dimers and spherical anions at different electrolyte concentrations and asymmetric ion valencies, using the grand canonical Monte Carlo simulation. The density profiles of neutral spheres have two maxima corresponding to parallel and perpendicular orientations of the dimer against the electrode surface. The height of the second maximum relative to the height of the first one increases with a decreasing electrolyte concentration. Thus, we have concluded that the probability of the perpendicular orientation increases with a decreasing electrolyte concentration.

The variation in the electrolyte concentration and ion valency leads to a diversity of shapes of differential capacitance curves. At low concentrations, the capacitance curves have a minimum surrounded by two asymmetric maxima. With increasing concentrations, the minimum transforms into a distorted maximum. For the 1:2 electrolyte, the maximum is located at positive σ, while for the 2:1 electrolyte, at a negative σ value.

Acknowledgements

Monika Kaja and Stanisław Lamperski acknowledge the financial support from the Faculty of Chemistry, Adam Micklewicz University of Poznań.

References

1. Gouy M., J. Phys. (Paris), 1910, **9**, 457; doi:10.1051/jphystap:019100090045700.
2. Chapman D.L., Philos. Mag., 1913, **25**, 475; doi:10.1080/14786440408634187.
3. Outhwaite C.W., Bhuiyan L.B., Levine S., J. Chem. Soc., Faraday Trans. 2, 1980, **76**, 1388; doi:10.1039/F29807601388.
4. Lamperski L., Outhwaite C.W., Bhuiyan L.B., J. Phys. Chem. B, 2009, **113**, 8925; doi:10.1021/jp900037h.
5. Henderson D., Lamperski S., Outhwaite C.W., Bhuiyan L.B., Collect. Czech. Chem. Commun., 2010, **75**, 303; doi:10.1135/cccc2009094.
6. Henderson D., Blum L., J. Electroanal. Chem., 1980, **111**, 217; doi:10.1016/S0022-0728(80)80041-6.
7. Henderson D., Blum L., Smith W.R., Chem. Phys. Lett., 1979, **63**, 381; doi:10.1016/0009-2614(79)87041-4.
8. Pizio O., Sokołowski S., J. Chem. Phys., 2013, **138**, 204715; doi:10.1063/1.4807777.
9. Bešter M., Vlachy V., J. Chem. Phys., 1992, **96**, 7656; doi:10.1063/1.462366.
10. Torrie G.M., Valleau J.P., J. Chem. Phys., 1980, **73**, 5807; doi:10.1063/1.440065.
11. Lamperski S., Bhuiyan L.B., J. Electroanal. Chem., 2003, **540**, 79; doi:10.1016/S0022-0728(02)01278-0.
12. Zhou S., Lamperski S., Zydorczak M., J. Chem. Phys., 2014, **141**, 064701; doi:10.1063/1.4892415.
13. Tang Z., Scriven L.E., Davis H.T., J. Chem. Phys., 1992, **97**, 494; doi:10.1063/1.463595.
14. Lamperski S., Outhwaite C.W., Bhuiyan L.B., Mol. Phys., 1996, **87**, 1049; doi:10.1080/00268979600100721.
15. Płuciennik M., Outhwaite C.W., Lamperski S., Mol. Phys., 2014, **112**, 165; doi:10.1080/00268976.2013.805847.
16. Grimson M.J., Rickayzen G., Chem. Phys. Lett., 1982, **86**, 71; doi:10.1016/0009-2614(82)83119-9.
17. Outhwaite C.W., Lamperski S., Condens. Matter Phys., 2001, **4**, 739; doi:10.5488/CMP.4.4.739.
18. Sokołowski S., J. Chem. Phys., 1991, **95**, 7513; doi:10.1063/1.461377.
19. Bhuiyan L.B., Lamperski S., Wu J., Henderson D., J. Phys. Chem. B, 2012, **116**, 10364; doi:10.1021/jp304362y.
20. Lamperski S., Kaja M., Bhuiyan L.B., Wu J., Henderson D., J. Chem. Phys., 2013, **139**, 054703; doi:10.1063/1.4817325.
21. Wu J., Jiang T., Jiang D., Jin Z., Henderson D., Soft Matter, 2011, **7**, 11222; doi:10.1039/C1SM06089A.
22. Fedorov M.V., Georgi N., Kornyshev A.A., Electrochem. Commun., 2010, **12**, 296; doi:10.1016/j.elecom.2009.12.019.
23. Breitsprecher K., Košovan P., Holm C., J. Phys.: Condens. Matter, 2014, **26**, 284108; doi:10.1088/0953-8984/26/28/284108.
24. Breitsprecher K., Košovan P., Holm C., J. Phys.: Condens. Matter, 2014, **26**, 284114; doi:10.1088/0953-8984/26/28/284114.
25. Silvestre-Alcantara W., Kaja M., Henderson D., Lamperski S., Bhuiyan L.B., Mol. Phys., 2015, **114**, 53; doi:10.1080/00268976.2015.1083132.
26. Lanning O.J., Madden P.A., J. Chem. Phys. B, 2004, **108**, 11069; doi:10.1021/jp048102p.
27. Górniak R., Lamperski S., J. Phys. Chem. C, 2014, **118**, 3156; doi:10.1021/jp411698w.
28. Fedorov M.V., Kornyshev A.A., J. Phys. Chem. B, 2008, **112**, 11868; doi:10.1021/jp803440q.
29. Předota M., Bandura A.V., Cummings P.T., Kubicki J.D., Wesolowski D.J., Chialvo A.A., Macheskyr M.L., J. Phys. Chem. B, 2004, **108**, 12049; doi:10.1021/jp037197c.
30. Kaja M., Silvestre-Alcantara W., Lamperski S., Henderson D., Bhuiyan L.B., Mol. Phys., 2015, **113**, 1043; doi:10.1080/00268976.2014.968651.
31. Lamperski S., Zydor A., Electrochim. Acta, 2007, **52**, 2429; doi:10.1016/j.electacta.2006.08.045.
32. Chialvo A.A., Simonson J.M., Condens. Matter Phys., 2011, **14**, 33002; doi:10.5488/CMP.14.33002.
33. Jiménez-Ángeles F., Lozada-Cassou M., J. Phys. Chem. B, 2004, **108**, 7286; doi:10.1021/jp036464b.
34. Frenkel D., Smit B., Understanding Molecular Simulation: From Algorithms to Applications, Academic Press, San Diego 1996, p. 126.
35. Lamperski S., Outhwaite C.W., J. Colloid Interface Sci., 2008, **328**, 458; doi:10.1016/j.jcis.2008.09.050.
36. Allen M.P., Tildesley D.J., Computer Simulation of Liquids, Oxford University Press, Oxford, 1987, p. 349.
37. Lamperski S., Mol. Simul., 2007, **33**, 1193; doi:10.1080/08927020701739493.
38. Bhuiyan L.B., Outhwaite C.W., Phys. Chem. Chem. Phys., 2004, **6**, 3467; doi:10.1039/b316098j.
39. Lamperski S., Sosnowska J., Bhuiyan L.B., Henderson D., J. Chem. Phys., 2014, **140**, 014704; doi:10.1063/1.4851456.

15

Double layer for hard spheres with an off-center charge

W. Silvestre-Alcantara[1], L.B. Bhuiyan[1]*, S. Lamperski[2], M. Kaja[2], D. Henderson[3]

[1] Department of Physics and Laboratory for Theoretical Physics, University of Puerto Rico,
San Juan, Puerto Rico 00936-8377, USA

[2] Department of Physical Chemistry, Adam Mickiewicz University in Poznań,
Umultowska 89b, 61-614 Poznań, Poland

[3] Department of Chemistry and Biochemistry, Brigham Young University, Provo UT 84602-5700, USA

Simulations for the density and potential profiles of the ions in the planar electrical double layer of a model electrolyte or an ionic liquid are reported. The ions of a real electrolyte or an ionic liquid are usually not spheres; in ionic liquids, the cations are molecular ions. In the past, this asymmetry has been modelled by considering spheres that are asymmetric in size and/or valence (viz., the primitive model) or by dimer cations that are formed by tangentially touching spheres. In this paper we consider spherical ions that are asymmetric in size and mimic the asymmetrical shape through an off-center charge that is located away from the center of the cation spheres, while the anion charge is at the center of anion spheres. The various singlet density and potential profiles are compared to (i) the dimer situation, that is, the constituent spheres of the dimer cation are tangentially tethered, and (ii) the standard primitive model. The results reveal the double layer structure to be substantially impacted especially when the cation is the counterion. As well as being of intrinsic interest, this off-center charge model may be useful for theories that consider spherical models and introduce the off-center charge as a perturbation.

Key words: *electrical double layer, simulations, density functional theory, off-center charged spheres*

This article is dedicated to our colleague and friend, Stefan Sokołowski, in commemoration of his 65th birthday. DH first met and collaborated with "Don Esteban" in Mexico City but had admired his work long before that. Stefan has been our good friend and frequent collaborator since that time. We wish him a happy birthday and continued good health and productivity.

1. Introduction

The electrical double layer (EDL) has long fascinated experimentalists and theorists. An EDL is formed when the charged particles in a Coulomb fluid are attracted (or adsorbed) by a charged surface or electrode. The charge within or on the charged surface forms one layer and the charge of the adsorbed fluid forms the second layer that is of opposite sign to the charge of the surface but is equal in magnitude to the electrode charge; hence, the name double layer. In reality, these two layers of the EDL often consist of sublayers, so the term double layer is not quite accurate. Generally, the structure of the (usually metal) electrode is ignored and the electrode is considered to be a classical metal with the (electrode) charge being distributed uniformly on the electrode surface. The fluid layer, on the other hand, can be thought of as an extended diffuse layer whose net charge is equal in magnitude but opposite in sign to the electrode charge. The fluid layer may consist of sublayers, sometimes with alternating charge.

---

*E-mail: beena@beena.uprrp.edu

The EDL has been of great practical importance in electrochemistry and analytical chemistry. In recent years, the EDL has been found to be promising for new batteries, fuel cells, and supercapacitors [1] as well as for studies in biology [2]. The selectivity of a physiological channel can be thought of as an EDL problem with the channel playing the role of an electrode and the adsorbed ions playing the role of the diffuse layer [3]. The adsorption of ions by DNA is another example of an EDL (see for example, reference [4]).

The importance of the EDL is attested to in some recent reviews [2, 5, 6]. Despite this importance, the acceptance of the recent theoretical developments has been somewhat hampered by the fact that deviations from the predictions of the classical and often inadequate Poisson-Boltzmann (PB) theory of Gouy [7], Chapman [8], and Stern [9] (GCS) occur at high concentrations and/or high electrode charge and can be difficult to observe in aqueous systems. The popularity of the GCS theory stems from its intuitive simplicity and ease of use. Thus, it might be useful in quick and qualitative analysis of experimental results but it is unsatisfactory if one wishes to understand these results at a fundamental level. In passing, we note that the classical theory for bulk electrolytes that parallels the GCS theory is the more well-known Debye-Hückel (DH) theory (see for example, reference [10]). Historically, the GCS theory predates the DH theory, however, unlike the former theory, the latter theory is a linearized version of the bulk PB theory.

Until recently, a broad interest in the theory of the EDL has not been helped by the fact that it is difficult to achieve experimentally the electrode charges and/or high electrolyte concentrations with aqueous electrolytes where the deficiencies of the GCS theory are most apparent. By contrast, computer simulations provide a means of testing theories for such difficult situations. They are the gold standard against which theories can be compared since the theory and simulation are based on the same assumed model for the inter-particle interactions. For a given model of the electrode and ions, simulations provide exact results, apart from statistical uncertainties, against which a theory may be compared. Torrie and Valleau [11] found in their seminal simulations that the capacitance near the point of zero charge (pzc) and at high concentration was greater than what was predicted by the GCS theory. Blum [12], and Henderson, Saavedra-Barrera and Lozada-Cassou [13] showed that the sophisticated mean spherical approximation (MSA) and hypernetted chain theory (HNC) also exhibited this behavior. The MSA and HNC are closely related; the MSA is a linearized version of the HNC. A careful examination [5, 14] of experimental results confirms this behavior but the effect is small. In their simulations, Torrie and Valleau also found that at high electrode charge the capacitance is smaller than the GCS predictions. One of us recalls a meeting at which a prominent experimentalist ridiculed this result as being outside experimental range. However, the effect is still real. Ions occupy space and their charge cannot be crowded indefinitely. They cannot continue to form a monolayer on the electrode as the electrode charge increases. Regrettably, the MSA and the HNC fail to predict this latter effect. The MSA is a linear response theory and makes no prediction about the behavior of the capacitance at large electrode charge, while the HNC proves inadequate and predicts a greater capacitance at a large charge [15].

The modified Poisson-Boltzmann theory (MPB) [16] also predicts an increasing capacitance with an increasing concentration at small electrode charge and seems consistent with the continued decline in the capacitance as a function of electrode charge at high electrode charge. Although the numerical solutions of the MPB equations lack convergence at sufficiently high electrode charge, within the range of the surface charge for which solutions exist, the capacitance curve passes from a double hump curve at small concentrations to a single hump curve at high concentrations consistent with simulations [17, 18]. In these works [17, 18], the MC and MPB were applied to a restricted primitive model (RPM) (charged hard spheres with a common diameter). The density functional theory (DFT)[19] seems to be satisfactory in a similar vein. Here, too the DFT applied to the RPM double layer gave very good agreement with the corresponding MC simulation results for the capacitance [20]. The capacitance at small electrode charge continues to increase with an increasing concentration but decreases with an increasing electrode charge at large electrode charge, and shows the aforementioned double hump to single hump transition.

Interest in the EDL at high concentrations and electrode charge is increasing because EDLs can be formed in ionic liquids. Ionic liquids are organic electrolytes. Due to the fact that a solvent is not present, ionic liquids are, in essence, room temperature molten salts. They can exist at high concentrations and can support higher electrode charges. As a result, they are ideal for the testing of modern theories of the EDL. As Kornyshev has aptly stated [20], an ionic electrolyte gives a new paradigm for EDL studies. As Kornyshev observes, ionic liquid capacitance curves can show double and single hump shapes. There have been a variety of recent experimental studies in which the capacitance exhibits both single and double hump shapes [21–25] as the ion concentration is varied.

Summarizing the situation, the differential capacitance of the EDL of an ionic liquid at small electrode charge increases, apparently without limit, and the differential capacitance, at large electrode charge decreases as the electrode charge increases in magnitude. This latter effect is due to the fact that the ions cannot be adsorbed into a monolayer as the GCS theory suggests. These two features have the effect of causing the capacitance to pass from a double hump shape with a minimum at or near the pzc at low concentrations to a single hump shape at higher concentrations. By contrast, the GCS theory predicts only a minimum in the capacitance that gradually fills in with an increasing concentration with the capacitance gradually becoming independent of the concentration and electrode beyond some high concentration result.

Ionic liquids, in the real world, are formed from asymmetric ions. We have considered EDLs in a wide range of models that approximate ionic liquids [26]. We have modelled the asymmetry of ionic liquids by means of size asymmetry [27], through the use of dimers formed by tangentially touching spheres [27–34], and dimers of fused spheres [35]. Another method of introducing asymmetry is to keep spherical ions but allow the charge of one species of the ions to be off-center. We do this by considering the charged sphere of a dimer cation to be smaller than the diameter of the uncharged sphere of the dimer and fusing the smaller charged sphere of the dimer cation so that it should be entirely within the larger neutral sphere. This results in an electrolyte consisting of spheres but with one species (the anions) having a charge at the sphere center and the other species (the cations) having an off-center charge. In this paper, we will report some results for the density and potential profiles of this off-center ion model double layer. It is to be hoped that as well as its intrinsic interest, this model of off-center charges might lend itself to theories that use an expansion in powers of the degree to which the charge is off-center. In the event of such theories being developed, our simulations will be of value for testing these theories.

In addition to our work using spheres, there is another body of work that uses simulations with fairly realistic complex ionic liquids [36–39]. The investigators using complex ionic liquid models seek specificity whereas we seek generality. Such complex ionic liquid models do not lend themselves to theory. At present, only simulations are possible for these complex models. Both approaches have their merit.

2. Model; simulation method

We employ a fluid electrolyte model that consists of anions and cations that are represented by charged hard spheres in a uniform dielectric background. The charged spheres have a charge whose magnitude is equal to the proton charge e and have an equal diameter, $d = 4.25 \times 10^{-10}$ m. This model is similar to that of the restricted primitive model electrolyte but we allow the charge of the cations to be off-center, while the charge of the anions continues to be located at the sphere center. The electrode is considered to be a non-polarizable, hard planar wall of uniform (surface) charge density, σ, that is located on the electrode surface. To yield cation spheres with an off-center charge, we use dimer cations consisting of an uncharged hard sphere and a smaller charged hard sphere, which can fuse into each other. Thus, when the charged sphere is encapsulated entirely within the uncharged sphere, we get a cation with an off-center charge. The anion is a charged sphere whose diameter is equal to that of the uncharged or neutral sphere of the dimer. Specifically, if d_+, d_-, and d_0 are the diameters of the positive, negative, and neutral spheres, respectively, then we have, $d_- = d_0 = d$ and $d_+ = d/2$.

In the Hamiltonian, the various interaction potentials occurring in the system are as given below. The interaction between the spheres is

$$u_{ij}(r) = \begin{cases} \infty, & r < d_{ij}^c, \\ \dfrac{Z_i Z_j e^2}{4\pi\epsilon_0\epsilon_r r}, & r \geq d_{ij}^c, \end{cases} \tag{1}$$

while the interaction of a sphere with the electrode is given by

$$w_s(x) = \begin{cases} \infty, & x < d_{ws}^c, \\ -\dfrac{\sigma Z_s e x}{\epsilon_0\epsilon_r}, & x \geq d_{ws}^c. \end{cases} \tag{2}$$

Here, Z_s is the valency of particle species s, ϵ_0 and ϵ_r are the vacuum and relative permittivities, r is the distance between the centers of two spheres, and x is the perpendicular distance of a sphere from the electrode plane.

The quantity d_{ij}^c is the distance of closest approach between two particles of type i and j, respectively, while d_{ws}^c is the distance of closest approach of a particle of species s to the electrode. In the present study, the following three cases may be distinguished:

(i) for the dimer cation case (no fusion), we get as before (see for example, reference [34]) $d_{ij}^c = (d_i + d_j)/2$ and $d_{ws}^c = d_s/2$, d_s being the diameter of a particle of type s,

(ii) for the off-centered charged cation case, we have $d_{++}^c \in [d/2, 3d/2]$, $d_{+0}^c (= d_{+-}^c) \in [3d/4, 5d/4]$, and $d_{ij}^c = d$ where $ij = 00, 0-, --$. Also, $d_{w+}^c \in [d/4, 3d/4]$ and $d_{w0}^c = d_{w-}^c = d/2$, and finally

(iii) when the smaller positively charged sphere is completely fused inside the larger neutral sphere such that the centers of the two spheres are coincident, we have $d_{ij}^c = d$ and $d_{ws}^c = d/2$.

The degree to which the charge of the cations is off-center is controlled by the separation of, or distance between, the *centers* of the spheres in the dimer, s_{+0}. Thus, when there is no fusion, and the spheres in the dimer are merely tangentially touching, the separation between them is $s_{+0} = 3d/4$. We have $s_{+0} = d/4$ when the positively charged sphere is completely inside the neutral sphere and an off-center charged cation results. This is the case of most interest in the present work. The $s_{+0} = 0$ when the centers of the positive and neutral spheres coincide, and the standard RPM ensues. It ought to be mentioned, however, that although the sizes of the cation and the anion are different, viz., $d/2$ and d, respectively, the fact that the positive sphere is symmetrically embedded within the neutral sphere, makes the composite sphere an effective hard sphere cation of the same size as the hard sphere anion, and hence the RPM. For the diameters considered here, the off-center charge lies between the center of the neutral sphere and a distance of $d/4$ from its center (i.e., mid-way to the surface of the neutral sphere). Results for an off-center charge at a distance greater than mid-way to the surface, $d/4$, could be obtained by using a smaller value for the diameter of the positive sphere. However, we expect that a value of s_{+0} that is less than, or equal to, $d/4$ would be of most interest, at least in a theory in which the degree to which the charge is off-center was treated as a perturbation. The relation between s_{+0} and the degree to which the small charged sphere is fused into the larger uncharged sphere is shown in figure 1.

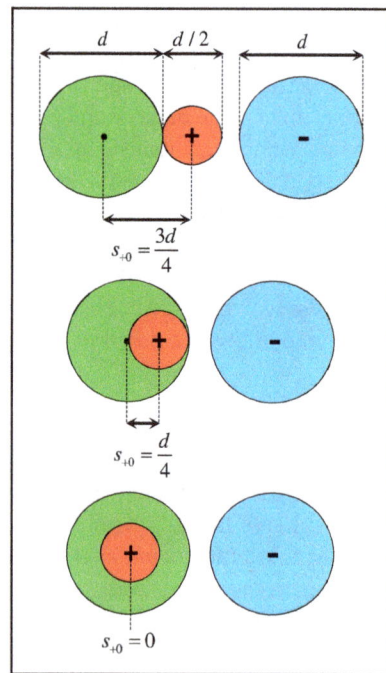

Figure 1. (Color online) Schematic diagram of an off-center cation containing RPM. The distance s_{+0} is between the centers of the positively charged sphere and the neutral sphere. From top to bottom: dimer model (no fusion), the off-center charge RPM, and standard RPM, respectively.

The MC simulations were performed in the canonical (NVT) ensemble following the standard Metropolis algorithm methods. The details of the simulations are the same as in our previous publications [28–35] and will not be repeated here. When $s_{+0} = 3d/4$ and $s_{+0} = 0$, that is the dimer and RPM situations, respectively, the simulations were checked against independent simulations of the dimer and RPM double layers using our previous programs. These results are useful tests of our present simulations and were found to be consistent with our earlier results within statistical errors.

3. Results

We consider a system of anions, cations, and uncharged hard spheres with a background relative permittivity of $\epsilon_r = 78.5$ at a temperature of $T = 298.15$ K. The hard sphere anions and the neutral spheres have a diameter of $d = 4.25 \times 10^{-10}$ m, while the hard sphere cations are smaller (half the anion or the uncharged sphere diameters). The electrolyte concentration considered here is $c = 1$ mol/dm^3. In this paper we will restrict ourselves to the symmetric valency cases of monovalent and divalent charged ions, viz., $Z_+ = -Z_- = Z = 1$ or 2, for the 1:1 or 2:2 cases, respectively.

In our numerical calculations, it is convenient to use reduced or dimensionless units that are denoted by an asterisk. However, in reporting results we will also give the equivalent physical units. The reduced temperature is $T^* = 4\pi\epsilon_0\epsilon_r d k_B T / Z^2 e^2$, whose values in the present study are $T^* = 0.595$ for the monovalent case ($Z = 1$), and $T^* = 0.149$ for the divalent case ($Z = 2$), respectively. The reciprocal of the T^* is the plasma coupling constant Γ of the literature, viz., $\Gamma = 1/T^*$. The bulk reduced density of the *free* particles of species s is defined as $\rho_s^* = \rho_s d^3$ with ρ_s being the corresponding bulk number density. Thus, for $c = 1$ mol/dm^3, we have $\rho_+^* = 0.00578$, $\rho_-^* = 0.0462$, and $\rho_0^* = 0.0462$, respectively. The reduced surface charge density on the electrode is defined $\sigma^* = \sigma d^2 / e$, while the reduced electrode potential is $\psi^* = e\psi / k_B T$, where ψ is the electrode potential in volts.

The double layer structure is described principally by the electrode–particle singlet distribution function $g_s(x) = \rho_s^*(x)/\rho_s^* = \rho_s(x)/\rho_s$, where $\rho_s^*(x)$ (or $\rho_s(x)$) is the local value of the corresponding quantity. In addition to reporting the values for the $g_s(x)$ of the particles of species s, we will also report the values of the reduced potential, $\psi^*(x)$. The potential profile $\psi(x)$ is a weighted integral of the singlet distributions, viz.,

$$\psi(x) = \frac{e}{\epsilon_0\epsilon_r} \sum_s Z_s \rho_s \int_x^\infty dx' (x - x') g_s(x'). \tag{3}$$

It is a useful indicator of the overall charge distribution in the system besides being the relevant quantity in characterizing the capacitance behaviour of the double layer. All of the profiles are calculated as functions of the perpendicular distance, x from the electrode, which is located at $x = 0$. The reduced unit for x is, of course, x/d.

In figures 2–4, results are presented for $Z = 1$ for $\sigma = \pm 0.1$ C/m^2, which corresponds to $\sigma^* = \pm 0.113$, whereas the results shown in figures 5–7 are for these same values of σ, or σ^*, but for $Z = 2$. Our main interest in this work is the case of spheres with an off-center charge, $s_{+0} = d/4$. These are shown in the middle panels of the density profiles in figures 2, 3, 5, 6, and in both panels of the potential profiles in figures 4 and 7. Results are also given for $s_{+0} = 3d/4$, that is, a dimer consisting of tangentially touching spheres, and $s_{+0} = 0$, which is the well studied RPM. The results for these two cases are not new, but the fact that these results represent the starting point and the end point in our quest for an off-center charged ion model, is a valuable internal consistency test of the new code. Furthermore, comparison of the results for various values of s_{+0}, especially as the degree of fusion increases, makes the effect of fusing the cation into the neutral sphere physically more apparent.

Considering first figure 2, the case of a negative surface charge, we notice in general that being counterions, the cations are attracted to the negatively charged electrode and bring the neutral spheres with them. On the other hand, the anions are the coions and are repelled from the electrode. In the no fusion situation, $s_{0+} = 3d/4$ (top panel), the profile of the neutral sphere of the dimer has a peak for x slightly less that d. The charged head of the dimer can approach the electrode more closely than can the neutral tail of the dimer since the positive sphere has a smaller diameter than the neutral one and the dimer can rotate. At the other extreme we have the case of complete fusion such that $s_{+0} = 0$ (bottom panel), that is, the RPM case, and the distributions of the positive and neutral spheres become identical because their centers are coincident so that the neutral spheres have effectively disappeared. The middle panel of this figure corresponds to $s_{+0} = d/4$ when the cation has just disappeared completely into the neutral sphere leading to a RPM with off-center charged cations, the central theme of the present study. The profiles for the center of this off-center charged sphere has a peak for x somewhat greater than $d/2$, whereas the profile of the charge of this off-center charged sphere has its peak for x nearer $3d/4$. These profiles also indicate that the distance of the charge of the off-center sphere from the electrode plane can be less than the corresponding distance of the center of the sphere itself since the sphere can rotate to bring

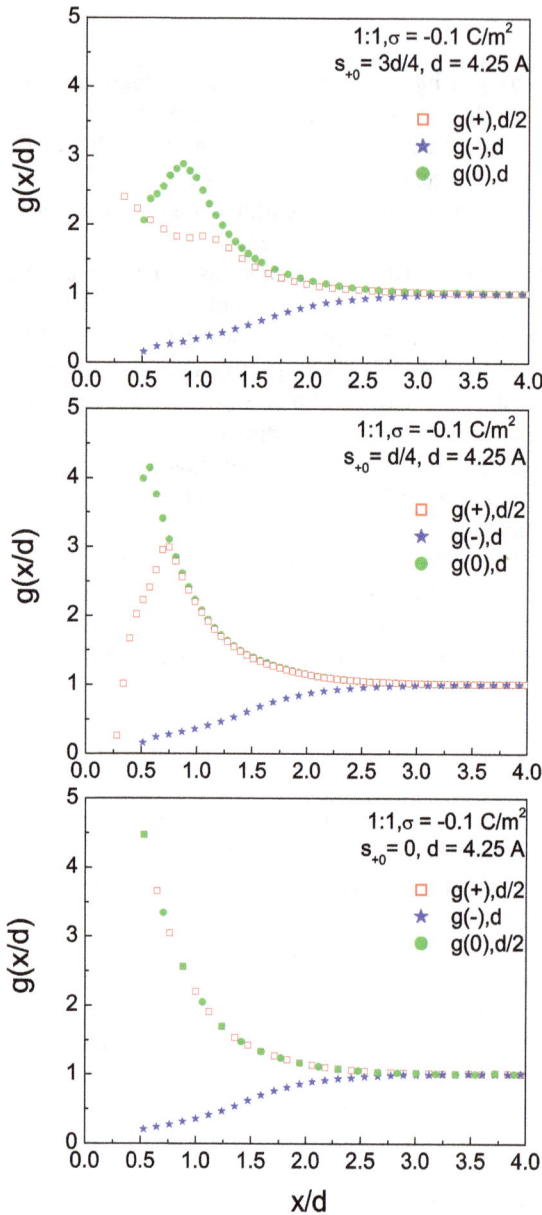

Figure 2. (Color online) Electrode–particle singlet distributions g_s for a 1:1 valency electrolyte system at concentration $c = 1$ mol/dm^3, reduced electrode surface charge density $\sigma^* = -0.113$ ($\sigma = -0.1$ C/m^2), and for $s_{+0} = 3d/4$ (top panel), $s_{+0} = d/4$ (middle panel), and $s_{+0} = 0$ (bottom panel).

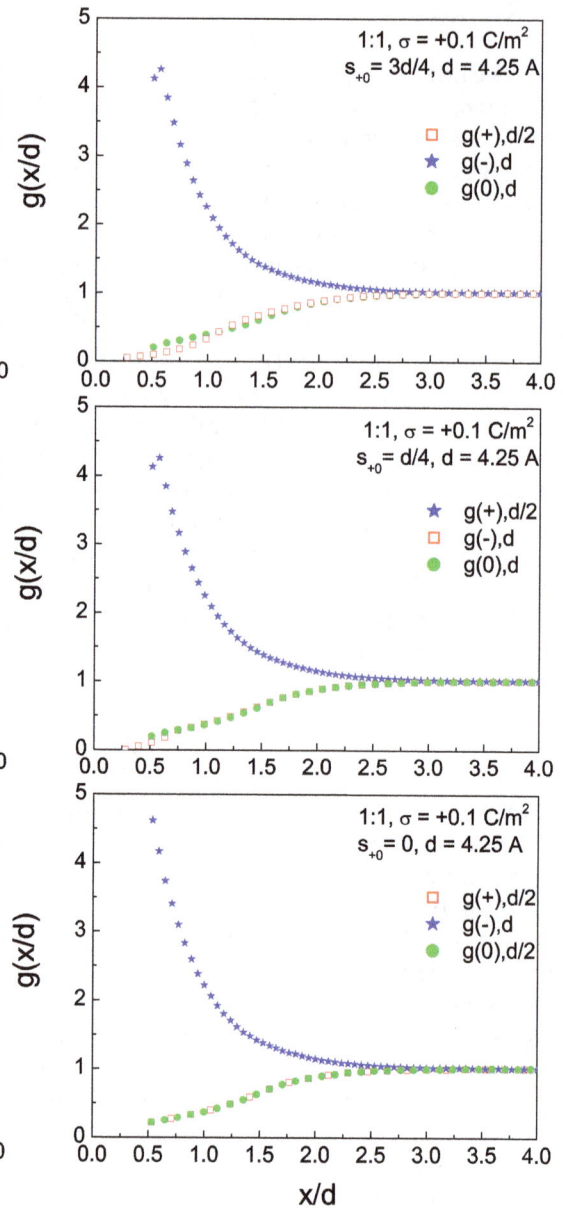

Figure 3. (Color online) Electrode–particle singlet distributions g_s for a 1:1 valency electrolyte system at concentration $c = 1$ mol/dm^3 and reduced electrode surface charge density $\sigma^* = +0.113$ ($\sigma = +0.1$ C/m^2. Rest of the notation and legend as in figure 2.

the charge center closer to the wall. The profiles of the monomer anions are only slightly affected by the value of s_{+0}; this is because the anion population near the negatively charged electrode is relatively less. This profile seems to be monotonous for all cases.

The case of a positively charged electrode is considered in figure 3. The striking feature of the results here is the qualitative similarity of the various profiles across the panels. This can be understood from the fact that now it is the negatively charged monomer spheres that are the counterions and are attracted to the electrode, while the positively charged spheres and their attached neutral spheres are repelled.

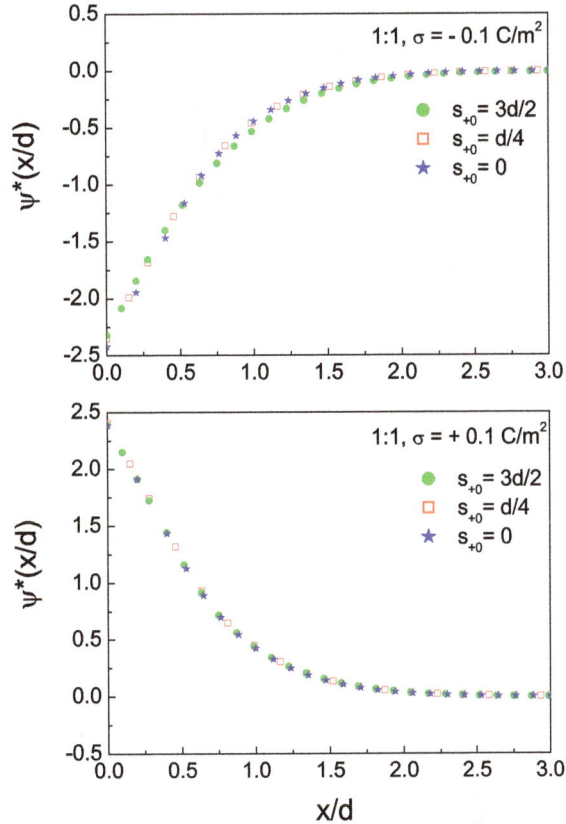

Figure 4. (Color online) Reduced mean electrostatic potential $\psi^*(x/d)$ for a 1:1 valency electrolyte system at concentration $c = 1$ mol/dm^3 with reduced electrode surface charge density $\sigma^* = -0.113$ ($\sigma = -0.1$ C/m^2) (upper panel), and $\sigma^* = +0.113$ ($\sigma = +0.1$ C/m^2) (lower panel). The three plots in each panel correspond to (i) solid circles , $s_{+0} = 3d/4$), (ii) open squares, $s_{+0} = d/4$), and (iii) stars, $s_{+0} = 0$.

The latter feature, in turn, implies that the internal geometry of the cation has little influence on the double layer structure in the immediate vicinity of the electrode. The profiles are thus more akin to that for a PM and having relatively less features than when the electrode is negatively charged. Analogous observations were made in some of our previous studies (see for example, reference [34]). It is interesting that the profile of the counterion has a small peak near the contact. In the course of these calculations we have also examined the cases with a higher surface charge density where this peak tended to become less conspicuous so that the peak may well be related to hard core effects. At any rate, it should have little affect on the potential profile since the potential profile is a first moment. The potential profile is also strongly influenced by the charges further from the electrode. The profiles for the positive and neutral spheres of the dimers are very nearly indistinguishable and are monotonous. These particles reside far away from the electrode and seemingly behave as a single entity. It is noted again that in the bottom panel $s_{+0} = 0$ for the RPM case, the g_+ and g_0 are identical as expected.

The 1:1 mean electrostatic potential profiles for the negatively and positively charged walls are illustrated in figure 4. In either situation the s_{+0} dependence is generally weak. For the monovalent case and hence relatively low coupling, the geometry of the ions gets masked in taking the weighted average of the g_s to get the ψ^*. We further note that the neutral species spheres have no bearing on the potential. These potential profiles are also monotonous.

The results for $Z = 2$, displayed in figures 5–7, show the effect of higher valency on the structure. The singlet distributions reveal oscillations, not seen with the 1:1 case, and the double layer is more compact. For a negative electrode (figure 5), for the no fusion dimer case (top panel) and the off-center charge fused sphere (middle panel) case, we note again that the positive spheres can approach the electrode more closely than the neutral spheres for reasons stated earlier. In figure 6 (positive electrode), the profiles

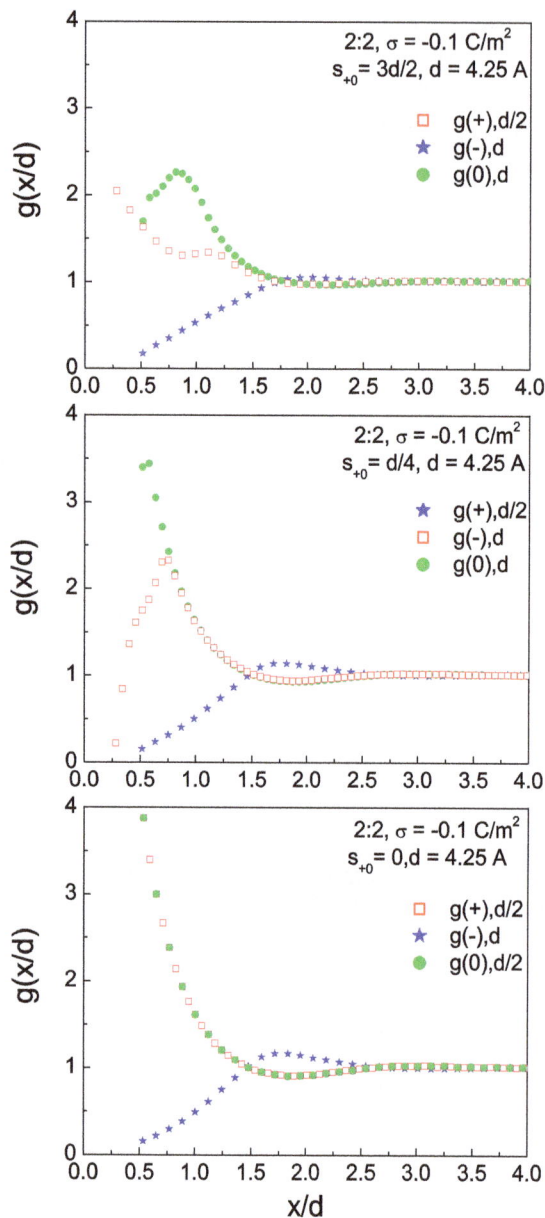

Figure 5. (Color online) Electrode–particle singlet distributions g_s for a 2:2 valency electrolyte system at concentration $c = 1$ mol/dm^3 and reduced electrode surface charge density $\sigma^* = -0.113$ ($\sigma = -0.1$ C/m^2). Rest of the notation and legend as in figure 2.

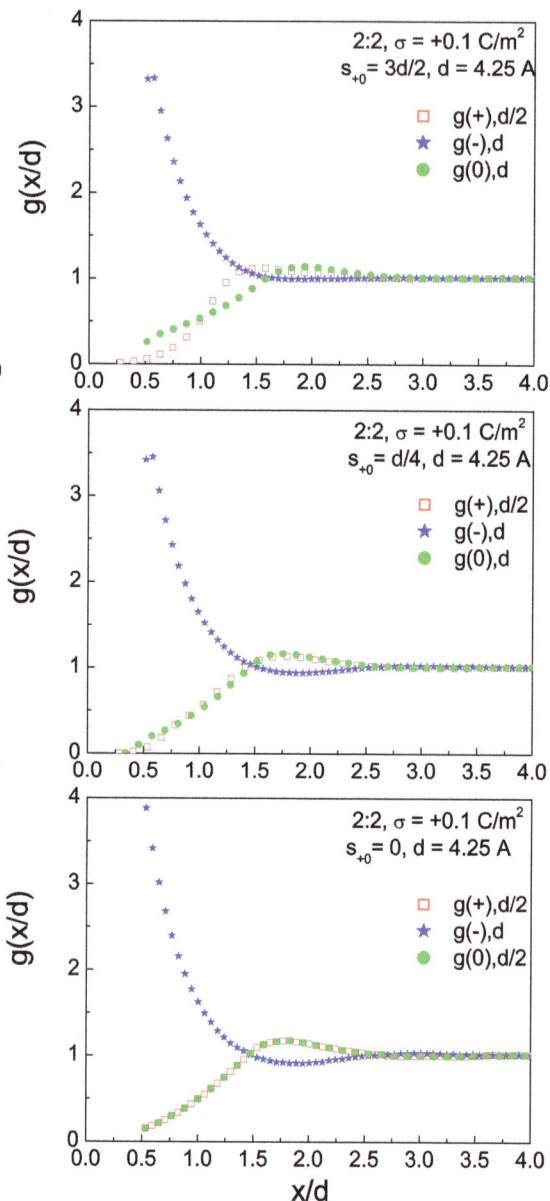

Figure 6. (Color online) Electrode–particle singlet distributions g_s for a 2:2 valency electrolyte system at concentration $c = 1$ mol/dm^3 and reduced electrode surface charge density $\sigma^* = +0.113$ ($\sigma = +0.1$ C/m^2). Rest of the notation and legend as in figure 2.

for positively charged and neutral ends of the dimer (top panel) or the profiles for the center of the off-center charge sphere and the charge-center of this sphere (middle panel) are very nearly the same as seen earlier in the 1:1 case. The general shape of the profiles is determined by the nature of counterions, which are spheres in all cases. However, rotation of the dimer coions is clearly evident. The dimer rotates so that the charged end is further from the electrode. This is still true for the off-center charged sphere but to a lesser extent. In the bottom panels (RPM case) of these two figures, the g_+ and g_0 are identical as before for the 1:1 case.

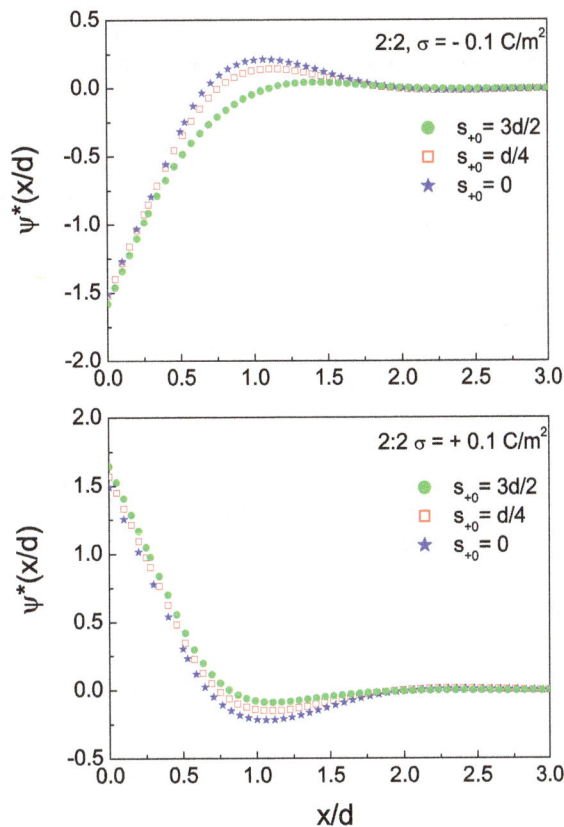

Figure 7. (Color online) Reduced mean electrostatic potential $\psi^*(x/d)$ for a 2:2 valency electrolyte system at concentration $c = 1$ mol/dm^3 with reduced electrode surface charge density $\sigma^* = -0.113$ ($\sigma = -0.1$ C/m^2) (upper panel), and $\sigma^* = +0.113$ ($\sigma = +0.1$ C/m^2) (lower panel). Rest of the notation and legend as figure 4.

The ψ^* profiles depicted in figure 7 are of interest. Unlike the monovalent cases in figure 5, the potential profiles for the divalent ions are no longer monotonous. Near the electrode, there is a broad maximum or a shallow minimum in ψ^* for the negative and positive surface charges, respectively. This is the phenomenon of *charge inversion* or *overscreening*, which refers to the physical situation where a layer of counterions predominates near the electrode and over-compensates the charge on the electrode. This is a common occurrence when divalent ions are present. In the present case, this feature manifests for all s_{+0}.

4. Summary

In this paper we have explored through MC simulations an asymmetric variation of the RPM planar double layer where asymmetry is imparted to the electrolyte by having an off-center charge for the cations. This is a logical extension of the familiar RPM. In the model introduced here, the repulsive hard, non-electrostatic, part of the inter-ionic interaction is spherical and central, and is represented by a hard sphere potential, which in many ways is the most difficult part of the interaction but which has been studied extensively and is now quite well understood. However, there is no reason to expect that the center of charge of an ion should be located at the center of mass of the ion. This would be especially true for the ions in an ionic liquid. One way of deriving a theory is to accept the hard sphere portion of a charged hard sphere fluid as the starting point and to introduce the electrostatic contribution as a perturbation. For example, it is possible to obtain the mean spherical approximation by means of a 'ring sum' of electrostatic terms. It may well be possible to formulate an analogous theory when the electrostatic interaction has a center of charge that differs from the geometric center, or center of mass, of the ion, especially when the difference between the center of charge and center of mass is small enough that it can be regarded as a perturbation.

Another pertinent question in line with the above concerns the generalization of the Henderson-Blum-Lebowitz (HBL) contact value theorem for PM double layers [40,41] to include the present off-center charge RPM double layer. Such an extension will certainly aid in the theoretical development as has been the case with the HBL sum rule in the double layer literature. But an expression for the osmotic pressure for a bulk charged hard sphere fluid with an off-center charge does not yet exist so that again a perturbation approach would be preferable. We hope to take up the issue in the near future.

A cation with an off-center charge was conceived from a dimer cation formed by a tangentially tethered pair of a large hard sphere and a smaller charged hard sphere such that the latter can fuse entirely into the uncharged sphere. The interesting changes that occur as a result in the structural pattern of the double layer relative to that of the standard RPM makes the off-center charge RPM promising for the future.

Variations of the present off-center charge model can be contemplated. For instance, one could have off-center charges for both the anion and the cation. Still further asymmetry could be incorporated by having in addition, asymmetric ionic valencies and/or sizes. Work on such models is in progress.

Acknowledgements

Monika Kaja and Stanisław Lamperski acknowledge the financial support from the Faculty of Chemistry, Adam Micklewicz University in Poznań.

References

1. Winter M., Brodd R.J., Chem. Rev., 2004, **104**, 4245; doi:10.1021/cr020730k.
2. Cherstvy A.G., Phys. Chem. Chem. Phys., 2011, **13**, 9942; doi:10.1039/C0CP02796K.
3. Nonner W., Catacuzzeno L., Eisenberg B., Biophys. J., 2000, **79**, 1976; doi:10.1016/S0006-3495(00)76446-0.
4. Gromelski S., Brezesinski G., Phys. Chem. Chem. Phys., 2004, **6**, 5551; doi:10.1039/B410865E.
5. Henderson D., Boda D., Phys. Chem. Chem. Phys., 2009, **22**, 3822; doi:10.1039/B815946G.
6. Hayes R., Warr G.G., Atkin R., Chem. Rev., 2015, **115**, 6357; doi:10.1021/cr500411q.
7. Gouy M., J. Phys. (Paris), 1910, **9**, 457; doi:10.1051/jphystap:019100090045700.
8. Chapman D.L., Philos. Mag., 1913, **25**, 475; doi:10.1080/14786440408634187.
9. Stern O., Z. Elektrochem. Angew. Phys. Chem., 1924, **30**, 508; doi:10.1002/bbpc.192400182.
10. McQuarrie D.A., Statistical Mechanics, University Science Books, Sausalito, 2000.
11. Torrie G.M., Valleau J.P., J. Chem. Phys., 1980, **73**, 5807; doi:10.1063/1.440065.
12. Blum L., J. Phys. Chem., 1977, **81**, 136; doi:10.1021/j100517a009.
13. Lozada-Cassou M., Saavedra-Barrera R., Henderson D., J. Chem. Phys., 1982, **77**, 5150; doi:10.1063/1.443691.
14. Blum L., Henderson D., Parsons R., J. Electroanal. Chem., 1984, **161**, 389; doi:10.1016/S0022-0728(84)80196-5.
15. Woelki S., Henderson D., Condens. Matter Phys., 2011, **14**, 43801; doi:10.5488/CMP.14.43801.
16. Outhwaite C.W., Bhuiyan L.B., J. Chem. Soc., Faraday Trans. 2, 1983, **79**, 707; doi:10.1039/F29837900707.
17. Lamperski L., Outhwaite C.W., Bhuiyan L.B., J. Phys. Chem. B, 2009, **113**, 8925; doi:10.1021/jp900037h.
18. Outhwaite C.W., Lamperski L., Bhuiyan L.B., Mol. Phys., 2011, **109**, 21; doi:10.1080/00268976.2010.519731.
19. Rosenfeld Y., J. Chem. Phys., 1993, **98**, 8126; doi:10.1063/1.464569.
20. Lamperski S., Henderson D., Mol. Simul., 2011, **37**, 264; doi:10.1080/08927022.2010.501973.
21. Kornyshev A.A., J. Phys. Chem. B, 2007, **111**, 5545; doi:10.1021/jp067857o.
22. Alam M.T., Islam M.M., Okajima T., Oshaka T., Electrochem. Commun., 2007, **9**, 2370; doi:10.1016/j.elecom.2007.07.009.
23. Islam M.M., Alam M.T., Oshaka T., J. Phys. Chem. C, 2008, **112**, 16568; doi:10.1021/jp8058645.
24. Alam M.T., Islam M.M., Okajima T., Oshaka T., J. Phys. Chem. C, 2008, **112**, 16600; doi:10.1021/jp804620m.
25. Alam M.T., Islam M.M., Okajima T., Oshaka T., J. Phys. Chem. C, 2009, **113**, 6596; doi:10.1021/jp810865t.
26. Lockert V., Horne M., Sedev R., Rodopoulis T., Ralston J., Phys. Chem. Chem. Phys., 2010, **12**, 12499; doi:10.1039/C0CP00170H.
27. Wu J., Jiang T., Jiang D., Jin Z., Henderson D., Soft Matter, 2011, **7**, 11222; doi:10.1039/C1SM06089A.
28. Lamperski S., Sosnowski J., Bhuiyan L.B., Henderson D., J. Chem. Phys., 2014, **140**, 014704; doi:10.1063/1.4851456.

29. Henderson D., Wu J., J. Phys. Chem. B, 2012, **116**, 2520; doi:10.1021/jp212082k.
30. Bhuiyan L.B., Lamperski S., Wu J., Henderson D., J. Phys. Chem. B, 2012, **116**, 10364; doi:10.1021/jp304362y.
31. Henderson D., Lamperski S., Bhuiyan L.B., Wu J., J. Chem. Phys., 2013, **138**, 144704; doi:10.1063/1.4799886.
32. Lamperski S., Kaja M., Bhuiyan L.B., Wu J., Henderson D., J. Chem. Phys., 2013, **139**, 054703; doi:10.1063/1.4817325.
33. Kaja M., Silvestre-Alcantara W., Lamperski S., Henderson D., Bhuiyan L.B., Mol. Phys., 2015, **113**, 1043; doi:10.1080/00268976.2014.968651.
34. Silvestre-Alcantara W., Henderson D., Wu J., Kaja M., Lamperski S., Bhuiyan L.B., J. Colloid Interface Sci., 2015, **449**, 175; doi:10.1016/j.jcis.2014.11.070.
35. Silvestre-Alcantara W., Kaja M., Henderson D., Lamperski S., Bhuiyan L.B., Mol. Phys., 2015, **114**, 53; doi:10.1080/00268976.2015.1083132.
36. Vatamanu J., Borodin O., Smith G.D., J. Am. Chem. Soc., 2010, **132**, 14825; doi:10.1021/ja104273r.
37. Vatamanu J., Borodin O., Bedrov D., Smith G.D., J. Phys. Chem. C, 2012, **116**, 7940; doi:10.1021/jp301399b.
38. Hu Z., Vatamanu J., Borodin O., Bedrov D., Phys. Chem. Chem. Phys., 2013, **15**, 14234; doi:10.1039/C3CP51218E.
39. Feng G., Jiang D., Cummings P.T., J. Chem. Theory Comput., 2012, **8**, 1058; doi:10.1021/ct200914j.
40. Henderson D., Blum L., J. Chem. Phys., 1978, **69**, 5441; doi:10.1063/1.436535.
41. Henderson D., Blum L., Lebowitz J., J. Electroanal. Chem., 1979, **102**, 315; doi:10.1016/S0022-0728(79)80459-3.

16

Cavity-ligand binding in a simple two-dimensional water model*

G. Mazovec, M. Lukšič, B. Hribar-Lee

University of Ljubljana, Faculty of Chemistry and Chemical Technology,
Večna pot 113, SI-1000 Ljubljana, Slovenia

By means of Monte Carlo computer simulations in the isothermal-isobaric ensemble, we investigated the interaction of a hydrophobic ligand with the hydrophobic surfaces of various curvatures (planar, convex and concave). A simple two-dimensional model of water, hydrophobic ligand and surface was used. Hydration/dehidration phenomena concerning water molecules confined close to the molecular surface were investigated. A notable dewetting of the hydrophobic surfaces was observed together with the reorientation of the water molecules close to the surface. The hydrogen bonding network was formed to accommodate cavities next to the surfaces as well as beyond the first hydration shell. The effects were most strongly pronounced in the case of concave surfaces having large curvature. This simplified model can be further used to evaluate the thermodynamic fingerprint of the docking of hydrophobic ligands.

Key words: *cavity-ligand binding, water confinement, surface hydration, potential of mean force, Monte Carlo computer simulation, two-dimensional water model*

1. Introduction

Due to the important influence of the confinement on the structural, thermodynamical, and phase equilibrium properties of simple fluids, these systems have been extensively studied recently [1–3]. The topic is also of great interest in the studies of biological systems in which many unanswered questions are concerned with hydration and dehydration of molecules of arbitrary shapes where water as a solvent finds itself confined close to the molecular surface.

One of the important issues that have begun to emerge is the effect of the shape, curvature, and roughness of a surface on its interaction with a ligand [4–10]. In order to understand the phenomena such as binding of the ligand to the receptor, or drugs to proteins, it is crucial to know the potential of the mean force between two biomolecules in question. Virtually all the binding sites in biology have a concave shape which imposes very particular geometrical constraints to the solvated water [11] and, therefore, the findings referring to the potential of the mean force between two spherical surfaces cannot be generalized to these problems.

To summarize the recent molecular dynamics studies on model systems of purely hydrophobic cavities [10, 12–14], it has been shown that water appears to be an active component in cavity-ligand association. An important impact of changes in water structure during the binding process has been noticed [10, 14]. While the cavity hydration can be altered by changing its radius, in weakly hydrated cavity regions the reorganization of water molecules and suppression of the solvent fluctuations leads to enthalpy driven association [15]. A different interpretation of the computer simulation results have been given by Graziano [17]. His analysis shows that the Gibbs free energy gain upon association of a ligand in a concave hydrophobic cavity is mainly due to the decrease in the solvent-excluded volume, that translates in a gain of configurational-translational entropy of water molecules. This entropic driving force is masked by a

---

*Dedicated to the 65th birthday of Prof. Dr. Stefan Sokołowski.

large enthalpy gain associated with the reorganization of water-water hydrogen bonds upon association of the two nonpolar objects [17].

In this work we have focused on the investigation of the potential of the mean force between a hydrophobic ligand and a planar, concave, and convex purely hydrophobic surface with different curvatures. Our main interest was in the interpretation of the potential of the mean force through the water microstructural changes due to the confinement. To better visualize these effects we have chosen a simple two-dimensional Mercedes-Benz (MB) water model [16].

The paper is organized as follows: After this short introduction, the model and method are described. Next, the results are presented and discussed, and the conclusions are given in the end.

2. Model and method

The Mercedes-Benz model [16] was used to describe water molecules. In this model, the water molecule is represented as a two-dimensional Lennard-Jones (LJ) disk with three equally separated hydrogen bonding arms, and interacts with another water molecule through the potential U_{ww}:

$$U_{ww}(\mathbf{X}_i, \mathbf{X}_j) = U_{LJ}(r_{ij}) + U_{HB}(\mathbf{X}_i, \mathbf{X}_j). \tag{2.1}$$

$\mathbf{X}_i$ denotes the vector with coordinates and orientation of the i-th particle, and r_{ij} is the distance between the centres of molecules i and j. Lennard-Jones 12-6 potential is:

$$U_{LJ}(r_{ij}) = 4\epsilon_{LJ}\left[\left(\frac{\sigma_{LJ}}{r_{ij}}\right)^{12} - \left(\frac{\sigma_{LJ}}{r_{ij}}\right)^{6}\right], \tag{2.2}$$

where ϵ_{LJ} represents the depth of the potential well, and σ_{LJ} is the Lennard-Jones diameter.

The hydrogen bond strength is a Gaussian function of the intermolecular distance and the angle between two hydrogen bonding arms:

$$U_{HB}(\mathbf{X}_i, \mathbf{X}_j) = \epsilon_{HB} G(r_{ij} - r_{HB}) \sum_{k,l=1}^{3} G(\hat{\mathbf{i}}_k \cdot \hat{\mathbf{u}}_{ij} - 1) G(\hat{\mathbf{j}}_l \cdot \hat{\mathbf{u}}_{ij} + 1), \tag{2.3}$$

where $G(x)$ is an unnormalized Gaussian function:

$$G(x) = \exp\left(-x^2/2\sigma^2\right). \tag{2.4}$$

The unit vector $\hat{\mathbf{i}}_k$ represents the k-th arm of the i-th molecule ($k = 1, 2, 3$). Unit vector $\hat{\mathbf{u}}_{ij}$ is the direction vector of the line joining the centres of the i-th and j-th molecule. Parameters $\epsilon_{HB} = -1$ and $r_{HB} = 1$ determine the energy and the length of the optimal hydrogen bond. Values for the model parameters are as follows: $\sigma_{LJ} = 0.7$ $r_{HB} = 0.7$, $\epsilon_{LJ} = 0.1 |\epsilon_{HB}| = 0.1$, and $\sigma = 0.085$.

The hydrophobic ligand was represented as a Lennard-Jones disk of the same size ($\sigma_{LJ} = 0.7$) as the water molecule.

To model a hydrophobic surface of a given curvature, a molecular wall was formed from Lennard-Jones discs (see figure 1). One surface was planar, while the others were bent. The interaction of the test water or hydrophobic ligand with the surfaces having positive curvatures (concave), and with negative curvatures (convex) was tested. The two radii used for concave surfaces were $R^* = 4.55$ and 1.40, while for the convex surfaces the radii were 2.80 and 5.95. The hydrophobic particle and the water molecules interacted with the particles forming the hydrophobic surface through the Lennard-Jones potential [equation (2.2)]. The LJ parameters (σ_{LJ} and ϵ_{LJ}) were the same for all particles in question.

Reduced units were used throughout the paper: $r^* = r/r_{HB}$, $T^* = k_B T/|\epsilon_{HB}|$, $p^* = p r_{HB}^2/|\epsilon_{HB}|$.

Isobaric-isothermal Monte Carlo computer simulations [18] were performed for systems with 120 MB water molecules and a single hydrophobic ligand (test particle). The hydrophobic surface was placed in the centre of the simulation box. Reduced temperature, T^*, and pressure, p^*, of the systems were 0.20 and 0.19, respectively. Initial configuration of water molecules was chosen randomly and then equilibrated in 10^7 cycles long simulation. Statistics were then collected over 10^8 cycles long production run

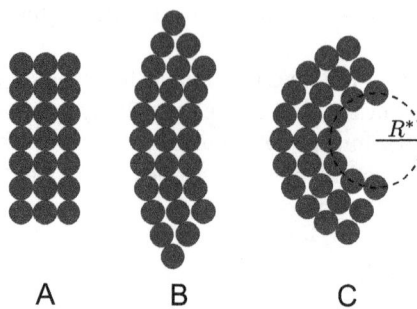

Figure 1. Schematic representation of the surfaces used in simulations. All surfaces were made of three layers of hydrophobic particles with Lennard-Jones (LJ) diameter $\sigma_{LJ} = 0.7$. Planar surface (A) was composed of $3 \cdot 7 = 21$ LJ discs, while curved surfaces (B and C) were made of $7 + 9 + 11 = 27$ LJ discs. Radii of the inner and outer layer of the surface B were $R^* = 4.55$ and 5.95, respectively, while for the surface C the inner and the outer radii were 1.40 and 2.80, respectively. The surfaces were placed in the middle of the simulation box.

which started from equilibrated configuration. In one cycle, either rotation or displacement (probabilities for rotation and displacement being equal) of a water molecule was attempted. Maximal displacement and rotation were adjusted throughout the simulation, so that approximately one half of Monte Carlo moves were accepted. After every 120th cycle, an attempt to change the volume of the simulation box was made, where the maximal change was adjusted in the same way as maximal displacement/rotation.

The pair correlation and angular distribution functions of water next to the surface were calculated using the histogram method, while the potential of the mean force between the hydrophobic ligand and the surface was calculated using the Widom's insertion technique [19].

3. Results and discussion

Here, we present the results of the Monte Carlo simulations of interaction of the hydrophobic ligand with the hydrophobic surface in water. All the results are given for $T^* = 0.2$ and $p^* = 0.19$. Figure 2 (a) shows the potentials of the mean force (PMFs) between the hydrophobic particle and the central particle

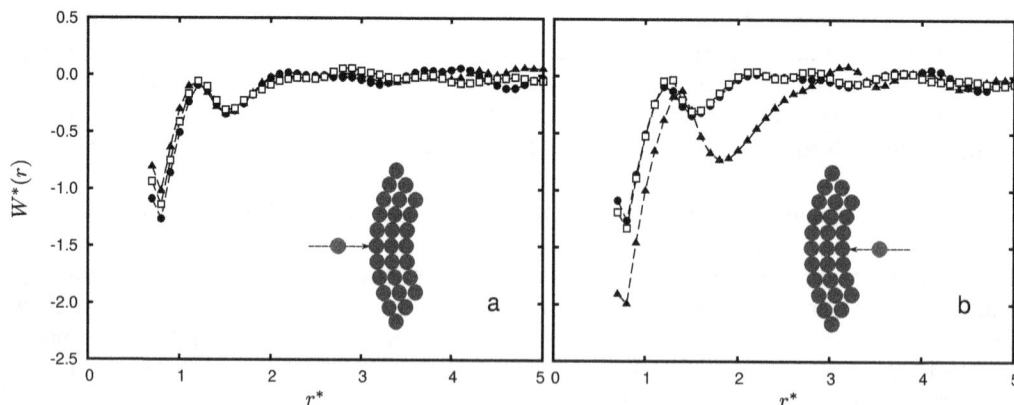

Figure 2. Potentials of the mean force (PMFs) between a hydrophobic ligand (test particle) and the central particle of the hydrophobic surface. Centre-to-centre distance between test particles is given by r^* (in the direction of the x-axis; see the insets). Results for the convex shaped surfaces are shown in panel (a), and for the concave case in panel (b). The reduced PMF, $W^*(r) = W(r)/k_B T$, for the case of linear surface is shown in both panels (circles). Two curvatures of the surface were tested — panel (a): 5.95 (squares) and 2.80 (triangles), panel (b): 4.55 (squares) and 1.40 (triangles). $T^* = 0.20$ and $p^* = 0.19$.

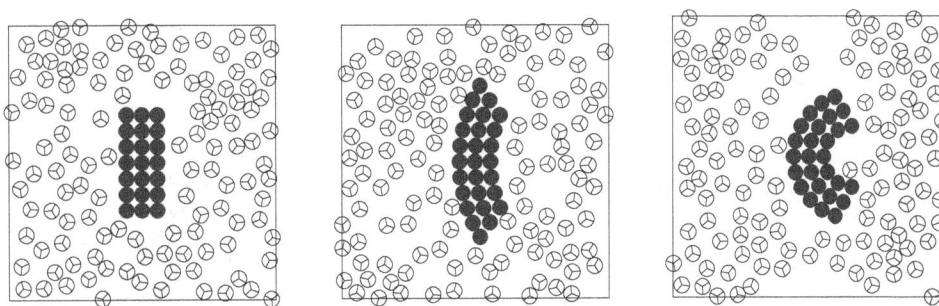

Figure 3. Representative simulation snapshots for various surfaces. White circles represent MB water molecules with three hydrogen bonding arms (Mercedes-Benz logo). The grey circles are hydrophobic particles forming the surfaces. $T^* = 0.20$ and $p^* = 0.19$.

of the surface for convex surfaces, while the PMFs for concave surfaces are shown in figure 2 (b). The PMF with planar surface is shown for comparison in both panels. An important difference from the PMF between two hydrophobic discs in water (figure 1 of reference [6]) is observed. There is a larger difference in the values of the peaks belonging to the contact minimum (CM) and solvent separated minimum (SSM) compared to the homogeneous system (two hydrophobic discs); CM state is here much more stable than the SSM state. This suggests that the hydrophobic particle would prefer to stay close to the surface. Not much difference between the PMFs for different surface curvatures is observed in the convex cases, while in the case of strongly concave surface, the contact state is additionally stabilized and the position of the SSM is shifted further away from the surface.

To interpret these results in view of water microstructure, we analysed the hydration of the surfaces more in detail. Characteristic simulation snapshots are shown in figure 3. A dehydration of the hydrophobic surface is noticed for all surfaces studied. The water molecules are pushed away and cavities are formed at the surface contact as well as beyond the first layer of water molecules. This is further confirmed by the water-surface radial distribution functions shown in figure 4. All $g(r)$ show a layered structure of water molecules away from the surface. There is no significant qualitative difference in the $g(r)$ between the planar surface case and convex bent surfaces. The value of the contact peak slightly decreases with an increasing curvature [see panel (a) of figure 4].

On the other hand, a qualitatively different behaviour is observed in the case of strongly concave surface. Here, water is completely pushed away from the cavity, forming the first hydration layer approximately two hydrogen bonds away from the surface [figure 4 (b)]. This is in agreement with the pre-

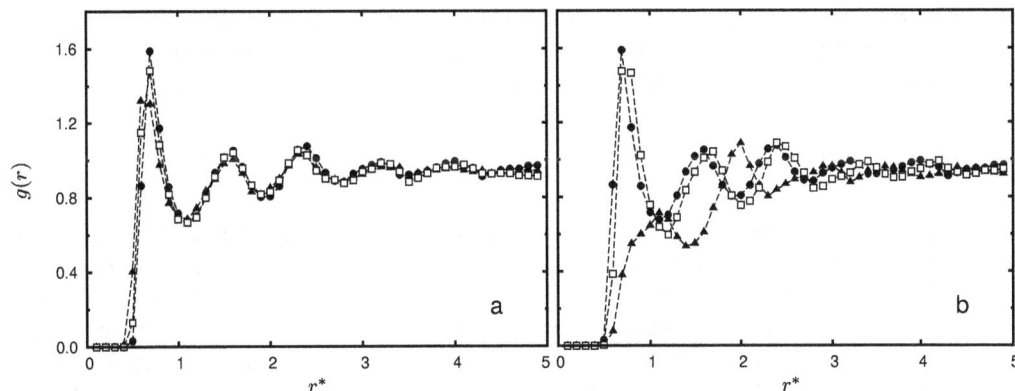

Figure 4. Radial distribution functions between a test water molecule and the central particle of the hydrophobic surface. Distance r^* has the same meaning as in figure 2. Surface parameters, notations and conditions (T^*, p^*) are the same as in figure 2.

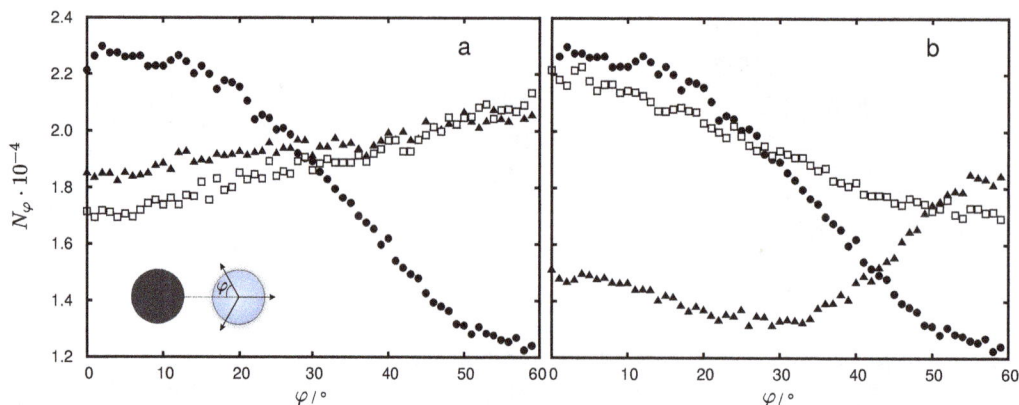

Figure 5. Angular distribution of the first hydration shell water with respect to the hydrophobic surface. Central particle of the surface closest to the test water was chosen to gather statistics. The angle ϕ is defined $0°$ when one of the water's hydrogen bonding arms points towards the centre of the surface's particle [see the sketch in panel (a)]. Results for the convex shaped surfaces are shown in panel (a), and for the concave case in panel (b). For comparison, the unnormalized distribution, N_φ, for the case of linear surface is shown in both panels (circles). Surface parameters, notations and conditions (T^*, p^*) are the same as in figure 2.

vious studies by Baron et al. and Graziano [10, 12, 13, 15, 17]. Due to the incapability of forming hydrogen bonds, a water molecule in such a state is entropically stabilized.

Further, the orientation of the surface water molecules belonging to the first hydration shell was examined. Figure 5 (a) shows the results for the convex surfaces, and panel (b) for the concave surfaces. The case of planar wall is shown in both panels for comparison. As already noticed by Southall et al. [20], a model water molecule close to the planar surface can no longer form all three hydrogen bonds with neighbouring water molecules. To compensate for this loss, it "wastes" one of the hydrogen bonds by pointing it towards the surface. This is seen from the angular distribution where the most probable orientation of the water molecule at the planar surface is the one pointing the hydrogen bond towards the surface ($\varphi = 0°$). The two hydrogen bonds pointing away from the surface participate in the formation of cavities facilitating a favourable SSM in the PMFs.

As the surface adopts the convex shape, the angular preference of the water to the surface is much less expressed. For the case in panel (a) of figure 5, we can conclude that angular preferences of water have a negligible contribution to the shape of the PMFs.

The situation gets reversed in the case of strongly concave surface [triangles in figure 5 (b)]. The preferential angle of water molecule in this case is the one where the hydrogen bonding arm points at an angle $60°$ with respect to the centre-centre coordinate [see the insert in panel (a)], suggesting the strongly obstructed hydrogen bond formation due to the lack of space in the cavity. This way a water molecule was stabilized by an increase of the rotational entropy.

4. Conclusions

The results for the potential of the mean force between hydrophobic solute and hydrophobic surface of various shapes in the model water solutions were presented. The results were analysed in view of hydration water structure and orientation. All the hydrophobic surfaces studied showed dewetting which was most pronounced for the bent concave surface. The angular orientation of the water molecules facilitated the formation of cavities which stabilized the non-covalent direct and water separated binding of the hydrophobic molecule to the surface. In the future work, temperature dependence of the phenomena will be investigated and the results will be used for the enthalpy-entropy decomposition of the free energy of ligand-surface interaction.

Acknowledgements

This work was supported by the Slovenian Research Agency through grant P1-0103-0201 and Slovene-Korean bilateral grant BI-KR/13-14-003. M.L. and B.H.-L. acknowledge the support of the NIH Grant 2R01GM063592-14.

References

1. Mansoori G.A., Rice S.A., Adv. Chem. Phys., 2014, **156**, 197; doi:10.1002/9781118949702.ch5.
2. Pizio O., Patrykiejew A., Sokołowski S., J. Chem. Phys., 2004, **121**, 11957; doi:10.1063/1.1818677.
3. Pizio O., Patrykiejew A., Sokołowski S., J. Phys. Chem. C, 2007, **111**, 15743; doi:10.1021/jp0736847.
4. Shavkat I., Mamatkulov S.I., Khabibullaev P.K., Netz R.R., Langmuir, 2004, **20**, 4756; doi:10.1021/la036036x.
5. Rudich Y., Benjamin I., Naaman R., Thomas E., Trakhtenberg S., Ussyshkin R., J. Phys. Chem. A, 2000, **104**, 5238; doi:10.1021/jp994203p.
6. Southall N.T., Dill K.A., Biophys. Chem., 2002, **101**, 295; doi:10.1016/S0301-4622(02)00167-9.
7. Wallqvist A., Berne B.J., J. Phys. Chem., 1995, **99**, 2885; doi:10.1021/j100009a052.
8. Pratt A.J.L., Chandler D., J. Chem. Phys., 1980, **73**, 3430; doi:10.1063/1.440540.
9. Chorny I., Dill K.A., Jacobson M.P., J. Phys. Chem. B, 2005, **109**, 24056; doi:10.1021/jp055043m.
10. Baron R., Setny P., McCammon J.A., J. Am. Chem. Soc., 2010, **132**, 12091; doi:10.1021/ja1050082.
11. Sharp K.A., Nicholls A., Fine R.F., Honig B., Science, 1991, **252**, 106; doi:10.1126/science.2011744.
12. Baron R., Molinero V., J. Chem. Theory Comput., 2012, **8**, 3696; doi:10.1021/ct300121r.
13. Baron R., Setny P., Paesani F., J. Phys. Chem. B, 2012, **116**, 13774; doi:10.1021/jp309373q.
14. Setny P., Baron R., McCammon J.A., J. Chem. Theory Comput., 2010, **6**, 2866; doi:10.1021/ct1003077.
15. Baron R., Setny P., McCammon J.A., In: Protein-Ligand Interactions, Gohlke H. (Ed.), Wiley-VCH, Weinheim, 2011, 145–169.
16. Dill K.A., Truskett T.M., Vlachy V., Hribar-Lee B., Annu. Rev. Biophys. Biomol. Struct., 2005, **34**, 173; doi:10.1146/annurev.biophys.34.040204.144517.
17. Graziano G., Chem. Phys. Lett., 2012, **533**, 95; doi:10.1016/j.cplett.2012.03.020.
18. Allen M.P., Tildesley D.J., Computer Simulation of Liquids, Oxford University Press, New York, 1991.
19. Frenkel D., Smith B., Understanding Molecular Simulation: From Algorithms to Applications, Academic Press, San Diego, 2002.
20. Southall N.T., Dill K.A., J. Phys. Chem. B, 2000, **104**, 1326; doi:10.1021/jp992860b.

17

Gyroidal nanoporous carbons — Adsorption and separation properties explored using computer simulations*

S. Furmaniak[1], P.A. Gauden[1], A.P. Terzyk[1], P. Kowalczyk[2]

[1] Faculty of Chemistry, Physicochemistry of Carbon Materials Research Group, Nicolaus Copernicus University in Toruń, Gagarin St. 7, 87–100 Toruń, Poland

[2] School of Engineering and Information Technology, Murdoch University, Murdoch, Western Australia 6150, Australia

Adsorption and separation properties of gyroidal nanoporous carbons (GNCs) — a new class of exotic nanocarbon materials are studied for the first time using hyper parallel tempering Monte Carlo Simulation technique. Porous structure of GNC models is evaluated by the method proposed by Bhattacharya and Gubbins. All the studied structures are strictly microporous. Next, mechanisms of Ar adsorption are described basing on the analysis of adsorption isotherms, enthalpy plots, the values of Henry's constants, α_s and adsorption potential distribution plots. It is concluded that below pore diameters ca. 0.8 nm, primary micropore filling process dominates. For structures possessing larger micropores, primary and secondary micropore filling mechanism is observed. Finally, the separation properties of GNC toward CO_2/CH_4, CO_2/N_2, and CH_4/N_2 mixtures are discussed and compared with separation properties of Virtual Porous Carbon models. GNCs may be considered as potential adsorbents for gas mixture separation, having separation efficiency similar or even higher than activated carbons with similar diameters of pores.

Key words: *gyroidal nanoporous carbons, adsorption, gas mixtures separation, Monte Carlo simulations*

1. Introduction

Adsorption of gas mixtures on solid surfaces has been attracting great interest for many years [1–7]. In the last decades there has been observed an increased interest to the development of new separation and purification techniques. It is evident that adsorption using various adsorbents is still a versatile tool for these purposes. On the other hand, the basic problem appearing in experimental studies is caused by difficulties in the synthesis of nanomaterials possessing desired properties. It is still not simple to control porosity and/or the chemical nature of the surface, and the both parameters at the same time. Moreover, experimental data on mixed systems are very limited, i.e., in the case of mixtures consisting of two volatile components the problem of the surface coverage determination has not been fully solved yet. Predicting adsorption behaviour of mixtures from pure component data is very important, from both the theoretical and practical viewpoints [8–14]. It is well known that the theoretical calculations provide additional opportunities for studies to better understand the separation processes. However, despite the intensive experimental and theoretical studies, our knowledge of the properties and the structure of mixed adsorbed layers is rather sparse, especially, on new generations of nanoporous adsorbents.

Computer simulation is an efficient method for resolving the above mentioned problems, since it is capable of modelling the processes of interest at the required level of detail in a controllable environment, providing the necessary tools for establishing the connections between the observed phenomena

---

*This paper is dedicated to Prof. Stefan Sokołowski on the occasion of his 65th birthday.

and their molecular-level physical background. Prof. Sokołowski's (and co-workers) research topics of interest have also been concerned with the issue of mixtures using Monte Carlo simulations [15–22], Density Functional Theory [23–28], and Dissipative Particle Dynamics [29]. Their inspiring articles discuss, for example, the following problems: (i) adsorption from mixtures of monomers [15], dimers [15], the chain molecules [30, 31], and even polymer mixtures [27], (ii) adsorption from mixtures on homogeneous [25, 32] and heterogeneous surfaces [15, 33, 34], (iii) layering transition, capillary condensation, wetting phenomena, and multilayer adsorption of binary ideal mixtures, systems exhibiting negative deviations from ideal mixing or positive one, binary mixture with partially miscible components, etc. [16, 17, 34–38], (iv) interaction of charged chain particles and spherical counterions in contact with charged hard wall [31], (v) analysis of the properties of two-dimensional symmetrical mixtures in an external field of square symmetry [39, 40], (vi) demixing and freezing in two-dimensional symmetrical mixtures [21], and (vii) the behaviour of mixed two component submonolayer films (Ar and Kr [41, 42] or Kr and Xe [22] on graphite). The majority of the analysed adsorbents have an ideal geometry of pores, for example a slit-like [25, 26, 32, 34, 43]. However, more complex models have also been studied, for example, pillared slit-like pores [28] and slit-like pores with walls decorated by tethered polymer brushes in the form of stripes [29].

In the last decades, novel exotic porous carbon nanostructures (such as carbon nanotubes (CNTs), single-walled carbon nanohorns, graphene and graphitic nanoribbons, ordered porous carbons, worm-like nanotubes and graphitic nanofibers, stacked-cup carbon nanofibers, cubic carbon allotropes, carbon onions, carbyne networks, and others) have been projected to be among the most useful materials for selective adsorption and separation of fluid mixtures [44–48]. However, in the theoretical studies, different carbon adsorbents are studied, such as: carbon nanotubes [13, 49–53], carbon nanohorns [13, 51, 54], 2D and 3D ordered carbon networks [55], hydrophobic virtual porous carbons (VPCs) [12, 14, 56–62], oxidized VPCs [12, 14, 60, 62, 63], and triply periodic carbon minimal surfaces (Schwarz's primitive and Schoen's gyroid) [45, 59, 64–70]. Recently, scientists have paid attention to the next generation of porous carbon molecular sieves materials, i.e., crystalline exotic cubic carbon allotropes: cubic carbon polymorphs (CCPs) [45, 71–73], diamond-like super structures of CNTs (super diamonds) [74], diamond-like frameworks [75], porous aromatic frameworks (PAFs) [76, 77], diamond-like carbon frameworks (i.e., diamondynes, also named D-carbons) [78], tetrahedral node diamondyne [79], carbon allotropes proposed by Karfunkel and Dressler [45, 80], compressed carbon nanotubes [45, 81], sodalite-like nanostructures [45, 82], folding of graphene slit-like pore walls [52, 83], gyroidal nanoporous and mesoporous carbons (GNCs and GMCs, respectively) [84–86].

One of the most interesting and promising adsorbent from the above mentioned is GNC. In the current study, we consider nine different GNC structures having surface built in a way ensuring connection of each carbon atom with exactly three neighbours, similarly as "schwarzites". Nicolaï et al. [84] confirmed that the curvature and the rigidity do not play a crucial role in the performance of GNC structures for ionic conduction. The major role, however, is played by the pore size and pore volume. Indeed, the larger the pore is, the larger is the ionic transport. Finally, the mentioned authors stated that GNC structures with tunable properties can be widely applied to water filtration or energy storage.

2. Simulation details

We used the structures of nine gyroidal nanocarbons (denoted as GNC-04, GNC-07, GNC-09, GNC-11, GNC-12, GNC-13, GNC-15, GNC-18 and GNC-21) published previously by Nicolaï et al. [84] (see figure 1 in [84]). In the case of the first six systems, original boxes generated by Nicolaï et al. [84] were multiplied (eightfold) in order to obtain box size at least two times greater than the cut-off distances used during simulations described below. The porosity of all the studied carbonaceous adsorbents was characterised by a geometrical method proposed by Bhattacharya and Gubbins (BG) [87]. The implementation of the method was described in detail elsewhere [88, 89]. The BG method provided histograms of pore sizes (effective diameters — d_{eff}). These data were also used to calculate the average sizes of pores accessible for Ar atoms ($d_{\text{eff,acc,av}}$) [88]. In addition, the volume of pores accessible for Ar was determined using a combination of the BG method and Monte Carlo integration [88].

Argon adsorption isotherms at its boiling point (i.e., $T = 87$ K) on all the studied nanocarbons were simulated using the hyper parallel tempering Monte Carlo method (HPTMC) proposed by Yan and de Pablo [90]. The simulation scheme was the same as in previous work [88]. We considered 93 replicas corresponding to different relative pressure values (p/p_s, where p and p_s are equilibrium and saturated Ar vapour pressure, respectively) in the range $1.0 \times 10^{-10} - 1.0$. Other details of the performed HPTMC simulations are described in [88]. The average numbers of Ar atoms in the simulation box were used to calculate the adsorption amount of Ar per unit of mass of the adsorbent (a) [88]. The isosteric enthalpy of adsorption (q^{st}) was also determined from the theory of fluctuations [88, 91] to reflect the energetics of the process.

In order to analyse the mechanism of Ar adsorption, we constructed high resolution α_s-plots [92] based on simulated adsorption isotherms. We used Ar adsorption isotherm simulated in the ideal slit-like system composed of graphene sheets with effective pore width equal to 10 nm as the reference one. We also determined the values of Henry's constant (K_H) from the slope of the linear part of adsorption isotherms in the low-pressure range [83]. Finally, adsorption potential distribution (APD) curves [93–95] were calculated. The APD curve is the first derivative of the so-called characteristic curve, presenting adsorption amount as a function of the adsorption potential (A_{pot}) defined as:

$$A_{pot} = -RT \ln \frac{p}{p_s}, \tag{2.1}$$

where R is the universal gas constant. The differentiation was performed numerically by the approximation of the isotherms using some empirical functions and calculating their derivatives.

We also simulated the adsorption and separation of three binary gas mixtures (important from practical point of view): CO_2/CH_4, CO_2/N_2, and CH_4/N_2 on all the studied GNCs. The computations were performed for $T = 298$ K using the grand canonical Monte Carlo method (GCMC) [91, 96]. The simulation scheme was the same as in our previous works [60, 62]. Simulations were performed for the total mixture pressure $p_{tot} = 0.1$ MPa (i.e., atmospheric pressure) and for the following mole fractions of components in the gaseous phase (y): 0.0, 0.01, 0.025, 0.05, 0.1, 0.2, 0.3, 0.4, 0.5, 0.6, 0.7, 0.8, 0.9, 0.95, 0.975, 0.99, and

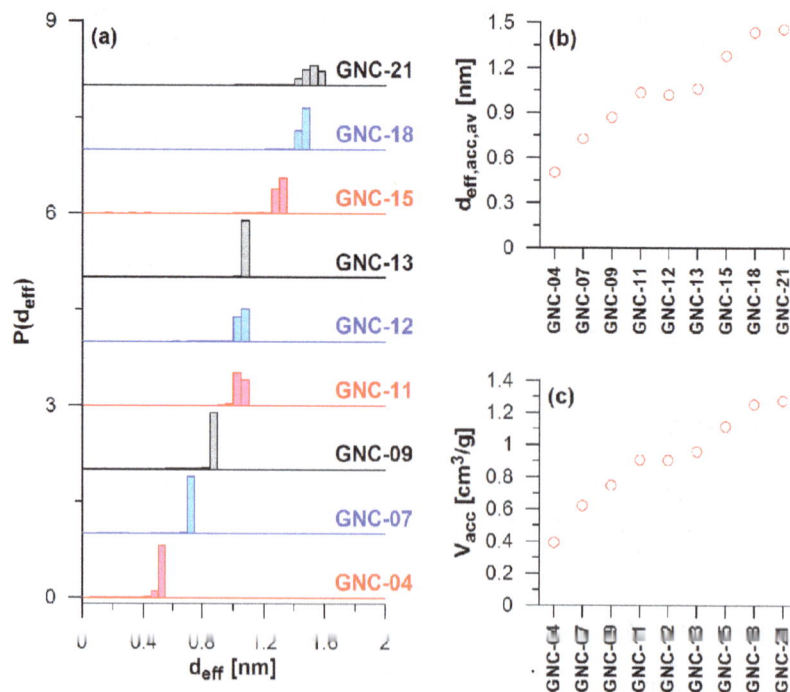

Figure 1. (Color online) Comparison of (a) pore size histograms for all the studied nanocarbons determined from the BG method (the subsequent histograms are shifted by 1 unit from the previous ones), (b) average sizes and (c) volume of pores accessible for Ar atoms, respectively.

1.0. For each point, the mole fractions of components in the adsorbed phase (x) were calculated from the average numbers of molecules present in the simulation box. The efficiency of the process of separation of mixtures was reflected by the value of equilibrium separation factor (the 1st component over the 2nd one):

$$S_{1/2} = \frac{x_1/x_2}{y_1/y_2}.$$ (2.2)

The adsorbed phase is enriched in the 1st component if $S_{1/2} > 1$.

3. Results and discussion

Figure 1 (a) collects histograms of effective pore sizes, determined from the application of BG method, for all the studied GNCs. All the structures are strictly microporous, i.e., the diameters of pores do not exceed 2 nm. Generally, the size of dominant pores increases in the considered series (from the GNC-04 up to GNC-21). However, there are two groups of nanocarbons having similar diameters of the main pores: (i) GNC-11, GNC-12 and GNC-13 — d_{eff} around 1 nm and (ii) GNC-18 and GNC-21 — d_{eff} around 1.5 nm. It should be noted that in the case of GNC-21, some amount of pores wider than in GNC-18 is also present. These regularities are reflected by the values of the average pore diameter [figure 1 (b)]. The increase in pore diameters is accomplished by the increase in pore volume from ca. 0.4 cm^3/g for GNC-04 up to ca. 1.3 cm^3/g for GNC-18 and GNC-21 [figure 1 (c)].

Figure 2 shows the comparison of Ar adsorption isotherms simulated for all the studied GNCs. The

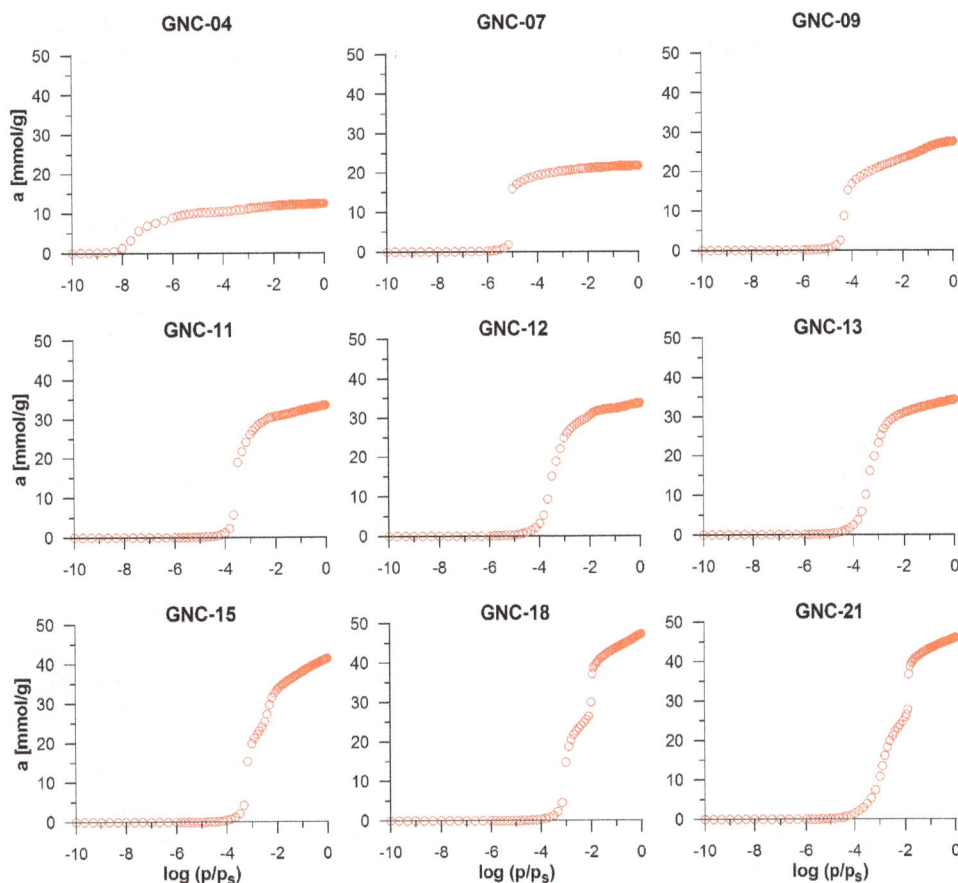

Figure 2. (Color online) Comparison of Ar adsorption isotherms (T = 87 K) simulated for all the considered nanocarbons.

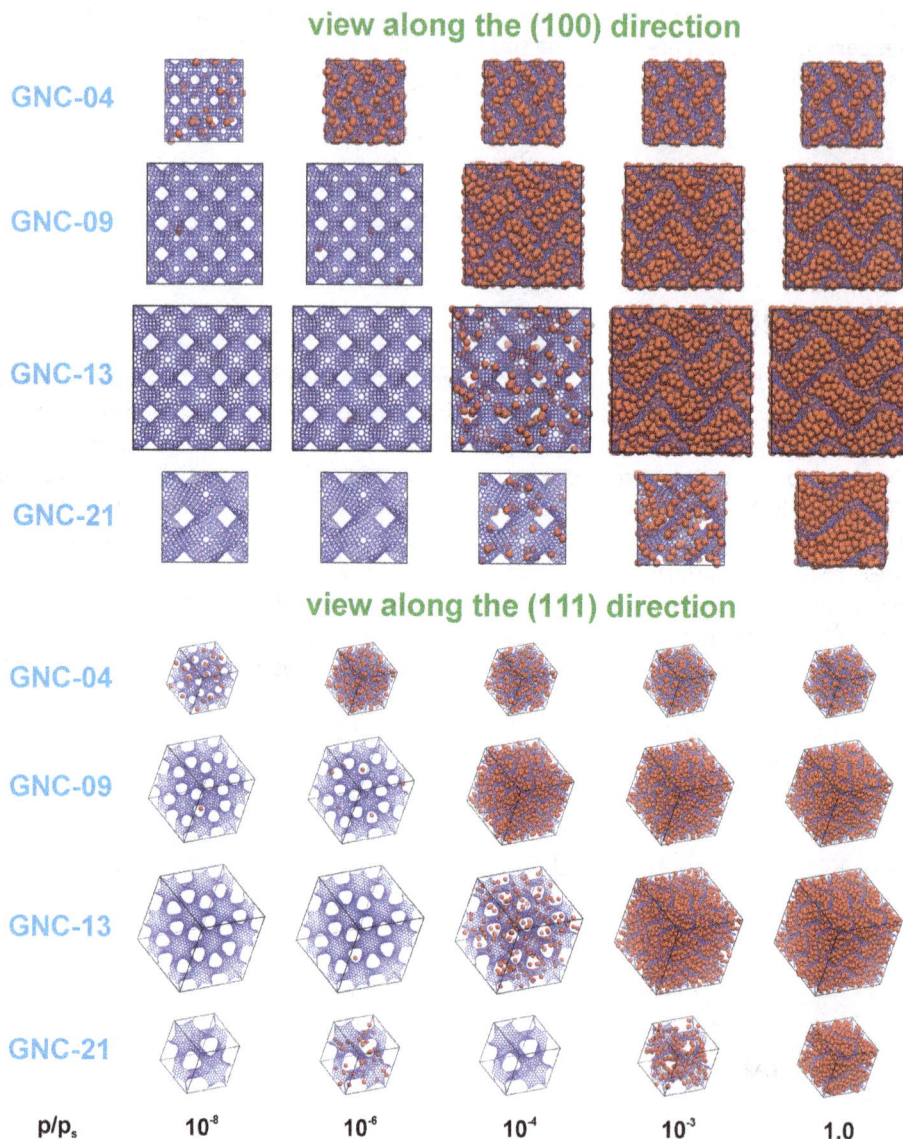

Figure 3. (Color online) Equilibrium argon configurations for selected nanocarbons and selected values of relative pressure (the frames reflect the size of simulation boxes, all the structures are in the same scale). It should be noted that this figure was created using the VMD program [97].

changes observed in the shape of isotherms reflect the differences in the properties of nanocarbons. Adsorption capacity varies from ca. 12 mmol/g for GNC-04 up to ca. 45 mmol/g for GNC-18 and GNC-21. These changes correspond to the differences in the pore volume [figure 1 (c)]. At the same time, the shift of the pore filling pressure toward higher values is observed. The pores of the GNC-04 structure are filled in the relative pressure range $10^{-8} - 10^{-6}$. However, the total filling of GNC-18 and GNC-21 occurs for similar values of relative pressure (around 10^{-2}). The middle carbons of the series (i.e., GNC-11, GNC-12 and GNC-13) are filled in the similar range of relative pressure ($p/p_s > 10^{-4}$). The differences in the pore filling are also clearly seen on equilibrium snapshots of Ar configurations in the simulation boxes shown in figure 3. These regularities are related to the differences in diameters of dominant pores present in the individual GNCs [figure 1 (a)]. Finally, one can observe that in the case of initial structures (up to GNC-13), the pore filling is a single-step process. However, the pores of GNC-15, GNC-18 and GNC-21 are filled in two steps. This is also caused by the rise in pore sizes. For pores wider than 1 nm, in the first step

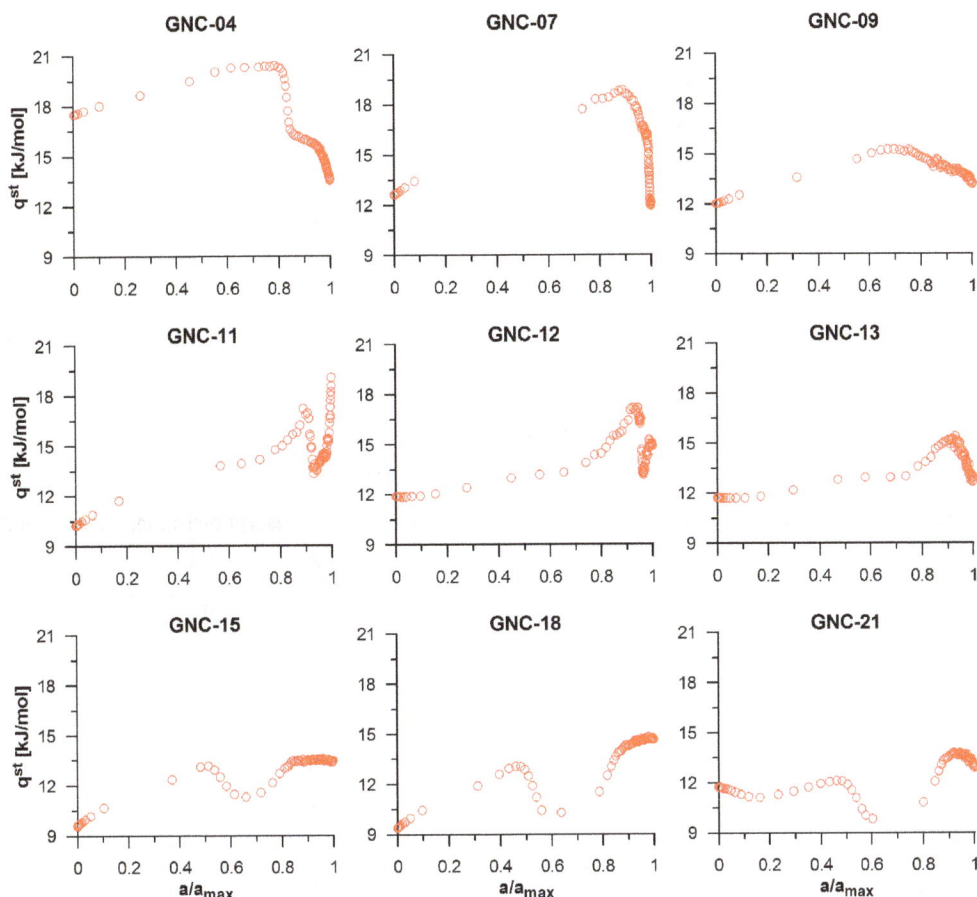

Figure 4. (Color online) Comparison of isosteric enthalpy of Ar adsorption [the data related to figure 2; for clarity adsorption amounts are normalised by the maximum observed value (a_{max})].

a monolayer is formed and next the remaining free volume is filled.

Figure 4 shows the plots of isosteric enthalpy of adsorption related to the isotherms shown in figure 2. For low loadings, the q^{st} values increase as the adsorption amount rises. In this range, Ar atoms are adsorbed mainly on high-energetic centres. The increasing adsorption amount causes other Ar atoms to appear in the vicinity of the initially adsorbed ones. This rises the contribution of fluid-fluid interactions to the total energy of a system. Here, the only exception is GNC-21 structure. There is observed a decrease in q^{st} for relative adsorption up to ca. 0.1. This system has probably got a heterogeneous surface. Consequently, the subsequent Ar atoms are adsorbed on centres having lower energy and this reduces the effects of the increase in lateral Ar-Ar interactions. In the intermediate range of adsorption, the enthalpy rises for all the structures until the entire adsorbent surface is covered. Next, the values of q^{st} decrease since Ar is adsorbed at the places more distant from the surface and this is connected with lower solid-fluid contributions. In the case of the structures having the widest pores (especially GNC-15, GNC-18 and GNC-21), the second peak on q^{st} is also observed. This peak is related to the total filling of pores.

Comparing the enthalpy at zero coverage for all the systems, one can observe that GNC-04 has a remarkably higher value of this parameter (ca. 17.5 kJ/mol) than the other systems. This is connected with the presence of the narrowest pores having high adsorption energy. The other structures may be divided into two groups. The first one includes GNC-07, GNC-09, GNC-12, GNC-13 and GNC-21 carbons. In this case, the enthalpy at zero coverage is close to 12 kJ/mol. For the second group (GNC-11, GNC-15 and GNC-18), this enthalpy value is in the range 9–10 kJ/mol. This may suggest some similarities in the surface nature of the group members (for example curvature, which is the main factor determining the energy of adsorption on the surface). The above-described differences in the energy of interactions with the surface of

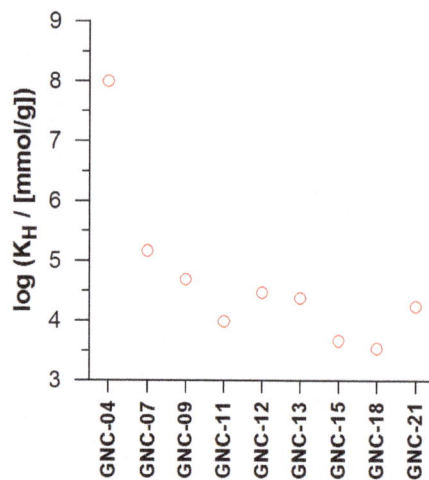

Figure 5. (Color online) Comparison of Henry's constants related to the Ar adsorption isotherms presented on figure 2.

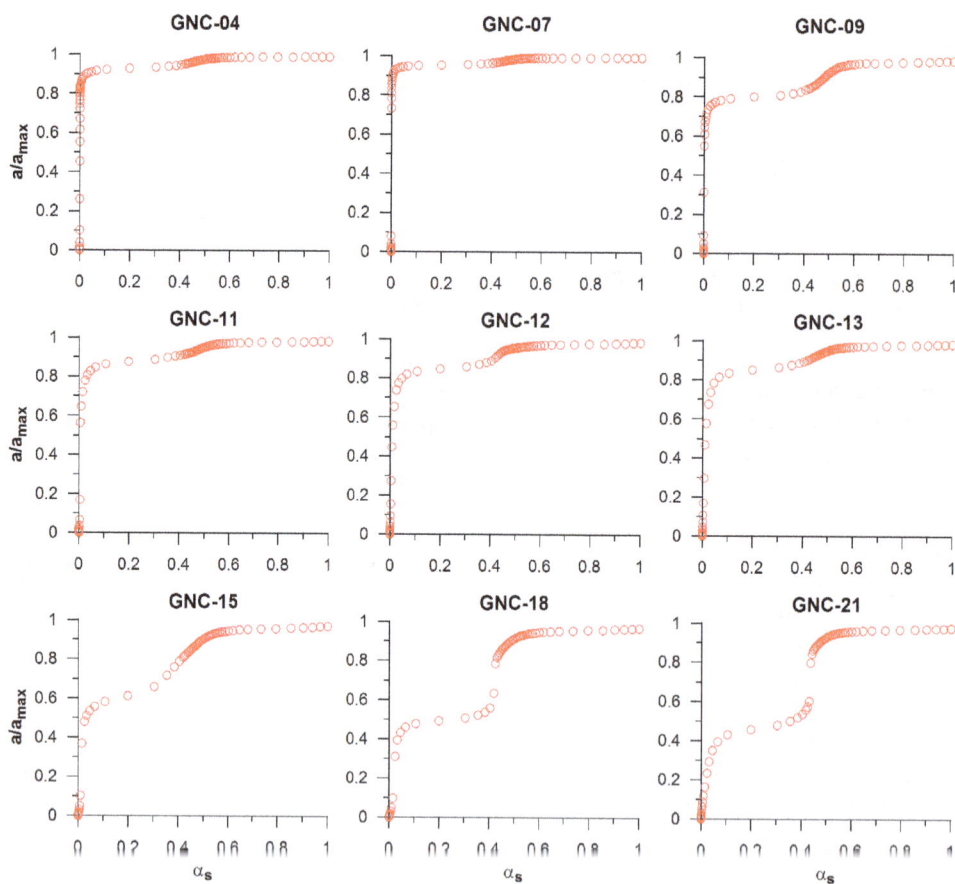

Figure 6. (Color online) Comparison of α_s-plots related to the Ar adsorption isotherms presented in figure 2 [for clarity the adsorption amounts are normalized by the maximum observed value (a_{max})]. The α_s is the normalized adsorption on the reference material, i.e., in the ideal slit-like system with the effective pore width equal to 10 nm.

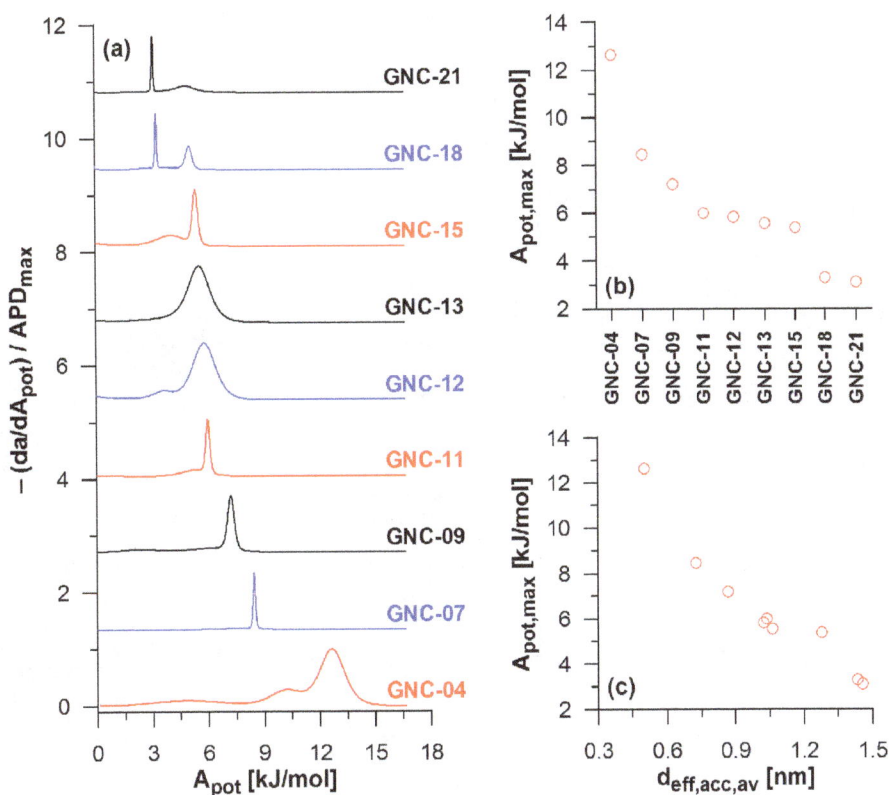

Figure 7. (Color online) (a) Comparison of APD curves [for clarity, all the curves are normalized by the maximum observed value (APD$_{max}$); subsequent curves are shifted by 1.35 units]. (b) Location of the main peak on the APD curves ($A_{pot,max}$). (c) Correlation between the location of the main peak and the average sizes of pores accessible for Ar atoms.

adsorbents fully correspond to the variation of Henry's constants shown in figure 5. This is not surprising since solid-fluid interactions are the main factor affecting the shape of the isotherm in the low pressure range. Hence, the value of K_H for GNC-04 system is at least 1000 times greater than for the other ones. For three of the remaining structures (i.e., GNC-11, GNC-15 and GNC-18), lower values of K_H ($< 10^4$ mmol/g) are recorded. The same carbons have the lowest enthalpy of adsorption at zero coverage.

Figure 6 presents comparison of α_s-plots related to the Ar adsorption isotherms. One can see that the adsorption process is dominated by a FS swings (GNC-04 and GNC-07) and the FS-CS swings (remaining structures) [98]. It can be noticed that with the rise in the pore diameter, the combination of primary and secondary micropore filling mechanism occurs. The boundary between those mechanisms is located for the structures with pore diameters around ca. 0.8 nm. It is also interesting that the range on α_s-plots connecting FS and CS swings is not linear as it is observed for the case of slit-like carbon micropores. This can be caused by the surface curvature.

Figure 7 (a) compares APD curves for all the systems studied. The presented data are complementary to adsorption isotherms shown in figure 2. On all the APD curves, at least one (dominant) peak is observed. It corresponds to the pore filling. Its location [$A_{pot,max}$, figure 7 (b)] is related to the pressure of the pore filling according to equation 2.1. Hence, this parameter may be correlated with the size of pores — see figure 7 (c). This figure quantitatively confirms the above-described qualitative differences in the pore filling process. The width of the main peak also provides some information on the process. The narrow peak means that condensation occurs in a narrow range of relative pressure. By contrast, a wide peak denotes a wide condensation range. For example, the pore filling in GNC-04 system occurs, as mentioned above, for $10^{-8} < p/p_s < 10^{-6}$ and this is reflected by a wide peak with the maximum located at ca. 12.6 kJ/mol. For this system, the other two peaks are also observed (the third one with the maximum

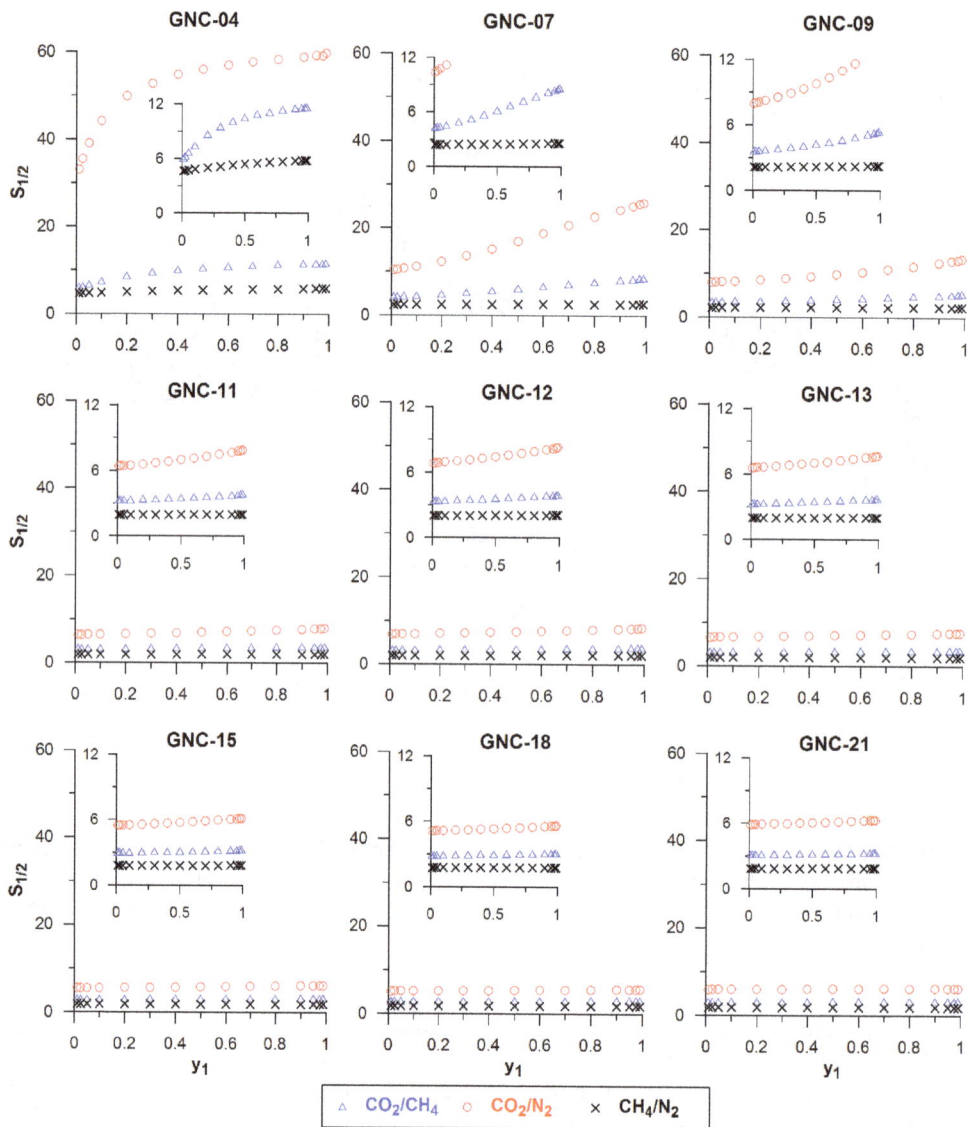

Figure 8. (Color online) Comparison of equilibrium separation factors [$S_{1/2}$, equation (2.2)] for the adsorption of all the mixtures ($T = 298$ K, $p_{tot} = 0.1$ MPa) on all the nanocarbons studied. The data plotted as the function of the 1st component mol fraction in the gaseous phase (y_1, the 1st component is CO_2 for CO_2/CH_4 and CO_2/N_2 mixtures and CH_4 for CH_4/N_2 mixture).

at ca. 45 kJ/mol is very broad). These peaks reflect the other sub-steps of the Ar density rise in pores of this structure. Similar interpretation also concerns the additional peaks observed for GNC-11 and GNC-12. In the case of GNC-18 and GNC-21 structures, the observed second peak is related to the above mentioned monolayer formation. A slightly different scenario occurs for GNC-15 carbon. Here, the dominant peak is connected with the monolayer formation and the second low (also wide) peak reflects the filling of the remaining pore volume. This fact explains why the location of the main peak for this structure deviates from the distinct trend visible for all the other GNCs in correlation shown in figure 7 (c).

Figure 8 presents a comparison of equilibrium separation factors for adsorption of all three studied mixtures for different compositions of gaseous phase. In addition, figure 9 directly compares the efficiency of separation of equimolar mixtures for all the studied systems. The separation is a consequence of differences in the adsorption of mixture components. Since the critical temperature for the studied gases decreases significantly in the sequence $CO_2 > CH_4 > N_2$, the adsorption affinity decreases in the

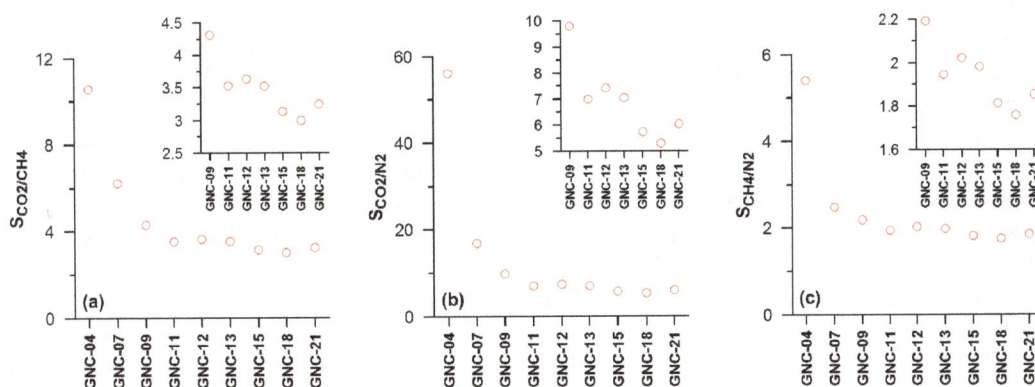

Figure 9. (Color online) Comparison of equilibrium separation factors for adsorption of equimolar ($y = 0.5$) mixtures: (a) CO_2/CH_4, (b) CO_2/N_2 and (c) CH_4/N_2 ($p_{tot} = 0.1$ MPa, $T = 298$ K) — see figure 8.

same sequence (for the subsequent gases, the process occurs for increasing value of reduced temperature). Consequently, for the given GNC, the highest equilibrium separation factor is observed for CO_2/N_2 and the smallest one for CH_4/N_2 mixture. The qualitative differences in separation efficiency between the structures studied are the same for all the mixtures studied. The highest values of equilibrium separation factor (remarkably higher than for the other systems) are observed for GNC-04 carbon. This is connected

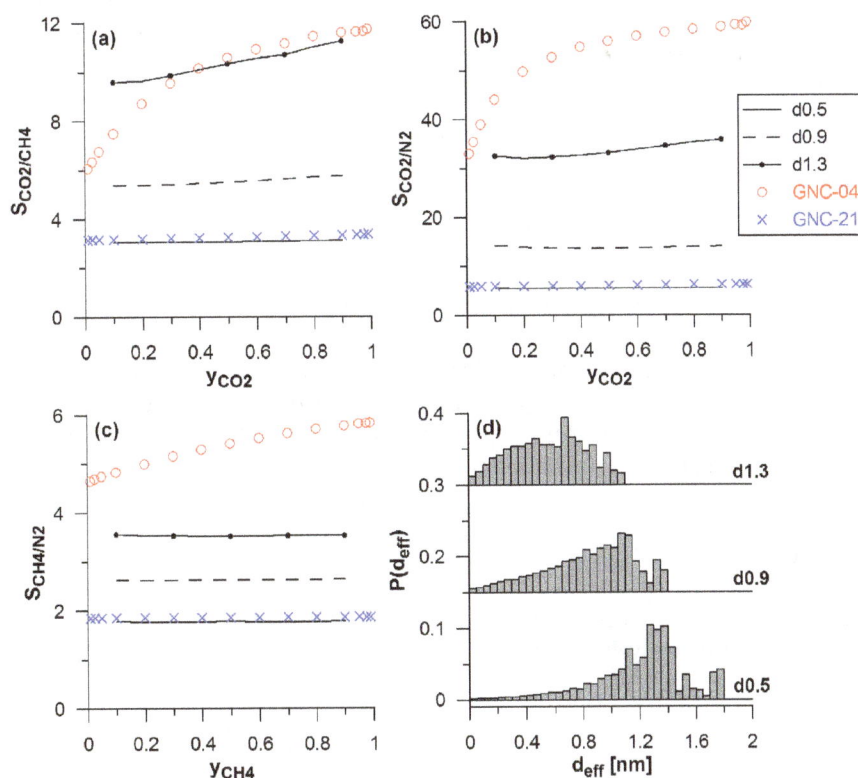

Figure 10. (Color online) Comparison of equilibrium separation factors [(a) CO_2/CH_4, (b) CO_2/N_2 and (c) CH_4/N_2] for mixture adsorption ($p_{tot} = 0.1$ MPa, $T = 298$ K) on GNC-04 and GNC-21 nanocrabons and on three virtual porous carbons (VPCs): d0.5, d0.9 and d1.3 [88] (the data for VPCs taken from our previous paper [60]). In addition, panel (d) presents the pore size histograms for the presented VPCs [88] (the subsequent histograms are shifted by 0.15 units from the previous ones).

with the presence of the narrowest pores. Among the other nanocarbons, GNC-07 and GNC-09 also exhibit higher values of $S_{1/2}$. However, they are lower than in the case of GNC-04 since these structures have wider pores. The efficiency of separation for the next GNCs is similar. Nevertheless, some small differences are also noticeable (see insets in figure 9). The GNC-11, GNC-15 and GNC-18 structures are less efficient in comparison with the adjacent carbons in the series. These regularities are analogous to that observed for Henry's constants shown in figure 5 and discussed above. These facts suggest that in the case of GNCs with pores wider than ca. 1 nm, the main factor affecting the efficiency of the separation process is the energetics of fluid interaction with the curved surface of the nanocarbons studied.

Finally, figure 10 compares the equilibrium separation factors (all three mixtures studied) for GNC-04 and GNC-21 structures and for three virtual porous carbons (VPCs) described in detail previously [88] and having different porosity — see figure 10 (d). As one can see, the GNC-21 structure exhibits a separation efficiency similar to the d0.5 carbon. The main pores of both adsorbents have a similar width. However, this VPC has also some narrower micropores which presence probably positively affects the separation efficiency. Such micropores are absent in the case of GNC-21 and, nonetheless, this nanocarbon exhibits similar values of $S_{1/2}$. This is the consequence of the adsorption energetics on a curved surface of this structure. On the contrary, the GNC-04 nanocarbon exhibits the efficiency of the CO_2/CH_4 mixture separation similar to the d1.3 carbon. This VPC has micropores distributed in the range up to ca. 1 nm [figure 10 (d)]. A large part of them has diameters similar to or lower than the GNC-04 structure. In the case of CO_2/N_2 and CH_4/N_2 mixtures, the values of equilibrium separation factors for GNC-04 are higher than for d1.3 carbon. This comparison (especially for CO_2/CH_4 mixture — similar efficiency for both adsorbents and for CH_4/N_2 mixture — higher efficiency for GNC-04) may suggest that a regularly curved surface of gyroidal carbons exhibits higher affinity to CH_4 molecules than a heterogeneous surface of activated carbons.

Summing up, the GNCs studied may be considered as potential adsorbents for gas mixture separation. The efficiency of this process is similar to or higher than for activated carbons with similar diameters of pores. The GNC-04 or similar structures seem to be quite promising materials for this purpose since this nanocarbon contains narrow and quite uniform pores (ca. 0.5 nm).

4. Conclusions

Adsorption and separation properties of GNCs — a new class of exotic nanocarbon materials, are studied for the first time using computer simulation technique. All the structures studied are strictly microporous. The mechanisms of Ar adsorption are described basing on the analysis of adsorption isotherms, enthalpy plots, the values of Henry's constants, α_s and adsorption potential distribution plots. Below the pore diameters ca. 0.8 nm, primary micropore filling process dominates. For structures possessing larger micropores, primary and secondary micropore filling mechanisms are observed. GNCs may be considered as potential adsorbents for gas mixture separation, having separation efficiency similar to or higher than this for activated carbons with similar diameters of pores.

Acknowledgements

The authors acknowledge the use of the computer cluster at Poznań Supercomputing and Networking Centre (Poznań, Poland) as well as the Information and Communication Technology Centre of the Nicolaus Copernicus University (Toruń, Poland).

References

1. Gas Separation by Adsorption Processes, Series on Chemical Engineering Vol. 1, Yang R.T. (Ed.), Imperial College Press, London, 1997.
2. Ruthven D.M., Farooq S., Knaebel K.S., Pressure Swing Adsorption, Wiley-VCH, New York, 1994.
3. Sinaiski E.G., Lapiga E.J., Separation of Multiphase, Multicomponent Systems, Wiley-VCH, Weinheim, 1997.
4. Jaroniec M., Madey R., Physical Adsorption on Heterogeneous Solids, Elsevier, New York, 1988.

5. Rudziński W., Everett D.H., Adsorption of Gases on Heterogeneous Surfaces, Academic Press, New York, 1992.

6. Rouquerol F., Rouquerol J., Sing K.S.W., Llewellyn P., Maurin G., Adsorption by Powders and Porous Solids. Principles, Methodology and Applications, 2nd Edn., Elsevier, Amsterdam, 2014; doi:10.1016/B978-0-08-097035-6.09991-5.

7. Bansal R.C., Goyal M., Activated Carbon Adsorption, CRC Press, Boca Raton, 2005.

8. Palmer J.C., Moore M.V., Roussel T.J., Brennan J.K., Gubbins K.E., Phys. Chem. Chem. Phys., 2011, **13**, 3985; doi:10.1039/c0cp02281k.

9. Liu L., Nicholson D., Bhatia S.K., J. Phys. Chem. C, 2015, **119**, 407; doi:10.1021/jp5099987.

10. Ozturk T.N., Keskin S., Ind. Eng. Chem. Res., 2013, **52**, 17627; doi:10.1021/ie403159c.

11. Lu X., Jin D., Wei S., Wang Z., An C., Guo W., J. Mater. Chem. A, 2015, **3**, 12118; doi:10.1039/C4TA06829G.

12. Billemont P., Coasne B., De Weireld G., Adsorption, 2014, **20**, 453; doi:10.1007/s10450-013-9570-z.

13. Furmaniak S., Terzyk A.P., Kaneko K., Gauden P.A., Kowalczyk P., Ohba T., Chem. Phys. Lett., 2014, **595-596**, 67; doi:10.1016/j.cplett.2014.01.031.

14. Di Biase E., Sarkisov L., Carbon, 2015, **94**, 27; doi:10.1016/j.carbon.2015.06.056.

15. Borówko M., Patrykiejew A., Rżysko W., Sokołowski S., Langmuir, 1997, **13**, 1073; doi:10.1021/la950940a.

16. Grabowski K., Patrykiejew A., Sokołowski S., Thin Solid Films, 1999, **352**, 259; doi:10.1016/S0040-6090(99)00291-6.

17. Grabowski K., Patrykiejew A., Sokołowski S., Thin Solid Films, 2000, **379**, 297; doi:10.1016/S0040-6090(00)01398-5.

18. Patrykiejew A., Sałamacha L., Sokołowski S., Pizio O., Phys. Rev. E, 2003, **67**, 061603; doi:10.1103/PhysRevE.67.061603.

19. Patrykiejew A., Sałamacha L., Sokołowski S., Dominguez H., Pizio O., Phys. Rev. E, 2003, **67**, 031202; doi:10.1103/PhysRevE.67.031202.

20. Patrykiejew A., Sokołowski S., Pizio O., J. Phys. Chem. B, 2005, **109**, 14227; doi:10.1021/jp048170b.

21. Patrykiejew A., Sokołowski S., Phys. Rev. E, 2010, **81**, 012501; doi:10.1103/PhysRevE.81.012501.

22. Patrykiejew A., Sokołowski S., J. Chem. Phys., 2012, **136**, 144702; doi:10.1063/1.3699330.

23. Bryk P., Sokołowski S., Pizio O., J. Phys. Chem. B, 1999, **103**, 3366; doi:10.1021/jp982028r.

24. Noworyta J.P., Henderson D., Sokołowski S., Chan K.-Y., Mol. Phys., 1998, **95**, 415; doi:10.1080/00268979809483175.

25. Patrykiejew A., Pizio O., Pusztai L., Sokołowski S., Mol. Phys., 2003, **101**, 2219; doi:10.1080/0026897031000099925.

26. Patrykiejew A., Pizio O., Sokołowski S., Sokołowska Z., Phys. Rev. E, 2004, **69**, 061605; doi:10.1103/PhysRevE.69.061605.

27. Bryk P., Sokołowski S., J. Chem. Phys., 2004, **120**, 8299; doi:10.1063/1.1695554.

28. Pizio O., Sokołowski S., Sokołowska Z., J. Chem. Phys., 2011, **134**, 214702; doi:10.1063/1.3597773.

29. Ilnytskyi J.M., Patsahan T., Sokołowski S., J. Chem. Phys., 2011, **134**, 204903; doi:10.1063/1.3592562.

30. Bryk P., Pizio O., Sokołowski S., J. Chem. Phys., 2005, **122**, 174906; doi:10.1063/1.1888425.

31. Pizio O., Bucior K., Patrykiejew A., Sokołowski S., J. Chem. Phys., 2005, **123**, 214902; doi:10.1063/1.2128701.

32. Sokołowski S., Rżysko W., Pizio O., J. Colloid Interface Sci., 1999, **218**, 341; doi:10.1006/jcis.1999.6392.

33. Kruk M., Patrykiejew A., Sokołowski S., Surf. Sci., 1995, **340**, 179; doi:10.1016/0039-6028(95)00681-8.

34. Martinez A., Patrykiejew A., Pizio O., Sokołowski S., J. Phys.: Condens. Matter, 2003, **15**, 3107; doi:10.1088/0953-8984/15/19/312.

35. Grabowski K., Patrykiejew A., Sokołowski S., Thin Solid Films, 1997, **304**, 344; doi:10.1016/S0040-6090(97)00214-9.

36. Kruk M., Patrykiejew A., Sokołowski S., Thin Solid Films, 1994, **238**, 302; doi:10.1016/0040-6090(94)90071-X.

37. Bucior K., Patrykiejew A., Pizio O., Sokołowski S., J. Colloid Interface Sci., 2003, **259**, 209; doi:10.1016/S0021-9797(02)00203-5.

38. Martinez A., Pizio O., Sokołowski S., J. Chem. Phys., 2003, **118**, 6008; doi:10.1063/1.1556850.

39. Sałamacha L., Patrykiejew A., Sokołowski S., J. Phys. Chem. B, 2009, **113**, 13687; doi:10.1021/jp901383v.

40. Patrykiejew A., Sokołowski S., J. Phys. Chem. B, 2010, **114**, 396; doi:10.1021/jp908710e.

41. Patrykiejew A., Sokołowski S., Condens. Matter Phys., 2014, **17**, 43601; doi:10.5488/CMP.17.43601.

42. Patrykiejew A., Rżysko W., Sokołowski S., J. Phys. Chem. C, 2012, **116**, 753; doi:10.1021/jp208323b.

43. Pizio O., Rżysko W., Sokołowski S., Sokołowska Z., J. Chem. Phys., 2015, **142**, 164703; doi:10.1063/1.4918640.

44. Suarez-Martinez I., Grobert N., Ewels C.P., Carbon, 2012, **50**, 741; doi:10.1016/j.carbon.2011.11.002.

45. Georgakilas V., Perman J.A., Tucek J., Zboril R., Chem. Rev., 2015, **115**, 4744; doi:10.1021/cr500304f.

46. Tomanek D., Guide Through the Nanocarbon Jungle, Morgan and Claypool Publishers, San Rafael, 2014.

47. Carbon Nanomaterials as Adsorbents for Environmental and Biological Applications, Carbon Nanostructures Series, Bergmann C.P., Machado F. (Eds.), Springer International Publishing, New York, 2015; doi:10.1007/978-3-319-18875-1.

48. Gogotsi Y., Presser V., Carbon Nanomaterials, CRC Press, Boca Raton, 2013.

49. Yeganegi S., Gholampour F., Acta Chim. Slov., 2012, **59**, 888.

50. Kowalczyk P., Chem. Phys. Phys. Chem., 2012, **14**, 2784; doi:10.1039/c2cp23445a.

51. Furmaniak S., Terzyk A.P., Kowalczyk P., Kaneko K., Gauden P.A., Chem. Phys. Phys. Chem., 2013, **15**, 16468; doi:10.1039/c3cp52342j.
52. Furmaniak S., Terzyk A.P., Gauden P.A., Kowalczyk P., Harris P.J.F., J. Phys.: Condens. Matter, 2014, **26**, 485006; doi:10.1088/0953-8984/26/48/485006.
53. Kowalczyk P., Terzyk A.P., Gauden P.A., Furmaniak S., Kaneko K., Miller T.F., J. Phys. Chem. Lett., 2015, **6**, 3367; doi:10.1021/acs.jpclett.5b01545.
54. Tanaka H., Kanoh H., Yudasaka M., Iijima S., Kaneko K., J. Am. Chem. Soc., 2005, **127**, 7511; doi:10.1021/ja0502573.
55. Romo-Herrera J.M., Terrones M., Terrones H., Dag S., Meunier V., Nano Lett., 2007, 7, 570; doi:10.1021/nl0622202.
56. Liu L., Nicholson D., Bhatia S.K., Chem. Eng. Sci., 2015, **121**, 268; doi:10.1016/j.ces.2014.07.041.
57. Terzyk A.P., Furmaniak S., Gauden P.A., Harris P.J.F., Wloch J., Kowalczyk P., J. Phys.: Condens. Matter, 2007, **19**, 406208; doi:10.1088/0953-8984/19/40/406208.
58. Terzyk A.P., Furmaniak S., Harris P.J.F., Gauden P.A., Wloch J., Kowalczyk P., Rychlickia G., Chem. Phys. Phys. Chem., 2007, **9**, 5919; doi:10.1039/b710552e.
59. Kowalczyk P., Gauden P.A., Terzyk A.P., Furmaniak S., Harris P.J.F., J. Phys. Chem. C, 2012, **116**, 13640; doi:10.1021/jp302776z.
60. Furmaniak S., Koter S., Terzyk A.P., Gauden P.A., Kowalczyk P., Rychlickia G., Chem. Phys. Phys. Chem., 2015, **17**, 7232; doi:10.1039/c4cp05498a.
61. Palmer J.C., Moore J.D., Roussel T.J., Brennan J.K., Gubbins K.E., Chem. Phys. Phys. Chem., 2011, **13**, 3985; doi:10.1039/c0cp02281k.
62. Furmaniak S., Kowalczyk P., Terzyk A.P., Gauden P.A., Harris P.J.F., J. Colloid Interface Sci., 2013, **397**, 144; doi:10.1016/j.jcis.2013.01.044.
63. Lu L., Wang S., Muller E.A., Cao W., Zhu Y., Lu X., Jackson G., Fluid Phase Equilib., 2014, **362**, 227; doi:10.1016/j.fluid.2013.10.013.
64. Kowalczyk P., Hołyst R., Terrones M., Terrones H., Phys. Chem. Chem. Phys., 2007, **9**, 1786; doi:10.1039/b618747a.
65. Mackay A.L., Nature, 1985, **314**, 604; doi:10.1038/314604a0.
66. Mackay A.L., Terrones H., Nature, 1991, **352**, 762; doi:10.1038/352762a0.
67. Lijma S., Ichihashi T., Ando Y., Nature, 1992, **356**, 776; doi:10.1038/356776a0.
68. Schwartz H.A., Gesammelte Mathematische Abhandlunge, Vol. 1 and 2, Springer, Berlin, 1980.
69. Townsend S.J., Lenosky T.J., Muller D.A., Nichols C.S., Elser V., Phys. Rev. Lett., 1992, **69**, 921; doi:10.1103/PhysRevLett.69.921.
70. Lee H.G., Kim J., Int. J. Numer. Methods Eng., 2012, **91**, 269; doi:10.1002/nme.4262.
71. Hu M., Tian F., Zhao Z., Huang Q., Xu B., Wang L.-M., Wang H.-T., Tian Y., He J., J. Phys. Chem. C, 2012, **116**, 24233; doi:10.1021/jp3064323.
72. Kowalczyk P., He J., Hu M., Gauden P.A., Furmaniak S., Terzyk A.P., Phys. Chem. Chem. Phys., 2013, **15**, 17366; doi:10.1039/c3cp52708e.
73. Kowalczyk P., Parsons D., Terzyk A.P., Gauden P.A., Furmaniak S., In: Carbon Nanomaterials Sourcebook: Nanoparticles, Nanocapsules, Nanofibers, Nanoporous Structures, and Nanocomposites, Vol. 2, Sattler K.D. (Ed.), CRC Press, London, 2016, ch. 6 (in press).
74. Tylianakis E., Dimitrakakis G.K., Martin-Martinez F.J., Melchor S., Dobado J.A., Klontzas E., Froudakis G.E., Int. J. Hydrogen Energy, 2014, **39**, 9825; doi:10.1016/j.ijhydene.2014.03.011.
75. Wang H., Cao D., J. Phys. Chem. C, 2015, **119**, 6324; doi:10.1021/jp512275p.
76. Huang L., Yang X., Cao D., J. Phys. Chem. C, 2015, **119**, 3260; doi:10.1021/jp5128404.
77. Yang Z., Peng X., Cao D., J. Phys. Chem. C, 2013, **117**, 8353; doi:10.1021/jp402488r.
78. Huang L., Xiang Z., Cao D., J. Mater. Chem. A, 2013, **1**, 3851; doi:10.1039/c3ta10292k.
79. Huang L., Zeng X., Cao D., J. Mater. Chem. A, 2014, **2**, 4899; doi:10.1039/C3TA15062C.
80. Karfunkel H.R., Dressler T., J. Am. Chem. Soc., 1992, **114**, 2285; doi:10.1021/ja00033a001.
81. Zhao Z.S., Xu B., Wang L.M., Zhou X.F., He J.L., Liu Z.Y., Wang H.T., Tian Y.J., ACS Nano, 2011, **5**, 7226; doi:10.1021/nn202053t.
82. Ribeiro F.J., Tangney P., Louie S.G., Cohen M.L., Phys. Rev. B, 2006, **74**, 172101; doi:10.1103/PhysRevB.74.172101.
83. Furmaniak S., Terzyk A.P., Gauden P.A., Wloch J., Kowalczyk P., Werengowska-Ciećwierz K., Wiśniewski M., Harris P.J.F., J. Phys.: Condens. Matter, 2015, **28**, 015002; doi:10.1088/0953-8984/28/1/015002.
84. Nicolaï A., Monti J., Daniels C., Meunier V., J. Phys. Chem. C, 2015, **119**, 2896; doi:10.1021/jp511919d.
85. Werner J.G., Hohelsel T.N., Wiesner U., ACS Nano, 2014, **8**, 731; doi:10.1021/nn405392t.
86. Choudhury S., Agrawal M., Formanek P., Jehnichen D., Fischer D., Krause B., Albrecht V., Stamm M., Ionov L., ACS Nano, 2015, **9**, 6147; doi:10.1021/acsnano.5b01406.
87. Bhattacharya S., Gubbins K.E., Langmuir, 2006, **22**, 7726; doi:10.1021/la052651k.
88. Furmaniak S., Comput. Meth. Sci. Technol., 2013, **19**, 47; doi:10.12921/cmst.2013.19.01.47-57.

89. Furmaniak S., Terzyk A.P., Gauden P.A., Kowalczyk P., Harris P.J.F., Koter S., J. Phys.: Condens. Matter, 2013, **25**, 015004; doi:10.1088/0953-8984/25/1/015004.

90. Yan Q., de Pablo J.J., J. Chem. Phys., 1999, **111**, 9509; doi:10.1063/1.480282.

91. Frenkel D., Smit B., Understanding Molecular Simulation, Academic Press, San Diego, 1996.

92. Setoyama N., Suzuki T., Kaneko K., Carbon, 1998, **36**, 1459; doi:10.1016/S0008-6223(98)00138-9.

93. Kruk M., Jaroniec M., Gadkaree K.P., Langmuir, 1999, **15**, 1442; doi:10.1021/la980789f.

94. Choma J., Jaroniec M., Colloids Surf. A, 2001, **189**, 103; doi:10.1016/S0927-7757(01)00572-6.

95. Choma J., Jaroniec M., Adsorpt. Sci. Technol., 2007, **25**, 573; doi:10.1260/0263-6174.25.8.573.

96. Tylianakis E., Froudakis G.E., J. Comput. Theor. Nanosci., 2009, **6**, 335; doi:10.1166/jctn.2009.1040.

97. Humphrey W., Dalke A., Schulten K., J. Mol. Graphics, 1996, **14**, 33; doi:10.1016/0263-7855(96)00018-5.

98. Kaneko K., Ishii C., Kanoh H., Hanzawa Y., Setoyama N., Suzuki T., Adv. Colloid Interface Sci., 1998, **77**, 295; doi:10.1016/S0001-8686(98)00050-5.

18

Critical point calculation for binary mixtures of symmetric non-additive hard disks

W.T. Góźdź, A. Ciach

Institute of Physical Chemistry Polish Academy of Sciences, Kasprzaka 44/52, 01-224 Warsaw, Poland

We have calculated the values of critical packing fractions for the mixtures of symmetric non-additive hard disks. An interesting feature of the model is the fact that the internal energy is zero and the phase transitions are entropically driven. A cluster algorithm for Monte Carlo simulations in a semigrand ensemble was used. The finite size scaling analysis was employed to compute the critical packing fractions for infinite systems with high accuracy for a range of non-additivity parameters wider than in the previous studies.

Key words: *phase coexistence, critical point, finite size scaling, Monte Carlo simulations*

1. Introduction

Two-dimensional fluid mixtures are quite common in soft-matter and in biological systems. Important examples are particles adsorbed at interfaces, on surfaces of pores in porous materials and biological membranes. When the adsorbed fluid forms a monolayer, it may be modelled as a two-dimensional system. The phenomenon of adsorption has been intensively studied, and one of the main contributors to the field is Stefan Sokolowski [1–6]. An interesting question that is not fully solved yet is the phase separation in a binary mixture on surfaces with different curvatures. This question is important not only for the adsorption on curved surfaces present in porous materials, but also for the properties of biological membranes surrounding organelle. In living organisms, the membranes are close to the critical point of the demixing transition [7, 8]. Therefore, the phase behavior of multicomponent two dimensional fluids and, particularly, the critical behavior may be of biological importance. During the phase transition, a small change of thermodynamic parameters causes a large change of the composition. It may result in significant shape transformations of the biological membranes since their shape is linked with their composition [9]. Especially interesting are the membranes which form triply periodic bilayers [10, 11], as well as porous materials whose internal surfaces are similar to minimal surfaces (the average curvature vanishes at every point, and the Gaussian curvature is negative).

The phase separation in binary fluid mixtures belongs to the Ising universality class, and the universal properties of the two-dimensional Ising model are well known from exact results [12]. The nonuniversal properties, however, should be determined for each particular system, and the only feasible method for a surface with arbitrary curvature is by computer simulations. Thus, it is important to develop a simulation procedure that is fast, efficient and accurate. Moreover, it is important to very accurately determine the critical parameters of a generic model on a flat surface, so that the results of approximate theories or the results of simulations on curved surfaces could be compared with reliable data. The lattice model is not appropriate for investigations of the properties of curved surfaces. Therefore, in this work we have chosen a generic continuous model of non-additive hard core mixtures. Some real phenomena, which can be modelled by a mixture of non-additive hard disks, are discussed, for example, in reference [13] and in references cited herein. The behavior of the hard core mixtures has been studied in bulk [14–21] and in restricted geometry [3, 22–24] in three dimensional systems. Much less attention was paid to the two dimensional systems [13, 20, 24–27].

We study the mixture of symmetric non-additive hard disks with the interaction potential defined by:

$$U_{\alpha\gamma}(r) = \begin{cases} \infty & \text{if} \quad r < \sigma_{\alpha\gamma}, \\ 0 & \text{if} \quad r > \sigma_{\alpha\gamma}, \end{cases} \tag{1.1}$$

where r is the distance between the centers of two disks, indexes $\alpha \in \{A, B\}$ and $\gamma \in \{A, B\}$ describe the species. The length scale is set by the A-component hard-disks diameter, $\sigma_{AA} = 1$. For symmetric non-additive mixtures,

$$\sigma_{BB} = \sigma_{AA} \tag{1.2}$$

and

$$\sigma_{AB} = \frac{1}{2}(\sigma_{AA} + \sigma_{BB})(1 + \Delta), \tag{1.3}$$

where Δ is the non-additivity parameter. We study the mixtures of positive non-additivity with different values of the parameter Δ. This potential is an idealization of interactions in a mixture of identical colloid particles with surfaces covered with polymeric brushes of two types, A and B. The polymeric brushes of different type effectively repel each other, but the polymers of the same type can interpenetrate. For this reason, the separation between the like particles can be shorter than between different particles.

Quite evidently, this is an athermal model, because all the allowed configurations are of the same energy. Nevertheless, such mixtures are capable of separating into two phases, i.e., one phase rich in component A and the other one rich in component B. The phase separation is not induced by competition between the internal energy and the entropy as in standard systems, but rather by competition between the entropy of mixing and the entropy associated with the area available for the particles. An increase of the packing fraction defined as

$$\eta = \eta_A + \eta_B = \frac{\pi}{4}\frac{N_A\sigma_{AA}}{S} + \frac{\pi}{4}\frac{N_B\sigma_{BB}}{S}, \tag{1.4}$$

where S is the surface area of the system and N_A and N_B are the numbers of the A and B particles, leads to a larger decrease of the available area in a homogeneous mixture than in a phase-separated system. This effect starts to dominate over the decrease of the entropy of mixing in a phase-separated system for $\eta > \eta_c$. Thus, η_c plays the role analogous to the critical temperature T_c, and $(\eta_c - \eta)/\eta_c$ plays a role analogous to $(T - T_c)/T_c$ in standard mixtures with interacting particles. Simulation results [19] show that this model system belongs to the Ising universality class. The universal properties of the model are known from the exact solution of the Ising model in two dimensions, but the nonuniversal properties, such as η_c, should be obtained by simulations.

At the absence of an external field, the symmetry of interactions implies, for the two coexisting phases I and II, the following relations:

$$x_A^{\mathrm{I}} = x_B^{\mathrm{II}}, \qquad x_B^{\mathrm{I}} = x_A^{\mathrm{II}}, \tag{1.5}$$

and

$$\mu_A^{\mathrm{I}} = \mu_A^{\mathrm{II}} = \mu_B^{\mathrm{I}} = \mu_B^{\mathrm{II}}, \tag{1.6}$$

where μ_α^{I} and x_α^{I} are the chemical potential and the composition of the component α in the i-th phase, respectively [19, 28]. Here, the difference of the chemical potentials $h = \mu_A - \mu_B$ plays the role of an external field.

We are interested in the critical properties of a mixture when the external field is zero. Along the symmetry line $h = 0$, the composition of the coexisting phases is symmetric, and therefore the critical point is at the concentration $x_c = N_A/(N_A + N_B) = 0.5$.

2. The method

We model an open system in contact with a reservoir by using the Semigrand [19, 29] Monte Carlo [30, 31] simulation method. Using this method, the system is simulated under a constant total number of particles N, total volume V, temperature T, and the difference of the chemical potential of one species with respect to an arbitrarily chosen species $\Delta\mu$. Thus, the number of molecules of each species fluctuates, while the total density remains constant. The semigrand ensemble is superior to the grand ensemble in simulating dense fluids because the particle insertion moves are inefficient for dense fluids. For a symmetric binary hard disks mixture, the internal energy and the chemical potential difference are both zero.

The realization of the semigrand ensemble Monte Carlo simulation for symmetric non-additive hard disks requires two kinds of moves: translation and identity change. The identity change moves can be implemented in the following way. A molecule can be chosen randomly from all the molecules, and the identity change move is always accepted if there is no overlap between the particles after the identity change. Such a procedure works quite well for a small number of molecules, up to a few hundred. The simulations near the critical point might be, however, time consuming. Therefore, we use a cluster algorithm to perform the simulations for larger system sizes [16, 24]. The idea of the cluster moves is as follows. The system is divided into a set of clusters. The molecules belong to a cluster if the distance from any molecule to any other molecule is smaller than $\sigma_{AA} + \Delta$. When the clusters are formed, the identity of all molecules in each cluster is randomly changed with the probability $p = 0.5$. We identify the clusters using the SLINK algorithm [32–34], which is fast and does not require a huge amount of computer memory. We have performed the calculations using either local MC moves or cluster moves. The results obtained by both methods were consistent, but the time of calculations was much shorter for the cluster algorithm.

For the translation moves, the maximum displacement is chosen to obtain 50% acceptance ratio. The identity-change and the translation moves are chosen randomly with the ratio of N translations per one cluster identity change move. We have performed calculations with a square or with a rectangular simulation box. The aspect ratio of the rectangular box is taken as $\sqrt{3} : 2$ to allow for the arrangement of the molecules into a triangular lattice. Periodic boundary conditions are employed. We have not observed any dependence of the results of calculations on the shape of the simulation box when the simulations were performed for fluid mixtures. The averages are taken over 10^6 Monte Carlo cycles, where a cycle consists of N translation and one cluster identity change move.

In molecular simulations, the results of calculations depend on a system size. The dependence is more pronounced for the calculations near the critical point since the correlation length becomes larger and larger when the system is closer and closer to the critical point. When the correlation length is larger than the size of the computational box, the results of the calculations become biased. That is why it may be very difficult to perform exact calculations of the critical point parameters.

When the system is in the two phase region, we do not always have only one phase in the simulation box during the simulation in the semigrand Monte Carlo method. It is possible that in the simulation box we will have either the first or the second phase. With some frequency, the first phase disappears and the second phase appears and vice versa. The higher is the frequency of this change, the closer the system is to the critical point. That is why it is hard to calculate the concentration at the coexistence as an ensemble average. One may try to overcome this problem by taking the most probable value of the concentration instead of the ensemble average, but near the critical point, this procedure may be problematic especially for two dimensional fluids. The shape of the coexistence curve for the two dimensional fluid is much flatter than for the three dimensional fluids. The critical exponent for the two dimensional fluids is $\beta = 1/8$ while for the three dimensional fluids, it is $\beta \approx 0.3258$. The distribution of the concentration is not sharply peaked and it is difficult to obtain the most probable value of the concentration with a sufficiently high accuracy.

Fortunately, it is possible to use the calculations performed for the systems of finite size to get the data on the infinite systems by using the concept of finite-size scaling [19, 35–37]. For each value of the non-additivity parameter Δ, the critical packing fraction of the infinite system, $\eta_c(\infty)$, can be calculated from the set of apparent critical packing fractions in systems with N particles, $\eta_c^*(N)$, according to the relation [19, 36, 37]:

$$\eta_c^*(N) - \eta_c(\infty) \propto N^{-1/(dv)}, \tag{2.1}$$

where the critical exponent v is equal to 1 for the 2D Ising universality class, and $d = 2$ for the two dimensional systems. The apparent critical point in the finite-size system can be determined from the distribution of the order parameter, $P_N(m)$ calculated in the simulations as a histogram. The order parameter in the case of the non-additive hard disks system is the concentration, $m = x - x_c$, where $x = N_A/(N_A + N_B)$

and x_c is the critical concentration which is exactly $1/2$ for symmetric mixtures. The distribution $P_N(m)$ is rescaled in such a way as to have a unit norm and a unit variance, and the rescaled distribution is denoted by $P_N^*(y)$, where the rescaled order parameter is $y = a_m^{-1} N^{\beta/(d\nu)} m$ with a_m denoting a proportionality constant. For the apparent critical packing fraction $\eta_c^*(N)$, $P_N^*(y)$ has an universal shape $P^*(y)$ [19, 37]. Since this model belongs to the Ising universality class, the universal function $P^*(y)$ is known [38, 39].

3. Results

The order-parameter distribution $P_N(m)$ is calculated in the simulations as a histogram, with the number of bins equal to the number of particles N. The function $P_N(m)$ obtained in simulations for fixed N and various η is then rescaled in such a way as to have a unit norm and a unit variance. The rescaled function $P_N^*(y)$ is compared with $P^*(y)$ for several values of η, and the best fit gives us $\eta_c^*(N)$. Figure 1 shows the order-parameter distribution function $P_N^*(y)$ calculated in the simulations for the system with $\Delta = 0.2$ and $N = 324$ disks, compared with the distribution function for the 2D Ising model [39]. $P_N^*(y)$

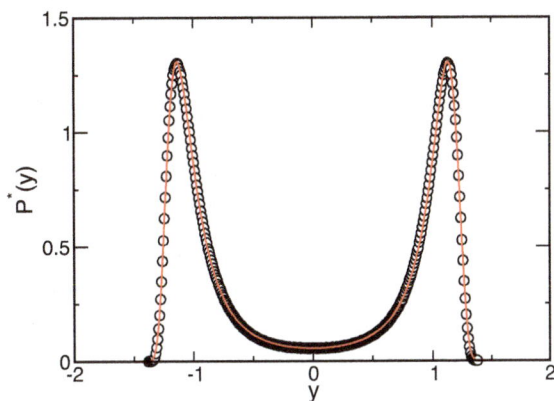

Figure 1. (Color online) The normalized distribution of the order parameter $P_N^*(y)$ (open circles) for $\Delta = 0.2$ and $\sqrt{N} = 18$, expressed as the function of the scaling variable $y = a_m^{-1} N^{\beta/(d\nu)} m$, at the packing fraction $\eta = 0.5384$. The solid line is the universal function $P^*(y)$ for the Ising model [39]. The critical exponent ν is equal to 1, and a_m is a proportionality constant.

agrees very well with the universal distribution $P^*(y)$ for the value of η that we identify with the apparent critical volume fraction $\eta_c^*(N)$. In the same way, we obtain apparent critical packing fractions for a set of systems with a different size. In figure 2, we present the plot of apparent critical densities for different system sizes calculated for the non-additivity parameter $\Delta = 0.2$. The apparent critical packing fractions are estimated for a set of finite systems with $\sqrt{N} = \{18, 20, 22, 24, 26, 28, 30\}$. The critical packing fraction as a function of $N^{-1/2}$ for an infinite system was obtained by fitting the set of apparent critical packing fractions to a straight line and extrapolating the value of the critical packing fraction at infinity. The same procedure was used to determine the critical packing fractions for all the values of the non-additivity parameter Δ.

In practice, the shape of the rescaled distribution function $P_N^*(y)$ is compared with the universal distribution of the order parameter, $P^*(y)$, for the first estimation of the apparent critical packing fraction. In order to determine the precise value of the apparent critical packing fraction, the fourth order cumulants,

$$U_N = 1 - \frac{\langle m^4 \rangle}{3 \langle m^2 \rangle^2}. \tag{3.1}$$

are calculated for the values of the packing fraction η for which the distribution functions are the most similar to the universal distribution $P^*(y)$. To calculate the matching point, the values of the cumulant $U_N(\eta)$ have been interpolated near the universal value U_N^*. We use the fourth order cumulant U_N [40], because its value at a fixed point has been already calculated for the 2D Ising universality class.

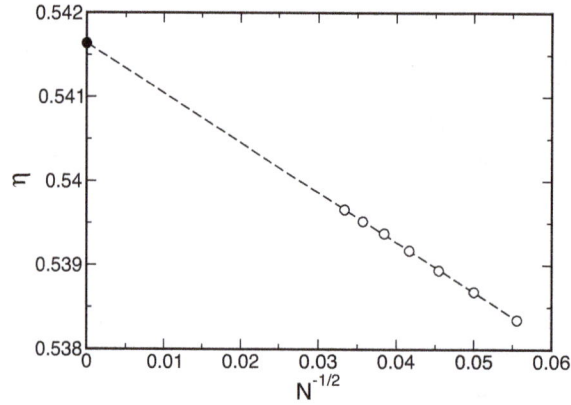

Figure 2. Scaling of the apparent critical packing fraction with the system size $\sqrt{N} = \{18, 20, 22, 24, 26, 28, 30\}$ for $\Delta = 0.2$. The open circles represent the apparent critical packing fractions for different system sizes obtained by matching $P_N^*(y)$ with the universal function as in figure 1. The solid circle is the critical packing fraction for the infinite system [see equation (2.1)]. The dashed line is the least square fit of the apparent critical packing fractions to a straight line.

The n-th moment $\langle m^n \rangle$ can be easily calculated from the distribution of the order parameter $P_N(m)$:

$$\langle m^n \rangle = \frac{\sum_m m^n P_N(m)}{\sum_m P_N(m)}. \tag{3.2}$$

The apparent critical packing fractions $\eta_c^*(N)$ satisfy the equation $U_N(\eta_c^*) = U^*$, and have been read off from the interpolated line for the value of $U^* = 0.61069$ [41]. In figure 3, we present the cumulants, $U_N(m)$, as functions of the packing fraction for different system sizes N. The curves cross at the values of the cumulant higher than the universal value U^*. Similar behavior was observed in references [20, 24].

In figure 4 we present the results of our calculation of the critical packing fractions for a set of the non-additivity parameter Δ, compared with the results already reported in the literature. The calculations reported in reference [26] significantly overestimate and in reference [27] significantly underestimate the results of other works [13, 20, 24, 25]. This is most probably the result of inappropriate simulation methods employed in those calculations. It can be easily noticed that the results of our calculations are in very good agreement with the results of the calculations reported in reference [20] and reference [24]. In all these works, the authors use cluster algorithms in the simulations and the critical packing fractions are calculated for infinite systems where different kinds of finite size scaling analysis were used.

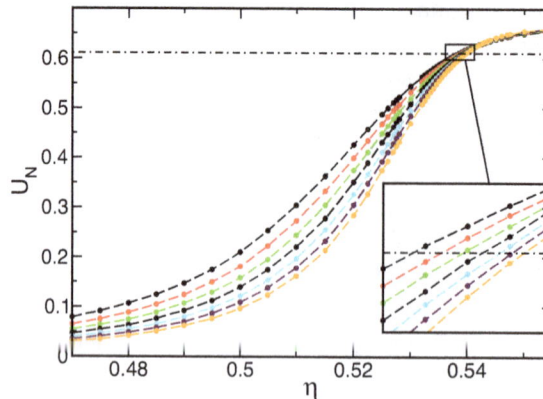

Figure 3. (Color online) Fourth order cumulant, $U_N(\eta)$, as a function of the packing fraction η for the non-additivity parameter $\Delta = 0.2$ and $\sqrt{N} = \{18, 20, 22, 24, 26, 28, 30\}$. The solid circles are the results of the simulations. The dashed line is just to guide the eye. The dash-dotted line is plotted at the universal value of the cumulant $U^* = 0.61069$.

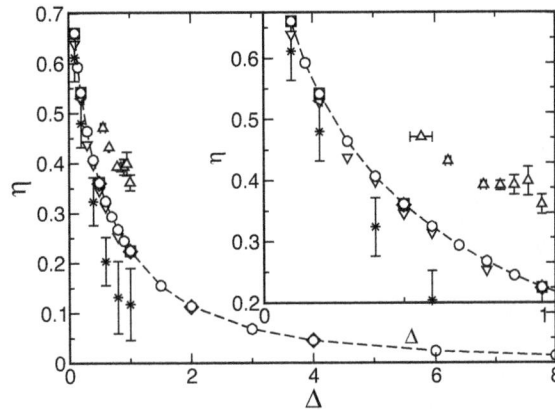

Figure 4. The critical packing fraction as a function of the non-additivity parameter Δ. The open circles show the calculations from this work, the triangles up — reference [26], the stars — [27], the triangles down — [13], the squares — [24], the diamonds — [20], the triangle right-hand — [25]. The dashed line is just to guide the eye.

In reference [25], the critical point packing fraction was obtained using the method of crossing the reduced second moment. This method is not as precise as the method of finite size scaling analysis. In reference [13], the critical packing fraction was estimated from the simulation of finite systems of the order of $N = 2000$ molecules. Thus, the values of the critical packing fractions should be lower than the values of the critical packing fractions obtained for infinite systems.

In our work, we employ the same cluster algorithm as in reference [24]. We use, however, a different numerical algorithm to identify the clusters, and different scaling procedure to obtain the critical packing fraction. In reference [20], a different cluster algorithm is used, where the clusters are built by superimposing two configurations and identifying the clusters by selecting the groups of overlapping molecules in the sense of hard core interactions. This method works very well for sufficiently large values of the non-additivity parameter Δ but is not as good for small Δ. In reference [20], calculations were performed for $Δ \in \{0.5, 1.0, 2.0, 4.0\}$. The cluster algorithm used in reference [24] and in our calculations works very well for any value of the non-additivity parameters. In reference [24], calculations were performed for $Δ \in \{0.1, 0.2, 0.5, 1.0\}$. In this work, we have performed calculations for all the values of the non-additivity parameter Δ for which the simulation results already existed, and we have extended the calculation to additional large and small non-additivities. We have performed the calculation for $Δ \in \{0.1, 0.15, 0.2, 0.3, 0.4, 0.5, 0.6, 0.7, 0.8, 0.9, 1.0, 1.5, 2.0, 3.0, 4.0, 6.0, 8.0\}$. The critical packing fractions for this set of parameters are presented in table 1. In figure 4, the results of our calculations are indistinguishable from the results of calculations in references [20] and [24]. An increase of the non-additivity parameter Δ results in a decrease of the critical packing fraction as shown in figure 4. In the mixture of non-additive hard disks, we have a competition between the entropy of mixing and the excluded volume effects. When the non-additivity parameter is large, the excluded volume effects are strong and the

Table 1. The critical packing fractions η_c for infinite systems for different values of the non-additivity parameter Δ.

Δ	0.1	0.15	0.2	0.3	0.4	0.5
η_c	0.6607(30)	0.5928(20)	0.5416(30)	0.4644(30)	0.4069(30)	0.3615(40)
Δ	0.6	0.7	0.8	0.9	1.0	1.5
η_c	0.3246(40)	0.2936(30)	0.2674(40)	0.2447(40)	0.2250(30)	0.1554(30)
Δ	2.0	3.0	4.0	6.0	8.0	
η_c	0.1140(40)	0.0686(30)	0.0455(30)	0.0241(20)	0.0148(20)	

demixing process takes place at a lower packing fraction. When the non-additivity parameter is small, the entropy of mixing dominates and the demixing takes place at a higher packing fraction.

4. Summary and conclusions

We have calculated the critical packing fractions with high accuracy for the non-additive hard disks mixtures for a wide range of the non-additivity parameter. We have used Monte Carlo simulations with cluster algorithm which resulted in rejection-free moves. We have used finite size scaling analysis based on the universal distribution of the order parameter to determine the critical packing fraction for infinite systems. The results of our calculations agree very well with the results of the previous calculations where the finite size scaling analysis was used. The proposed simulation method allows for accurate and unambiguous determination of the critical packing fraction for any value of the non-additivity parameter Δ. We hope that the results of our calculations will be the reference point for testing the results of approximate theories for hard disks systems. They should also allow one to compare the critical properties of the particles adsorbed at flat or at curved surfaces, and to determine in this way the effect of the curvature of the underlying surface on the critical properties of the adsorbed fluid mixture.

Acknowledgement

The support from NCN grant No $2012/05/B/ST3/03302$ is acknowledged. We would like to thank Noe Almarza and Paweł Rogowski for helpful discussions.

References

1. Góźdź W.T., Sokołowska Z., Sokołowski S., Z. Phys. Chem., 1991, **173**, No. 1, 95; doi:10.1524/zpch.1991.173.Part_1.095.
2. Bucior K., Patrykiejew A., Pizio O., Sokołowski S., Sokołowska Z., Mol. Phys., 2003, **101**, No. 6, 721; doi:10.1080/0026897021000021859.
3. Duda Y., Pizio O., Sokołowski S., J. Phys. Chem. B, 2004, **108**, No. 50, 19442; doi:10.1021/jp040340x.
4. Patrykiejew A., Sokołowski S., J. Phys. Chem. C, 2007, **111**, No. 43, 15664; doi:10.1021/jp0728813.
5. Boriówko M., Sokołowski S., Staszewski T., J. Phys. Chem. B, 2012, **116**, No. 10, 3115; doi:10.1021/jp300114y.
6. Patrykiejew A., Sokołowski S., J. Chem. Phys., 2012, **136**, No. 14, 144702; doi:10.1063/1.3699330.
7. Veatch S.L., Soubias O., Keller S.L., Gawrisch K., Proc. Natl. Acad. Sci. USA., 2007, **104**, No. 45, 17650; doi:10.1073/pnas.0703513104.
8. Honerkamp-Smith A.R., Veatch S.L., Keller S.L., Biochim. Biophys. Acta, 2009, **1788**, No. 1, 53; doi:10.1016/j.bbamem.2008.09.010.
9. Góźdź W.T., Bobrovska N., Ciach A., J. Chem. Phys., 2012, **137**, No. 1, 015101; doi:10.1063/1.4731646.
10. Góźdź W.T., Hołyst R., Phys. Rev. E, 1996, **54**, No. 5, 5012; doi:10.1103/PhysRevE.54.5012.
11. Dotera T., Kimoto M., Matsuzawa J., Interface Focus, 2012, **2**, No. 5, 575; doi:10.1098/rsfs.2011.0092.
12. Onsager L., Phys. Rev., 1944, **65**, 117; doi:10.1103/PhysRev.65.117.
13. Fiumara G., Pandaram O.D., Pellicane G., Saija F., J. Chem. Phys., 2014, **141**, No. 21, 214508; doi:10.1063/1.4902440.
14. Amar J.G., Mol. Phys., 1989, **67**, 739; doi:10.1080/00268978900101411.
15. Lomba E., Alvarez M., Lee L., Almarza N., J. Chem. Phys., 1996, **104**, 4180; doi:10.1063/1.471229.
16. Johnson G., Gould H., Machta J., Chayes L., Phys. Rev. Lett., 1997, **79**, No. 14, 2612; doi:10.1103/PhysRevLett.79.2612.
17. Saija F., Pastore G., Giaquinta P.V., J. Phys. Chem. B, 1998, **102**, No. 50, 10368; doi:10.1021/jp982202b.
18. Góźdź W.T., Pol. J. Chem., 2001, **75**, No. 4, 517.
19. Góźdź W.T., J. Chem. Phys., 2003, **119**, No. 6, 3309; doi:10.1063/1.1590746.
20. Buhot A., J. Chem. Phys., 2005, **122**, No. 2, 024105; doi:10.1063/1.1831274.
21. Pellicane G., Pandaram O.D., J. Chem. Phys., 2014, **141**, No. 4, 044508; doi:10.1063/1.4890742.
22. Góźdź W.T., J. Chem. Phys., 2005, **122**, No. 7, 074505; doi:10.1063/1.1844332.
23. De Sanctis Lucentini P.G., Pellicane G., Phys. Rev. Lett., 2008, **101**, 246101; doi:10.1103/PhysRevLett.101.246101.
24. Almarza N.G., Martín C., Lomba E., Bores C., J. Chem. Phys., 2015, **142**, No. 1, 014702; doi:10.1063/1.4905273.
25. Muñoz-Salazar L., Odriozola G., Mol. Simul., 2010, **36**, No. 3, 175; doi:10.1080/08927020903141027.

26. Nielaba P., Int. J. Thermophys., 1996, **17**, No. 1, 157; doi:10.1007/BF01448218.
27. Saija F., Giaquinta P.V., J. Chem. Phys., 2002, **117**, No. 12, 5780; doi:10.1063/1.1501126.
28. Góźdź W.T., Gubbins K.E., Panagiotopoulos A.Z., Mol. Phys., 1995, **84**, No. 5, 825; doi:10.1080/00268979500100581.
29. Kofke D.A., Glandt E.D., Mol. Phys., 1988, **64**, 1105; doi:10.1080/00268978800100743.
30. Metropolis N., Ulam S.M., J. Am. Stat. Assoc., 1949, **44**, No. 247, 335; doi:10.1080/01621459.1949.10483310.
31. Metropolis N., Rosenbluth A.W., Rosenbluth M.N., Teller A.H., Teller E., J. Chem. Phys., 1953, **21**, No. 6, 1087; doi:10.1063/1.1699114.
32. Florek K., Łukaszewicz J., Perkal J., Steinhaus H., Zubrzycki S., Colloquium Math., 1951, **2**, No. 3–4, 282.
33. Florek K., Łukaszewicz J., Perkal J., Steinhaus H., Zubrzycki S., Przegl. Antrop., 1951, **17**, 193.
34. Sibson R., Comput. J., 1973, **16**, No. 1, 30; doi:10.1093/comjnl/16.1.30.
35. Fisher M.E., Barber M.N., Phys. Rev. Lett., 1972, **23**, 1516; doi:10.1103/PhysRevLett.28.1516.
36. Phase Transitions and Critical Phenomena, 8th Edn., Domb C., Lebowitz J. (Eds.), Academic Press, London, 1983.
37. Wilding N.B., Phys. Rev. E, 1995, **52**, 602; doi:10.1103/PhysRevE.52.602.
38. Nicolaides D., Bruce A.D., J. Phys. A: Math. Gen., 1988, **21**, No. 1, 233; doi:10.1088/0305-4470/21/1/028.
39. Liu Y., Panagiotopoulos A.Z., Debenedetti P.G., J. Chem. Phys., 2010, **132**, No. 14, 144107; doi:10.1063/1.3377089.
40. Binder K., Phys. Rev. Lett., 1981, **47**, 693; doi:10.1103/PhysRevLett.47.693.
41. Kamieniarz G., Blote H.W.J., J. Phys. A: Math. Gen., 1993, **26**, No. 2, 201; doi:10.1088/0305-4470/26/2/009.

Permissions

List of Contributors

A. Patrykiejew and T. Staszewski
Department for the Modeling of Physico-Chemical Processes, Maria Curie-Skłodowska University, 20-031 Lublin, Poland

E. Núñez-Rojas
Departamento de Química, Universidad Autónoma Metropolitana-Iztapalapa, México, D.F. 09340

H. Dominguez
Instituto de Investigaciones en Materiales, Universidad Nacional Autónoma de México, México, D.F. 04510

M. Borówko and T. Staszewski
Department for the Modeling of Physico-Chemical Processes, Maria Curie-Skłodowska University, 20-031 Lublin, Poland

A. Malijevský
Department of Physical Chemistry, University of Chemistry and Technology Prague, 166 28 Praha 6, Czech Republic
Laboratory of Aerosols Chemistry and Physics, Institute of Chemical Process Fundamentals, Academy of Sciences, 16502 Prague 6, Czech Republic

A. Baumketner
Institute for Condensed Matter Physics, National Academy of Sciences of Ukraine, 1 Svientsistskii St., 79011 Lviv, Ukraine

W. Cai
Beijing Computational Science Research Center, Beijing 100094, China

A. Woszczyk and P. Szabelski
Department of Theoretical Chemistry, Maria-Curie Sklodowska University, Pl. M.C. Sklodowskiej 3, 20-031 Lublin, Poland

T. Bryk
Institute for Condensed Matter Physics of the National Academy of Sciences of Ukraine, 1 Svientsitskii St., 79011 Lviv, Ukraine
Institute of Applied Mathematics and Fundamental Sciences, Lviv Polytechnic National University, 12 Bandera St., 79013 Lviv, Ukraine

A.D.J. Haymet
Scripps Institution of Oceanography, UC San Diego, San Diego, California 92093, USA

M. Holovko, T. Patsahan and I. Kravtsiv
Institute for Condensed Matter Physics of the National Academy of Sciences of Ukraine, 1 Svientsitskii St., 79011 Lviv, Ukraine

D. di Caprio
Institut de Recherche de Chimie Paris, CNRS—Chimie ParisTech, 11 rue Pierre et Marie Curie, 75005 Paris, France

J.M. Ilnytskyi
Institute for Condensed Matter Physics of the National Academy of Sciences of Ukraine, 1 Svientsitskii St., 79011 Lviv, Ukraine

P. Bryk and A. Patrykiejew
Department for the Modeling of Physico-Chemical Processes, Maria Curie-Skłodowska University, 20–031 Lublin, Poland

T. Domański
Institute of Physics, M. Curie-Skłodowska University, 20-031 Lublin, Poland

P. Zeliszewska, A. Bratek-Skicki and Z. Adamczyk
J. Haber Institute of Catalysis and Surface Chemistry Polish Academy of Sciences, Niezapominajek 8, 30-239 Cracow, Poland

M. Ciesla
M. Smoluchowski Institute of Physics, Jagiellonian University, Łojasiewicza 11, 30-348 Cracow, Poland

Z.Ható, D. Boda and T. Kristóf
Department of Physical Chemistry, University of Pannonia, P. O. Box 158, Veszprém, H-8201, Hungary

D. Gillespie
Department Molecular Biophysics and Physiology, Rush University Medical Center, Chicago, IL 60612, USA

J. Vrabec and G. Rutkai
University of Paderborn, Laboratory of Thermodynamics and Energy Technology, 100 Warburger St., 33098 Paderborn, Germany

R. Tscheliessnig
Austrian Centre for Industrial Biotechnology (ACIB), Muthgasse 11, A-1190 Wien, Austria

L. Pusztai
Wigner Research Centre for Physics, Hungarian Academy of Sciences, Konkoly Thege út. 29-33, H-1121, Budapest, Hungary

M. Kaja and S. Lamperski
Department of Physical Chemistry, Adam Mickiewicz University of Poznan, Umultowska 89b, 61-614 Poznan, Poland

W. Silvestre-Alcantara and L.B. Bhuiyan
Laboratory of Theoretical Physics, Department of Physics, University of Puerto Rico, San Juan, 00931-3343, Puerto Rico

D. Henderson
Department of Chemistry and Biochemistry, Brigham Young University, Provo UT 84602-5700, USA

W. Silvestre-Alcantara and L.B. Bhuiyan
Department of Physics and Laboratory for Theoretical Physics, University of Puerto Rico, San Juan, Puerto Rico 00936-8377, USA

S. Lamperski and M. Kaja
Department of Physical Chemistry, Adam Mickiewicz University in Poznań, Umultowska 89b, 61-614 Poznań, Poland

G. Mazovec, M. Lukšič and B. Hribar-Lee
University of Ljubljana, Faculty of Chemistry and Chemical Technology, Vecna pot 113, SI-1000 Ljubljana, Slovenia

S. Furmaniak, P.A. Gauden and A.P. Terzyk
Faculty of Chemistry, Physicochemistry of Carbon Materials Research Group, Nicolaus Copernicus University in Torun, Gagarin St. 7, 87–100 Torun, Poland

P. Kowalczyk
School of Engineering and Information Technology, Murdoch University, Murdoch, Western Australia 6150, Australia

W.T. Góźdź and A. Ciach
Institute of Physical Chemistry Polish Academy of Sciences, Kasprzaka 44/52, 01-224 Warsaw, Poland

Index

www.ingramcontent.com/pod-product-compliance
Lightning Source LLC
Chambersburg PA
CBHW082027190326
41458CB00010B/3296